# THE
# Growth Warriors

# THE
# Growth Warriors

## Creating Sustainable Global Advantage for America's Technology Industries

Ronald Mascitelli

**Technology Perspectives**
**Northridge, CA**

*Editorial and Sales Offices:* Technology Perspectives
18755 Accra Street, Northridge, California 91326

Publisher's Cataloging-in-Publication

Mascitelli, Ronald
   The growth warriors : creating sustainable global
advantage for America's technology industries / Ronald
Mascitelli. — 1st ed.
   p. cm.
   Includes bibliographical references and index.
   Preassigned LCCN: 98-60147
   ISBN: 0-9662697-0-5
   1. High technology industries—United States. 2.
Technology—United States—Management. 3. Competition,
International. 4. Industrial productivity—United
States.   I. Title.

HC79.H53M37 1999              338'.064
                              QBI98-1063


1  3  5  7  9  10  8  6  4  2
First Edition

This book is printed on acid-free recycled paper meeting
the requirements of the American National Standard
for Permanence in Paper for Printed Library Materials.

Manufactured in the United States of America

*Dedication* –

To my wife, Renee,
whose love and support have
enabled me to pursue my dreams.

# Contents

# Acknowledgments

I would like to extend my sincere appreciation to the many individuals who have made constructive comments on early drafts of this book. In particular, the contributions of Jon Nadenichek and Yves Courbet were invaluable in refining my discussion of technology-driven economic growth. The production aspects of this project benefited from the editorial inputs of Virginia Iorio and Robert A. Juran, and the craftsmanship of Marc Bailey. My wife, Renee, deserves substantial credit for the accuracy and quality of the final text.

Finally, I would like to personally thank the hundreds of respected scholars and successful practitioners of competitive strategy upon whose works this book is based. Although the opinions expressed throughout this book are my own, the broad vision of the global technology enterprise described herein was first seen through their eyes.

Ron Mascitelli
October 1998

# List of Exhibits

## Figures

# Tables

# Boxes

# Acronym Dictionary

AAAS - American Association for the Advancement of Science
www.aaas.org/aaas.html
ABB - Asea Brown Bovari
AEA - American Electronics Association
www.aeanet.org/
AMD - Advanced Micro Devices
AME - Association for Manufacturing Excellence
www.ame.org/
ANSI - American National Standards Institute
web.ansi.org/default_ js.htm
APEC - Asia-Pacific Economic Cooperation
www.apecsec.org.sg/
APSI - Alliance to Promote Software Innovation
ARPA - Advanced Research Projects Agency (see DARPA)
ASEAN - Association of Southeast Asian Nations
www.asean.or.id/
ASIC - Application-Specific Integrated Circuit
ATM - Asynchronous Transfer Mode
ATP - Advanced Technology Program
www.atp.nist.gov/atp/atphome.htm
BPR - Business Process Reengineering
BSA - Business Software Alliance
www.bsa.org/
CAD - Computer Aided Design
CAM - Computer Aided Manufacturing
CD - Compact Disk
CEM - Contract Electronics Manufacturing
CFC - Chlorofluorocarbon
CIBER - Center for Industrial Business Education and Research
CNC - Computer Numerically Controlled
$CO_2$ - Carbon Dioxide
COTS - Commercial Off the Shelf
CRADA - Cooperative Research and Development Agreement
CT - Computed Tomography
DARPA - Defense Advanced Research Projects Agency
www.darpa.mil/index.html

DEC - Digital Equipment Corporation
DMS - Document Management Systems
DOC - Department of Commerce
www.doc.gov/
DRAM - Dynamic Random Access Memory
DVD - Digital Videodisk
EMC - Electromagnetic Compatibility
EPROM - Erasable Programmable Read-Only Memory
ERP - Enterprise Resource Planning
EU - European Union
www.europa.eu.int/index-en.htm
FCC - Federal Communications Commission
www.fcc.gov/
FCPA - Foreign Corrupt Practices Act
FDA - Food and Drug Administration
www.fda.gov/default.htm
FDI - Foreign Direct Investment
FFRDC - Federally Funded R&D Center
FGCP - Fifth Generation Computer Project
FTAA - Free Trade Area of the Americas
www.alca-ftaa.org/
FTWG - Focus Technology Working Group
GATT - Global Agreement on Tariffs and Trade
www.wto.org/
GDP - Gross Domestic Product
GII - Global Information Infrastructure
www.gii.org/
GNP - Gross National Product
GSM - Global System for Mobile Communications
HDTV - High-Definition Television
IEC - International Electrotechnical Commission
www.iec.ch/
IEEE - Institute of Electrical and Electronic Engineers
www.ieee.org/
ILM - Industrial Light and Magic
IMF - International Monetary Fund
www.imf.org/external/
IOS - Internet Operating System
IPC - Institute for Interconnecting and Packaging Electronic Circuits
www.ipc.org/index.html
IPR - Intellectual Property Rights
ISO - International Standards Organization
www.iso.ch/
IT - Information Technology
I/UCRC - Industry/University Cooperative Research Center
JEDI - Joint Environment for Digital Imaging
JETRO - Japan External Trade Organization
www.jetro.go.jp/top/index.html

| | |
|---|---|
| J&J - | Johnson and Johnson |
| LCD - | Liquid Crystal Display |
| LED - | Light-Emitting Diode |
| MFA - | Multifiber Arrangement |
| MFN - | Most-Favored Nation |
| MITI - | Ministry of International Trade and Industry<br>www.miti.go.jp/index-e.html |
| MNC - | Multinational Corporation |
| MOSS - | Market-Oriented, Sector-Specific |
| MOT - | Management of Technology |
| MSC - | Multimedia Supercorridor |
| MTC - | Manufacturing Technology Center |
| NACA - | National Advisory Committee for Aeronautics |
| NAFTA - | North American Free Trade Agreement<br>www.nafta.net/naftagre.htm |
| NASA - | National Aeronautics and Space Administration<br>www.nasa.gov/NASA_homepage.html/ |
| NCR - | National Cash Register |
| NEC - | Nippon Electric Company |
| NEMI - | National Electronics Manufacturing Initiative<br>www.nemi.org/ |
| NGM - | Next-Generation Manufacturing Project |
| NIH - | National Institutes of Health<br>www.nih.gov/ |
| NIH - | Not Invented Here |
| NIST - | National Institute of Standards and Technology<br>www.nist.gov/ |
| NREN - | National Research and Education Network<br>www.ccic.gov/pubs/blue94/section.3.2.html |
| NSF - | National Science Foundation<br>www.nsf.gov/ |
| NSI - | National System of Innovation |
| NTIS - | National Technical Information Service<br>www.ntis.gov/index.html |
| NTRS - | National Technology Roadmap for Semiconductors<br>www.sematech.org/member/docubase/techrpts.htm |
| NTT - | Nippon Telegraph and Telephone |
| OECD - | Organization for Economic Co-operation and Development<br>www.oecd.org/ |
| OEM - | Original Equipment Manufacturer |
| PAC - | Political Action Committee |
| PDA - | Personal Digital Assistant |
| PPP - | Purchasing Power Parity |
| PTO - | Patent and Trademark Office<br>www.uspto.gov/ |
| R&D - | Research and Development |
| ROI - | Return on Investment |

| | |
|---|---|
| SAE - | Society of Automotive Engineers |
| | www.sae.org/ |
| SBIR - | Small Business Innovation Research |
| | www.sbaonline.sba.gov/index.html |
| SEMATECH- | Semiconductor Manufacturing Technology |
| | www.sematech.org/public/home.htm |
| SEMI - | Semiconductor Equipment and Materials Institute |
| | www.semi.org/default.html |
| SGI - | Silicon Graphics Inc. |
| SIA - | Semiconductor Industry Association |
| | www.semichips.org/ |
| SME - | Small to Medium-Sized Enterprise |
| SOC - | System on a Chip |
| SPIN - | Scientific Performance Improvement Network |
| STA - | Semiconductor Trade Agreement |
| STC - | Science and Technology Center |
| TCI - | Tele-Communications Inc. |
| TFP - | Total-Factor Productivity |
| TQM - | Total Quality Management |
| TRACES - | Technology in Retrospect and Critical Events in Science |
| TRIPS - | Trade-Related Aspects of Intellectual Property Rights |
| | www.wto.org/wto/intellec/intellec.htm |
| TRP - | Technology Reinvestment Project |
| USDC - | United States Display Consortium |
| | www.semi.org/Focused/fpd/usdc1.html |
| USITC - | United States International Trade Commission |
| | www.usitc.gov/default.htm |
| USTR - | United States Trade Representative |
| | www.ustr.gov/ |
| VCR - | Videocassette Recorder |
| VER - | Voluntary Export Restraint |
| VLSI - | Very Large Scale Integration |
| VRA - | Voluntary Restraint Agreement |
| WIPO - | World Intellectual Property Organization |
| | www.wipo.org/eng/newindex/index.htm |
| WTO - | World Trade Organization |
| | www.wto.org/ |

# Introduction

In the battle for growth in high-technology industries, the greatest challenge lies in identifying the enemy. The days of epic contests among familiar champions have passed. Such protracted duels seem almost quaint in contrast to the chaotic free-for-alls that characterize today's technology markets. Likewise, it is no longer credible for firms to blame faceless foreign rivals for lagging competitiveness. Experience in global markets has demonstrated that competitors from all nations are struggling with the same turmoil and uncertainty.

Amid a roiling sea of opportunity, innovative firms burst forth, thrive, and die with shocking rapidity. Markets and technologies twist and turn in inscrutable ways, no longer driven by the traditional forces upon which business strategy has long been based. The behavior of rivals, the preferences of customers, and the structure of the value chain are now being shaped by the tectonic forces of technological progress and economic globalization. Evidently, in the battle for sustainable long-term growth, the true "enemy" is the changing competitive environment itself.

Until recently, there has been little need for the practitioners of technology enterprise (i.e., managers, executives, entrepreneurs) to understand the underlying nature of the competitive environment. In relatively stable markets, the "rules of engagement" form a familiar backdrop for narrowly focused, industry-specific strategic actions. But how does a firm formulate strategy when the rules, the players, and even the goals of the game change continuously? Under these disconcerting conditions, the ability to extend reasoning and experience into new and unfamiliar territory is essential. The survival of firms today depends on their ability to rapidly adapt to change in all aspects of the competitive environment.

This book is about global competitiveness in technology-intensive industries. Its approach to the topic, however, is unlike any other book on the subject. Incorporating dynamic elements into competitive strategy requires more than a minor adjustment to traditional thinking: We must seek an entirely new approach to strategy formation that is time-based, adaptable, decision-driven, and focused on the efficient creation of value. Moreover, such an approach must resonate with the forces that drive today's global technology markets, and be guided by a set of principles and frameworks for extending a firm's current knowledge and experience into uncharted territory.

The objective of this book is to explore how U.S. firms can achieve and sustain global competitiveness, based on fundamental principles of economics and international trade theory, and supported by the best available insights into technology markets and industrial innovation. To accomplish this ambitious goal I have enlisted collaborative help: The published works and personal experiences of many respected scholars, practitioners, and policymakers form the backbone of this book. This material was carefully filtered to identify those aspects of the global competitive environment that are common to all high-technology industries. To breathe life (and interest) into this conceptually challenging material, I have provided numerous current case examples.

Throughout this work I have emphasized simple visualization tools, models, and frameworks that can help executives and policymakers make sense out of chaos. Given my background as a physicist, it is not surprising that I have taken a "reductionist" approach, avoiding complex methodologies and detailed analyses in favor of revealing the underlying structures of technology enterprise. My motivations, however, are pragmatic: The evolution of complex systems is often based on relatively simple "organizing principles." From the regimented efficiency of ant colonies to the dynamism of capitalist economies, basic patterns of behavior and response can culminate in amazing levels of complexity. As the reader will soon discover, a similar set of organizing principles drives the patterns of global trade in technology markets, the creation of value through innovation, and the ability of firms to sustain competitive advantage in the face of blinding change.

This book is divided into three parts: First, a broad background is presented to establish basic relationships and perspectives within the global business environment. Second, the major factors impacting competitiveness in technology industries are developed in some detail. Finally, the focus narrows to the practical issues faced by decision-makers within high-tech firms. The reader might visualize a three-tiered pyramid, with an expansive base, a more tractable middle section, and a detailed apex.

In Part 1, a sweeping panorama of the current competitive environment for technology-intensive firms is presented. The first step toward achieving a sustainable advantage in dynamic markets is to understand the "lay of the land." Indeed, the competitive landscape is being transformed before our eyes from the last vestige of the postwar industrial age to an entirely new system for value creation. Information and telecommunications technologies have enabled an interlinked, real-time capitalist economy that encompasses nearly the entire planet, offering endless market opportunities.

According to the sociologist Manuel Castells, revolutions in information technology and communications enable a new level of economic development, based on the concept of the "network enterprise." The ability to create and share information throughout a global network consisting of many different types of businesses can potentially yield dramatic increases in productivity. To achieve these gains, however, we must embrace an appropriate social and business culture. As Castells observes, "There is an extraordinary gap between our technological overdevelopment and our social underdevelopment."[1]

The economic system of the industrial age was driven primarily by capital accumulation and mass production. In response to this environment, a feudal system of sorts evolved, in which firms grew powerful through vertical, command-and-control organizations. These insular structures placed fundamental limits on the levels of productivity that could be achieved. In particular, mass production of commodities is characterized by diminishing returns to scale.

Today, the concept of the vertical organization has been supplanted by an increasingly horizontal, networked global business environment. This shift in the structure of firms has been accompanied by a redefinition of competition, from head-to-head contests between vertical giants to highly collaborative partnerships in which the line between cooperation and competition is blurred. Higher levels of specialization and collaboration are enabling rapid productivity gains, while product and process innovations have taken center stage as the driving force behind economic growth.

In a real sense, the shift from mass production of physical products to an economy that is increasingly dominated by knowledge-based products has removed the limits to economic growth. Unlike their hardware counterparts, knowledge products are characterized by *increasing returns*: Once created, a piece of software, for example, can be replicated and sold indefinitely at virtually zero marginal cost. The combination of network effects, real-time information exchange, increased horizontal specialization, and increasing returns from knowledge products offers the potential of sustainable economic growth for firms and nations alike.

Chapter 1 forms a conceptual foundation for all that follows, by establishing that productivity represents the basis for competition in global technology markets. Firms that make the most efficient use of inputs to production (i.e., raw materials, capital, labor), or that achieve the highest value in the marketplace for their innovations, will tend to dominate their markets. Unfortunately, the productivity growth of American firms has lagged behind that of other advanced nations in the recent past, and despite our current economic resurgence, we are still at a global disadvantage in several critical technology industries.

The explosive growth of high-technology exports in recent years has forced American firms to confront the daunting complexities of global markets. In Chapter 2, an overview of economic globalization is presented, along with a survey of important topics in international trade. Throughout human history, the movement of goods between nations has been driven by the forces of comparative advantage: Unequal endowments of input factors such as raw materials and labor determined the patterns of trade. In our current technology-driven economy, however, trade is increasingly being driven by "created" advantages in productivity. A simple two-tiered model is presented that demonstrates the distinction between global trade in commodities and the international exchange of technology-intensive products.

In Chapter 3, industrial research and development is identified as a primary source of productivity growth in the postwar era. Massive R&D investments by the U.S. government on mission-specific defense technologies during the 1950s and 1960s served as the engine for technical progress throughout the capitalist world. In more recent times our domestic emphasis has shifted to commercial R&D investment by private industry. Throughout this period, American firms have demonstrated a frustrating dichotomy: Whereas the U.S. technology enterprise has led the world in both scientific research and industrial innovation, our firms have repeatedly failed to capitalize on that creative advantage in the global marketplace.

This relatively poor "end game" performance has been exacerbated by an insular attitude toward the technological advances of other nations. Despite significant progress in this regard, U.S. firms still lag behind their counterparts in Europe and Asia in ability to absorb external knowledge and in willingness to adopt new process technologies.

With the panoramic perspective of Part 1 serving as a foundation, Part 2 focuses on those aspects of the competitive environment that can impart a sustained advantage to technology-intensive firms. In Chapter 4, two important models are presented that represent organizing principles of the emerging technology-driven economy. The first model describes the market value of a product or service in terms of three dimensions: its physical content, the explicit knowledge (i.e., intellectual property) that it embodies, and the unique tacit skills and experience necessary for its creation. It is this last dimension that holds the key to sustained market leadership: Whereas physical artifacts and intellectual property can be imitated, pirated, or invented around, the talents and experiences carried in the heads of a firm's employees cannot be easily replicated by rivals. As the knowledge content of technology products rises in economic significance, it is this tacit aspect of innovation that can perpetuate a firm's market leadership.

The second framework described in Chapter 4 captures the evolution of technology enterprise over time. Three elements of industrial R&D are identified as constituting a self-reinforcing cycle of innovation: the rate at which firms can generate new innovations, their ability to appropriate economic returns from those innovations, and their capacity to absorb and accumulate technical knowledge. As the competitive environment evolves, each of these elements imparts a survival advantage to firms. Collectively, the positive feedback among these three factors can vault an individual firm, or an entire economic region, into a world-leading position.

The remaining four chapters of Part 2 explore the various facets of the competitive environment that can directly affect the innovation cycle of firms. In Chapter 5, industry and market forces are considered, including the effects of intense inter-firm rivalry, the global deployment of R&D activities, and the productivity-enhancing power of alliances and collaboration. A brief overview of basic market strategies is provided as background for the lay reader.

As global competition expands, the effects of international trade and competition policies are being felt by even the smallest of high-tech firms. Chapter 6 explores the impact of trade actions, subsidies, intellectual property protection, piracy, and antitrust policy on American high-tech industries. Research and development in the United States has long been subsidized by the federal government, under the guise of defense technology development and targeted military procurement. While our industries were focused on building the world's most sophisticated fighting machine, commercial development in Japan, Germany, and the newly industrialized nations was being "encouraged" through aggressive industrial policies. Does direct subsidy of firms by foreign governments justify the use of activist trade measures such as the imposition of antidumping duties to punish "unfair competition"?

Chapter 7 describes what has come to be known as the American "system of innovation." This term refers collectively to those factors within the domestic environment that enable rapid innovation and technical progress. Despite a common belief that globalization will tend to negate the impact of national factors, there is growing evidence of high levels of technological specialization within nations. It appears that national systems of innovation can become self-reinforcing over time, guiding domestic industries into becoming world-leading "poles" of specialized technical capability. Indeed, the American system of innovation will likely be the catalyst for our continued economic development throughout the twenty-first century.

At the heart of the American system of innovation is U.S. technology policy, a poorly integrated maze of programs and investments administered by a plethora of agencies. In the first section of Chapter 7, U.S. technology policy is compared to that of other nations in its ability to provide four key ingredients of technological leadership: 1) the generation of pre-competitive knowledge, 2) the creation of an enabling technology infrastructure, 3) the fostering of diffusion and adoption of new productive methods, and 4) the development of specialized human capital.

The final chapter of Part 2 may be the most important from the standpoint of future global competitiveness. Chapter 8 describes the advantages of geographic proximity to suppliers, rivals, customers, and sources of groundbreaking research and specialized human capital. Industrial clusters have been a recognized phenomenon since early in the industrial age. In recent years, the legendary rise of Silicon Valley has prompted the birth of technology clusters throughout the world. The potential for such regions to achieve a "critical mass" of knowledge spillovers and network externalities is the holy grail of competitiveness: Participation in a vital technology cluster may be the surest path to global leadership in innovation and productivity.

Once the spectrum of factors that impact evolving competition is understood, the final step toward achieving sustainable competitive advantage must be taken at the level of the firm. In Part 3, an integrated process for the development of dynamic strategies is presented that enables rapid adaptation to changing technological and market conditions. Rather than attempt to revise static methods for strategy formation, I describe an entirely new structure in Chapter 9, based on continuous decision-making at every level of the firm. As the pace of progress and the degree of uncertainty increase, the development of strategy is transformed into a real-time series of choices based on the best available current knowledge and insights into probable future events. The goal of this strategic decision-making process must be to enhance the three critical elements of the innovation cycle to achieve a persistent market advantage.

The successful execution of strategies under evolving conditions is determined by the quality of current information available to decision-makers and by their ability to visualize the possible outcomes of their choices. An overview of techniques for knowledge management and decision support is provided in Chapter 10, along with a survey of traditional methods for technology forecasting. While each of these forecasting techniques offers unique advantages, two methods provide the greatest benefit under conditions of rapid and unpredictable change: scenario-building and technology roadmapping. This pair of powerful, mutually reinforcing tools is discussed in considerable detail, including pragmatic advice on firm-level implementation and several current case examples.

The final chapter addresses the ultimate source of sustained competitive advantage: the market itself. A set of models for time-based strategy formation in high-technology markets is presented in Chapter 11, derived from current trends in the information technology, biotechnology, telecommunications, and semiconductor industries. These visualization tools offer insight into risk reduction and adaptability in product development, methods for enhancing customer loyalty, and the use of technology as a medium for expression and creativity.

In my closing remarks, a brief essay is presented on the issue of environmental sustainability. Technical progress can enable virtually unlimited wealth creation, but it cannot guarantee a better quality of life. If economic development continues to be based on the unregulated exploitation of the natural environment, the benefits of increasing income may be negated by the declining quality of our surroundings, our climate, and even the air we breathe. It is time for environmental issues to be subjected to market forces, thereby allowing the efficient, practical, and ultimately self-serving preservation of the natural world.

Before we begin our intellectual odyssey, I would like to offer the reader two suggestions. First, it should be clear from the scope of this work that detailed industry-specific insights will not be forthcoming. I have endeavored to build half of a bridge; it is up to the reader to construct the other half. This should be taken as a personal invitation to collaborate in adapting the general principles and models contained in this book to your specific needs.

Second, it is important to recognize that the point of this book is to alter your perceptions. Readers will engage this work at different levels, and will bring with them distinct views and opinions. My goal is not to change your mind, but rather to inform your mind. One cannot learn without changing one's opinions, and an informed opinion is clearly preferable to an uninformed one. It would be safe to say that if you haven't changed your mind recently on almost every aspect of technology enterprise, you probably have some learning to do.

As we begin our exploration of the global competitive environment for high-technology industries, it is heartening to recall the words of Sun Tzu from *The Art of War:* "Thus it is said that one who knows the enemy and knows himself will not be endangered in a hundred engagements."[2] In the spirit of achieving such enduring strategic insight, I encourage the reader to engage his or her mind, turn the page, and enjoy the journey.

# PART 1

## The Global Environment for Technology Enterprise

# 1
# Productivity and Economic Growth

## The Growth Imperative

What are the forces that drive economic growth? Why does modern society place such emphasis on wealth and prosperity? What roles do labor and capital play in achieving these goals? Is technology the cause, or the result, of this relentless pursuit?

From a pragmatic standpoint, such fundamental questions may seem to have little relevance to the strategic actions of firms. Yet nothing could be farther from the truth. The growth of companies is driven by the same underlying forces that drive macroeconomic growth, and is increasingly influenced by the actions of foreign and domestic governments. Is the resurgent growth of America's high-technology industries, for example, a direct result of activist public policy and enlightened government action, or should the credit be given to the irrepressible entrepreneur seeking his or her fortune through breakthrough innovation and increased specialization?[1]

The debate over whether economic growth stems from the pull of government policy or from the push of entrepreneurial ambition may never be resolved. As with any symbiotic relationship, it is difficult to distinguish the host from the beneficiary. The dismal performance of centrally planned economies during the latter half of the twentieth century suggests that government action alone cannot inspire a society to high levels of productivity and growth. The strong entrepreneurial spirit of the Chinese people, for example, was stifled until recently by the oppressive policies of the mainland, but has flourished within the generative climate of Taiwan.

The sections that follow explore the complex interdependencies between the global economic environment and the ambitions of technology-intensive firms.

Within this system, the desire for wealth is the common organizing principle; national governments, industries, and individual enterprises relentlessly pursue this goal, each sector dependent on the others for its success. Although their motivations are distinct, cooperation among these various actors is essential; economic expansion can be sustained only through a resonance of appropriate government policies, world-leading industrial productivity, and the creative spirit of the entrepreneur.

### Why Growth Matters

Among modern nation-states, there is no more ubiquitous a goal than rapid economic development. Indeed, there is a growing list of role models to inspire less-developed countries toward industrialization. Japan's legendary rise from the ashes of war to become an industrial powerhouse was achieved through sustained double-digit growth rates. The "Asian tigers" have transcended their developing-nation legacy through similar means, prompting the creation of an entirely new classification, the "newly industrialized country." More recently, a new generation of high-growth champions has taken center stage. Lowly Ireland has been invigorated by a steady stream of foreign direct investment to become known as the "Celtic tiger," while the prospect of China's newly liberalized economy overtaking that of the United States early in the twenty-first century has become palpably real.

Unfortunately, the dramatic political upheavals of the last decade have created some notable losers as well. The negative growth of Russia and the "economies in transition"[2] throughout much of the 1990s continues to be cause for global concern. Likewise, the prospect of a prolonged economic slowdown in Asia has evoked pathetic whimpers from the "tigers," and even mighty Japan has repeatedly stumbled in its attempts to re-ignite its domestic economy.

For all the attention given to short-term swings in gross domestic product (GDP) by policymakers and business leaders, the reality is that economic growth acts *slowly*. Those who are familiar with the effects of compound interest know that it is not the return earned on principal, but rather the cumulative effect of interest on top of interest, that creates wealth. Over the short term, an economic slowdown has little impact on standard of living. If slow growth persists, however, the compounded effects can be devastating.

Poorer countries with high GDP growth rates are, in a sense, gaining on richer countries with lower growth rates, as shown in Figure 1.1.[3] Like a great race of sovereign tortoises, the nations of the world struggle to sustain fractional increases in their rates of growth, hoping to reduce the economic gap between themselves and the leaders. Although these incremental gains are not immediately observable, the aggregate effects over even one lifetime can be enormous. At the turn of the twentieth century, for example, the United Kingdom and Argentina headed the list of the world's richest nations, while Japan was nothing more than an economic backwater.

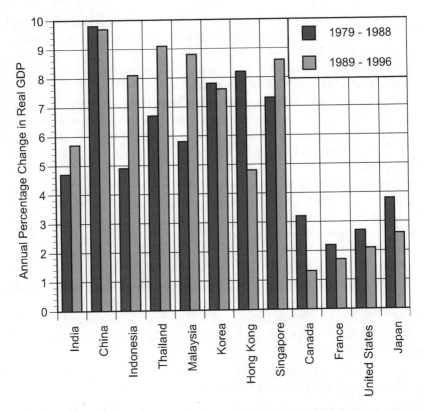

**Figure 1.1:** Average annual percentage change in real GDP for several developing and advanced economies from 1979 through 1988 and 1989 through 1996. Source: International Monetary Fund (1997, Statistical Appendix, Table A2, pg. 132).

Implicit in this recent global obsession with economic development is the tacit assumption that *growth is necessary.* In fact, this more accurately qualifies as a "deeply held belief." But is this seemingly obvious assumption valid? After all, there are some notable examples of cultures that have accepted extended periods of economic stagnation. Native Americans, aboriginal tribes in Australia and Africa, and even more "wealthy" societies such as the Balinese have achieved sustainable, zero-growth economies.[4] Why then is much of the modern world engaged in the desperate pursuit of growth?

The most obvious motivation for economic development is that the current wealth of many nations is inadequate to meet the basic needs of human existence. As of 1995, 76 percent of the world's population had an annual per capita income of under $3,020, and 53 percent subsisted on less than $730 per year.[5] With more than half of the world's people unable to enjoy even a minimal standard of living, the need for economic growth has become urgent from a global perspective.

Another driving force behind economic expansion is the continuing growth of the world's population.[6] Aggregate wealth must grow at a similar rate just to avoid backsliding. Unfortunately, high population growth rates are often characteristic of the same regions that suffer from crushing poverty, making their need for economic development all the more compelling.

The industrialized nations, on the other hand, enjoy a comparatively lavish standard of living. The basic needs of food and shelter have largely been met, and a large percentage of their populations live in comfort and safety. Could the more advanced economies not be satisfied with an acceptably high, but tediously stable, per capita GDP?

There is certainly no definitive answer as to why modern industrial societies hunger for economic growth. Sociologists suggest that our lust for wealth stems from a desire for increased control over our environment. Whatever the source of this hunger, innovations that increase the quality of life are coveted by almost every culture. A cure for a deadly disease, a labor-saving implement, or a new source of entertainment will carry a premium price in marketplaces the world over. Such opportunities for a richer and more comfortable life represent a strong incentive for people to work more efficiently, both to meet the demand for recent innovations and to earn enough money to purchase them. When these higher levels of productivity are aggregated, the result is an increase in the economic output of nations.

### Wealth Creation and Corporate Growth

There are two primary factors that currently shape the corporate strategies of most U.S. firms: the demands by shareholders for short-term profits, and the challenges of maintaining long-run competitiveness in rapidly changing markets. Such concerns over profitability and sustainability are by no means limited to relatively small start-ups. Over the last four decades, fully three-quarters of the firms that had been counted among the Fortune 500 have ceased to exist.[7]

The traditional model for U.S. corporate governance dictates that the shareholder comes first. In American capital markets, shareholders typically behave as investors rather than owners, showing little concern for the long-term welfare of the firms that utilize their funds. The narrow and generally short-term financial interests of shareholders tend to dominate over the concerns of other corporate stakeholders. Not surprisingly, executives have recognized that the only way to achieve a reasonable level of independent control over their firms is to appease these impatient interests through increased dividends and higher stock prices.[8]

Fewer than 5 percent of publicly traded American firms have a shareholder with greater than a 50 percent stake. Large mutual funds and pension funds are the dominant investors, and despite doing lip service to supporting long-term investment strategies, most fund managers are evaluated on their ability to generate quarterly gains. Hence, the prospect of capturing quick-turn profits remains attractive. Similarly, the executive compensation systems within many U.S. firms are strongly influenced by the impatience of shareholders, offering huge bonuses for short-term growth, and threatening unemployment for disappointing results.

In Germany and Japan, on the other hand, a large percentage of shares in publicly traded firms are held by banks and other financial institutions, or by firms in cross-shareholding agreements.[9] These arrangements accommodate a much longer time horizon for strategic action, as compared with the U.S. capital market.

In Japan, for example, half of the shares in public corporations are held by banks and insurance companies, with another quarter being held by firms. In Germany, the dominant shareholders are other businesses, which control 40 percent of all outstanding shares. The net result of this cross-shareholding ownership structure is that both Japanese and German firms can remain smaller and more focused than their American counterparts, and are free to pursue promising long-term investments in research and development.[10] The impact of this ownership structure carries through to the bottom line: American firms are criticized for achieving a shareholder return of "only" 10 to 12 percent, while Japanese firms can yield a measly few percent return for years without the slightest pressure to lay off workers.[11]

By comparison, more than half of the shares of publicly traded U.S. firms are held by individuals, either directly or through mutual funds, and another 25 percent are held by pension funds.[12] This ownership structure places firms in the untenable position of having to achieve rapid near-term growth at the expense of more promising investments with longer payback periods. Without the benefit of an extended planning horizon, U.S. companies are often forced into growth strategies involving either inappropriate acquisitions or unrestrained product-line expansion.

During the 1990s, American corporations have engaged in a frenzy of global mergers and acquisitions. These actions have been prompted in part by a positive economic climate and a belief that firms must be bigger and stronger to compete effectively in expanding world markets. A growing number of executives have become convinced that mergers will lead to improved efficiencies and greater shareholder value. This is a dangerous trend, particularly when acquisitions are based on a "hermit crab" mentality, in which smaller firms are appended to the acquiring firm without synergy or sense.[13]

Silicon Graphics Inc. (SGI), for example, has long been considered one of the shining stars of high technology, often being compared to Sun Microsystems as one of a new breed of highly innovative firms. Yet too-rapid growth and some ill-advised acquisitions have driven SGI's stock prices into a free fall. In 1997, SGI's market capitalization was $2.6 billion, compared with Sun's $12.3 billion. Just three years earlier, SGI had a higher market value than Sun. An insular attitude and questionable market strategies may deserve much of the blame for SGI's woes, but the ill-advised acquisition of Cray Research Inc. in 1996 was seen as a serious distraction from SGI's main lines of business.[14]

A similar fate awaited AT&T after its highly publicized purchase of National Cash Register (NCR). Under AT&T's questionable stewardship, the value of this formerly healthy enterprise was destroyed. On top of the roughly $5 billion that AT&T paid for NCR, a similar sum was reportedly lost to restructuring during its ownership. The result of this abortive merger was the eventual spinoff of the NCR unit, at half of its original value.[15]

Cisco Systems offers a more positive example of corporate growth through synergistic acquisition, having acquired fourteen leading-edge companies between 1993 and 1996. During that period Cisco more than doubled its sales and net income, and consolidated its lead in the data-networking market. Several of its acquisitions were designed to hedge its bets in the volatile arena of packet-switching standards. By purchasing both Granite Systems, a leader in gigabit ethernet switches, and StrataCom, an aggressive provider of products based on the asynchronous transfer mode (ATM) standard, Cisco was assured a technology base regardless of which standard won the market battle. These examples suggest that the key to success in a growth-by-acquisition strategy is identifying critical organizational synergies prior to purchase.[16]

Shareholder demands for short-term growth can also be a powerful driver for firms to enter international markets, often as an escape from stagnant domestic demand. In this context, global expansion may be undertaken either as an offensive action intended to inspire shareholders into taking a new look at a mature firm or as a defensive strategy to avoid being locked out of foreign markets in the future. The rapid expansion of firms into foreign markets often elicits a defensive response from global competitors, however, and may also catalyze nationalistic sentiment within importing nations. Given the powerful effects that foreign enterprise can have on developing economies, it is not surprising that host nations look upon a foreign invasion of their domestic markets with a wary eye.

In summary, the unrestrained growth of firms is not necessarily a good thing from the standpoint of long-term competitiveness. Growth that is forced in response to the pressures of entrepreneurial ambition or impatient capital, rather than pulled by market forces, will almost always lead to disaster. Despite overwhelming evidence that short time horizons have constrained the investment strategies of American firms, there has been little improvement in the corporate myopia that continues to limit the global competitiveness of American enterprise.

## The Limits to Input-Driven Growth

On a recent visit to Asia, I met with a representative of Singapore's Economic Development Board. During the course of the meeting, I asked a seemingly innocuous question: "What is the availability of technically skilled labor in Singapore?"

After an uncomfortable pause, he informed me that I would "experience difficulty" in finding appropriate technically skilled labor. Furthermore, I was told to expect difficulty in locating *unskilled* labor as well. When asked what the current unemployment rate was for Singapore, the representative responded, "Zero."

The discussion that followed was an object lesson in the limitations of input-driven growth.[17] Singapore was fresh out of labor, and labor is a primary input into what economists call the *production function*. As I will soon demonstrate, an inadequate supply of this vital ingredient could have ominous ramifications for Singapore's continued economic growth.

### Accounting for Growth

Basically, the production function can be thought of as a "black box." Inputs are loaded into the box, including such things as labor, raw materials, and capital. The box then transforms these inputs into an economic output that is of higher value than that of the collective inputs.[18] The difference between the value of the inputs and the value of the output is the result of *productivity*.

The relationship among inputs, outputs, and productivity is the focus of an arcane field of macroeconomics known as *growth accounting*. For a given production function, an equation can be constructed that describes the optimal proportions of labor and capital inputs needed to yield a specific change in output.[19] Adding a quantity of labor without an appropriate increase in the amount of capital, or vice versa, will yield a less-than-optimal result. This effect is known as *diminishing returns to scale*.

A brief example is in order. Suppose that you own a factory that produces widgets (what else?). Based on your current technology, manufacturing a widget requires a single worker toiling at a single widget-making machine for one day. Since the demand for widgets is high, you currently have one hundred employees working at an equal number of machines.

One day you receive a new order for five more widgets per day. Unfortunately, capital is scarce, so you decide to add five more workers, but purchase only a single additional widget machine. Will you be able to meet the new demand?

Since these new workers are lacking the proper proportion of capital equipment, they cannot be as productive as the original hundred. Similarly, if you happened to be flush with capital, but workers were scarce, you might have responded by buying five new machines and adding only a single worker. The result is the same; by not adding factor inputs in the proper proportions, you suffer diminishing returns to scale.

Now where does productivity enter into this? Suppose that you discover a new type of automated widget machine. Although the capital investment would be five times as great, the new machine can produce five widgets per day with only a single worker in attendance. By purchasing this machine, you would be able to meet your new demand even if workers are in short supply because, due to improved technology, *the productivity of the new worker is five times greater.*

The important thing about productivity is that, unlike factor inputs, which must be supplied continuously, the gains in output resulting from increased productivity are *sustainable.*[20] Once you have learned about the new automated widget machine, you will never go back to the outdated ones. This is the power of growth based on technical progress.

### The Asian Miracle Demystified?

How does diminishing returns apply to Singapore? As has been the case with the other Asian tigers, Singapore has experienced rapid economic growth over the last three decades. With an average growth rate for that period of more than 8 percent per year, it is representative of what has come to be known as the "Asian miracle." But how was this growth accomplished?

In a controversial series of papers, Alwyn Young suggests that the vast majority of Singapore's dramatic economic growth was due to huge increases in factor inputs, while productivity growth has been *essentially zero.*[21] Over the period from 1960 to 1990, Singapore's share of capital investment as a percentage of GDP rose from around 11 percent to more than 40 percent. Likewise, the participation of its population in the workforce rose from 27 to 51 percent. Young's analysis suggests that the primary engine for Singapore's economic growth has been increased factor inputs, rather than increased productivity.[22]

Assuming this analysis is correct (and there are strong indications that it is not), why could Singapore not continue to grow indefinitely using this economic model? Without an adequate labor force to support the introduction of new foreign investment, the marginal productivity of new capital will rapidly decline, and Singapore will suffer diminishing returns to scale. To put this in more practical terms, a firm might be reluctant to establish a headquarters or an R&D center in Singapore, knowing that they might have considerable difficulty in staffing it.[23]

The path to sustainable growth for Singapore is through increased productivity, a fact that is not lost on its Economic Development Board, as discussed in Box 1.1. But what about the other high-growth economies of Asia. Are they really "paper tigers," destined to experience slow or stagnant growth once they have absorbed all the capital they can efficiently deploy?

There is little doubt that increased inputs of labor and capital have played a significant role in the extraordinary development of the newly industrialized countries. It is unlikely, therefore, that the astonishing growth they have enjoyed in recent years can continue. A shift from rampant capital accumulation to balanced growth of both capital intensity and productivity is needed for sustained economic expansion.

## Box 1.1: Singapore's Quest for Productivity[35]

Over the last several decades, Singapore has positioned itself as a regional headquarters for multinational corporations. In fact, a significant proportion of its growth, particularly in the high-technology sector, has been fueled by foreign investment. The legacy of this high-growth strategy is that Singapore is struggling to develop domestic technology industries that can compete in global markets.

In an effort to sustain Singapore's impressive economic performance, its Economic Development Board has adopted a growth strategy that emphasizes the development of human capital and productive technology. Universities in Singapore currently graduate roughly five thousand engineers and scientists annually, and they plan to convert several schools to "all-technical" colleges. In addition to developing local talent, they are aggressively pursuing foreign experts by relaxing immigration rules and providing inexpensive housing. Offers of R&D funding for new product development under its Innovation Development Scheme represent an additional enticement to global entrepreneurs. Prime Minister Goh Chok Tong has announced that developing Singapore into a "learning nation" is his highest domestic priority.

The dilemma for Singapore is that they are trapped between two economic models. Recent labor shortages have driven wages beyond the point of competitiveness for semi-skilled manufacturing. Several companies, including Packard Bell, Dell Computer, and Iomega, have passed up Singapore in favor of the abundant low-cost labor available in Thailand and the Philippines. Meanwhile, Singapore's ability to expand into high-productivity sectors is limited by a shortage of trained technologists.

Singapore is struggling mightily to convert itself from a low-cost manufacturing hub into the world's biggest high-tech consultancy. For now, however, manufacturing is still Singapore's economic lifeblood. To remain competitive in the near term, they have relaxed levies on the use of unskilled foreign workers and formed a partnership with Indonesia to expand its manufacturing base onto two nearby islands, Bintan and Batam, that offer new sources of low-cost labor.

On the other hand, a large proportion of Japan's rise to economic prominence has been driven by productivity increases. Likewise, Hong Kong has shown an ability to increase productivity at substantial rates. Moreover, the recent merger of Hong Kong and China represents a formidable economic union: a nearly infinite reservoir of labor combined with a considerable indigenous technology base.

Although growth in China is not likely to continue at double-digit rates, this industrializing nation has not even begun to exhaust its potential for accumulating capital and expanding productivity. We should not expect to see the effects of diminishing returns limiting the economic growth of China anytime soon.

### Ramifications for the Growth of Firms

It is not surprising that diminishing returns to scale can also have a major impact on the growth of firms. Executives in high-growth companies are frequently confronted with the constraints of insufficient labor or capital. By applying the production-function concept to individual firms rather than nations, a number of more subtle input requirements rise in importance. Whereas macroeconomic models consider only the "generic" inputs of materials, capital, and labor, specialized inputs become significant when evaluating productivity at the level of the firm.

The most obvious among these specialized inputs is skilled labor. The production functions of technology-intensive industries often demand labor inputs from dozens of narrowly defined specialties. As the uniqueness of the required expertise grows, the availability of labor with the appropriate skill level diminishes. Thus, technology-intensive firms must search relentlessly for the talent they need to sustain growth. When desperation drives firms to hire inadequate talent, all other inputs to production (meaning all other areas of specialization) suffer the effects of diminishing returns to scale.

Along similar lines, the potential for increasing productivity through the implementation of new technologies may be limited by the availability of specialized inputs to production. Typically, advanced technologies carry a "skill premium," implying that there will be at least a temporary shift in the skill level of labor required to produce high-technology products or utilize new process innovations.[24]

Therefore, as an indirect result of diminishing returns to scale, smaller firms may be at a substantial disadvantage in the adoption of productive new technologies. Larger firms can more easily absorb the temporary drain on working capital and skilled labor required to implement these innovations. Small firms are often structured around finely tuned production functions in which every employee plays a vital role. With little excess capital or underutilized expertise to support technology upgrades, small firms may be reluctant to embrace productive innovations.[25]

A final way in which growth can be limited by the lack of proper inputs involves the rates of return achievable from investment in research and development. The stock market's valuation of R&D investment has changed considerably over the last three decades. In the period from 1973 to 1983, intangible R&D assets and tangible capital were perceived as being of equivalent value. By the mid-1980s, however, this relationship broke down, with R&D market valuation declining by more than a factor of three.[26]

What could cause such a decline in perceived value? There are several reasonable explanations, including reduced investor confidence during the restructuring wave of the 1980s. It is also possible, however, that the fall in market valuation of R&D investment was due to investors' recognition of diminishing returns to scale. For a research investment to be profitable, it must yield either increased pricing power in the marketplace or a reduction in the costs of production. If there are insufficient opportunities for R&D to be productive in this way, additional investment dollars will suffer diminishing returns to scale.

How does this explain the decline in stock market valuation of R&D investment? The 1980s were a time of defense-budget expansion and "Star Wars" technologies. Engineering schools were stretched to their limits to meet the demands of a burgeoning aerospace and defense sector. The opportunity to work on leading-edge technologies (under conditions of virtually unlimited funding) was irresistible to many of our nation's top scientists and engineers. This draw of technical talent into defense research may have left commercial firms with a paucity of this precious input to production. Although funds continued to be invested in research and development during the 1980s, the lack of adequate talent to exploit those investments may have contributed to a decline in R&D rates of return.

## A Brief History of American Productivity

Even the casual observer will note that there are substantial differences between the economies of Singapore and the United States. Yet we are subject to the same constraints of diminishing returns. Our recent extended growth cycle has been characterized by exceedingly low unemployment. With the percentage of unemployed workers falling below 5 percent, there have been continuing fears of inflation; as the labor pool dries up, the demand for scarce resources will exert upward pressure on wages, and ultimately prices. At least that's how it has worked in the past.

The robust business expansion of the 1990s, however, has been characterized by both price stability and rapid growth, despite sustained low unemployment. In fact, prices have been falling recently in many high-technology sectors, pulling down the total rate of inflation from 3.1 percent to 2 percent in 1997.[27] The popular explanation, extolled by economists and the media alike, is robust growth in productivity. Even the chairman of the Federal Reserve Board, Alan Greenspan, has hinted publicly at an "emerging digital economy" in which the productivity gains from information and communications technologies may finally be yielding sustainable, noninflationary growth.[28]

Since the advent of the personal computer in the late 1970s, business analysts have predicted a surge in productivity driven by the efficiency and power of information technology. Yet during the 1970s and 1980s the rate of American productivity growth fell to its lowest levels of the twentieth century. Is the long-anticipated productivity transformation finally upon us, or are we still trapped in an extended period of lackluster performance? Moreover, if information technology is finally paying productivity dividends, how can we ensure that this trend continues?

## The Measurement of Productivity

One would assume that productivity, a parameter that is so vital to our economic well-being, would be thoroughly understood. Unfortunately, this is not the case. Despite its ubiquitous use in economic and business literature, there are difficulties associated with the measurement of productivity that have plagued attempts to accurately quantify technology-driven growth.

The first and most famous study of the sources of growth in the United States was performed by Nobel laureate Robert Solow shortly after World War II.[29] By applying the methods of growth accounting to the period from 1909 to 1949, Solow sought to quantify the contributions of capital growth and technical progress to America's significant economic expansion.[30] His analysis suggested that roughly 80 percent of the growth in per capita income during that period could not be explained by increases in the capital stock. Although measurement error and other artifacts might have been present, this pioneering study supports the conclusion that the majority of real economic gains experienced in the United States during the first half of this century were due to technical progress.

This came as something of a surprise to many economists. Until Solow's work was published, it was widely believed that the primary driver for economic growth was the accumulation of capital. Although it was generally accepted that technological progress played an important role in growth, it was assumed that innovations occurred somewhat randomly, as an outgrowth of basic scientific research. That technology played such a central role in driving our economic expansion had not been anticipated.[31]

A considerably expanded study was subsequently performed by Edward Denison, covering the period from 1929 to 1982.[32] In this analysis, the percentage contribution of technical progress to economic growth was found to be somewhat lower (roughly 65 percent). Solow's conclusions were supported, however; productivity gains had made a significant contribution to America's economic expansion.

Despite this solid circumstantial evidence, there have been persistent problems in quantifying the contribution that technical progress has made to total-factor productivity. (Gains in *total-factor productivity* enhance the output potential of both capital and labor. Gains in *labor productivity*, on the other hand, refer to an increase in labor's contribution to output, which can result from either improved technology or capital accumulation.)[33] Empirical data suggest that capital accumulation and improved technology are closely interrelated. But which is the cause and which is the effect? Does improved technology motivate increased capital investment, or do higher levels of capital offer more opportunities to "learn by doing," as some have suggested?[34] Perhaps the causality works both ways.

More fundamentally, the methods of growth accounting may simply be incapable of quantifying the economic contributions of technical progress through innovation.[36] This analytical tool depends on some basic assumptions, such as constant returns to scale and perfect competition, that are not valid for high-technology enterprise, as I will demonstrate in Chapter 2.

Even the empirical data used in growth studies may grossly understate the contribution of technology to economic expansion.[37] The traditional analytical approach involves subtracting from economic output whatever can be attributed to accumulated inputs, such as capital and labor, and equating the residual to productivity.[38] The problem with this approach is that both the data and the choice of how to decompose them are highly subjective.

Government data on GDP growth fails to capture many of the benefits of technological innovation, such as increased product variety, performance, and quality. The current inflation-adjusted selling price of a Pentium II personal computer, for example, is roughly similar to that of an early 80286 model. Hence, these products would make equal contributions to real GDP in the years they were produced. The fact that the modern version can provide hundreds of times the power and productivity of the earlier model is not captured in government statistics.[39]

Although the quantitative measurement of productivity remains elusive, we can certainly agree on its qualitative contribution to our current standard of living. A brief look back into our economic history will demonstrate that technical progress has fundamentally altered both the quality of our lives and our prospects for the future.

### From Humble Beginnings

By the middle of the twentieth century, economists began to recognize that something extraordinary was happening to the standard of living throughout the industrialized world. Not only had per capita wealth reached unprecedented heights, but the rate of growth showed little sign of diminishing. This may seem routine from a modern viewpoint, but nothing of this sort had occurred before in all of history. To put our prosperous times in perspective, I will offer a few examples to illustrate how much our quality of life has improved over the last century as a result of technology and innovation.[40]

Most of us hold a romanticized view of what life was like prior to the mid-nineteenth century. Cinematic images of Elizabethan England or Colonial America belie the crushing poverty that was almost universal prior to the two industrial revolutions. In fact, the standard of living for much of the world had changed surprisingly little over the fourteen centuries from the fall of Rome to the eighteenth century. There were some important innovations, particularly during the Renaissance, including movable type, clocks, navigational instruments, and firearms. Unfortunately, these inventions failed to improve labor productivity, and were accessible only to the relatively rich.

During the eighteenth and nineteenth centuries, most people spent their lives in a desperate struggle against disease and starvation. Without means to preserve food, families were entirely dependent on the success of regional crops. As a result, poor weather, pestilence, and overworked soil threatened mass famine on a shockingly regular basis.

In 1850, the average life expectancy for an American was roughly forty years. During this brief lifetime, a worker would spend, on average, sixty hours per week at work. Farmers toiled from dawn until dusk using manual implements powered by animals or human sweat. In the rapidly growing cities, factory laborers were subjected to unsafe working conditions, poor ventilation, and repetitious drudgery. A standard workman's budget of the time allocated almost 50 percent of his earnings for food, and more than 90 percent was needed to provide the essentials of food, clothing, and shelter. There was no hope of vacation or retirement; most laborers worked themselves into the grave.

Prior to the Civil War, Americans endured a median standard of living roughly equal to that of China or the Philippines today. Beginning in about 1870, however, something remarkable began to happen. After fourteen centuries of virtual economic stagnation, the more industrialized countries of the world began to experience a steady increase in wealth that persists to this day. In the eleven decades from 1870 to 1980, America's real per capita output increased by more than 700 percent, as shown in Figure 1.2.

The source of this spectacular growth is attributable to that elusive concept of productivity. The advent of mass-production techniques, including interchangeable parts and the assembly line, provided the means to multiply labor and increase specialization. These breakthroughs brought the price of manufactured goods down to affordable levels. Henry Ford stated in the early 1900s that his goal was to "democratize the automobile." In fact, the contribution made by Ford and other great industrialists of the time was to harness *innovation* and make it *productive.*

The average rate of productivity growth in the United States over the last century has been roughly 1.5 percent.[41] Although this may not seem particularly impressive, compounded gains have resulted in a twentyfold increase in total-factor productivity and a dramatic improvement in the living standards of most Americans. The benefits of this surge in productivity have not been equally distributed, however. Even today, the poorest among us still live in conditions that are no better than those endured a century ago.

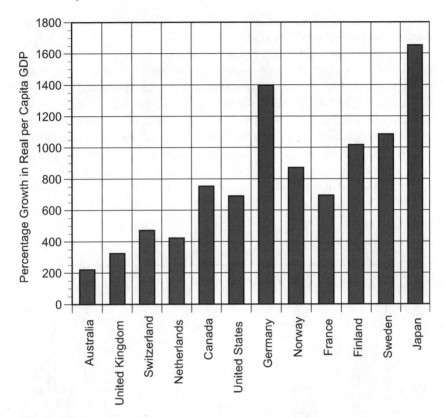

**Figure 1.2:** Percentage growth in real per capita GDP for several advanced economies over the period from 1870 to 1980. Source: Baumol, W. J. et al. (1989, pg. 13). Data used with permission.

### Is American Productivity in Decline?

There has been much debate in the popular press regarding America's declining competitiveness in world markets. These concerns are supported by evidence that the rate of productivity growth of U.S. industries slowed significantly over the last several decades. Although there is a consensus among economists that a slowdown did occur during the 1970s and 1980s, a careful analysis of the data demonstrates that there may be no cause for panic.

The first conclusion that can be drawn from a detailed analysis of long-run economic data is that total-factor productivity is cyclical.[42] This is largely a result of sensitivity to economic fluctuations; productivity tends to rise sharply as an economy emerges from a recession and to fall dramatically during a downturn.

Since the U.S. economy has been characterized by recurrent business cycles, productivity has tended to react in kind.[43]

A clearer picture of the long-run behavior of productivity can be formed by examining data in which the jagged cyclical fluctuations have been averaged out, as shown in Figure 1.3.[44] A sharp surge in productivity was experienced after World War II, followed by a decline beginning near the end of the 1960s. There was a pronounced slump during the 1970s, followed by an upturn, particularly in the manufacturing sector, beginning in the mid-1980s. Although the two-decade decline experienced during the postwar era is of some concern, there is evidence that this was a temporary phenomenon. Indeed, several recent studies have concluded that America now appears to have returned to its historical rate of productivity growth.[45]

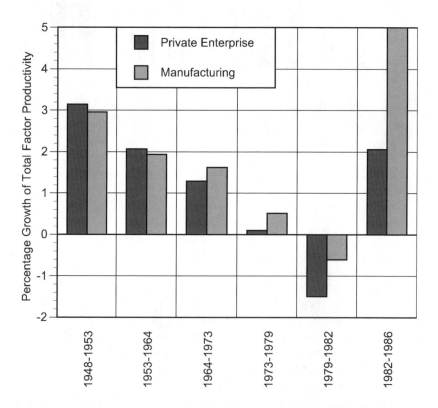

**Figure 1.3:** Annual growth rate of U.S. productivity for all private enterprise and for the manufacturing sector over the period from 1948 to 1986. Note that time periods differ in length due to the nature of the available data. Source: Denison, E. F. (1989, pg. 5), as compiled from U.S. Bureau of Labor Statistics, "Multifactor Productivity Measures - 1986."

A reasonable explanation for the behavior shown in Figure 1.3 might go as follows. The period from the 1930s to World War II was typified by very low productivity, resulting from the economic shock of the Great Depression. During and shortly after the war, there was a great surge of pent-up demand, coupled with a shortage of labor (which tends to bias productivity data upward). This productivity surge continued well into the 1950s, fueled by an unprecedented global demand for American manufactured goods. During this period, the United States became the undisputed world leader in virtually every industrial sector.

This postwar surge was something of an anomaly, which prompted a corrective backlash of relatively depressed productivity growth during the 1970s. The same period was marked by increased government consumption (which tends to crowd out private-sector capital investment), several severe recessions, oil-price shocks, and an increase in environmental regulation.[46] In addition, during this period as many as 20 percent of America's best and brightest technologists were deflected from the commercial sector into the military-industrial complex.[47] Without passing judgment on the value of this allocation of resources to our national security, it was clearly not the best policy from an industrial productivity standpoint.

In summary, although a downturn in productivity did occur in the 1970s and 1980s, we have experienced a significant rebound over the last several years. Output per workhour rose by 1.7 percent in 1997, following a 1.9 percent jump in 1996. Productivity in the manufacturing sector did especially well, registering a 4.4 percent gain in 1997, up from 3.7 percent in 1996.[48] These encouraging data suggest that America has escaped the productivity doldrums, and with a new generation of highly creative and motivated entrepreneurs taking center stage in American business, the future looks bright. Indeed, we may be at the threshold of a new renaissance period for American ingenuity and productivity.

### The Convergence Club

For most of the last half-century, the United States has enjoyed a commanding global lead in productivity, as measured by GDP per workhour. This remarkable record, however, could be nearing an end. The slowing of America's productivity growth during the 1970s and 1980s has eroded our position relative to other advanced economies. Even today, despite a strong recovery, our rate of productivity growth continues to lag behind that of several industrialized nations. If this trend persists, the United States could easily slide into economic mediocrity by early in the twenty-first century.

Fortunately, there may be some room for optimism. Much of the growth recorded in the newly industrialized nations, for example, has been driven by capital accumulation, as discussed previously. Once these countries catch up to global levels of capital intensity, we can expect diminishing returns to slow their progress. More important, there is general agreement that at least some of the gains in productivity experienced by other countries during the postwar period were a direct result of the imitation of American technology.

A close examination of the historical behavior of economic growth for both advanced and developing nations reveals a pattern that has tremendous importance for American competitiveness in world markets. Recent studies have observed that classical theories can no longer adequately explain the nature of global commerce. This new evidence suggests that the patterns of international trade and competition are increasingly determined by relative levels of total-factor productivity.[49]

Since 1870, productivity per worker has been steadily rising in virtually all nations. If we examine historical data after controlling for levels of capital investment, there is a general trend toward convergence in the productivity of all economies, from the richest to the poorest.[50] The relative rates of convergence, however, are not uniform. In fact, there appears to be a select group of the more-advanced economies that are converging rapidly, while the less-developed countries languish far behind. What factors determine membership in this privileged group? More important for American industry, can the United States continue to stay near the front of the pack?

Over the 110-year period from 1870 to 1980, the ratio of productivities between the highest and lowest of the leading industrial economies has steadily decreased, as shown in Figure 1.4. In 1870, Japanese labor was roughly one-eighth as productive as Australian labor (then the leader in this group). By 1980, this 8:1 ratio was reduced to 2:1, with the United States in the lead and Japan again bringing up the rear.[51]

One plausible explanation for this behavior suggests that industrialized nations are approaching similar levels of productivity due primarily to imitation and technology transfer. Presumably, countries that lag relative to the leader would tend to show higher growth rates, since they have the most to gain from imitation. This so-called *convergence hypothesis*, however, provides no explanation for the exclusion of all but a select few countries from the convergence club.

If the hypothesis of convergence through imitation and technology transfer is to explain the observed data, we must assume that the transfer of technical information has become increasingly more efficient over the last century. This seems reasonable, given dramatic improvements in communication and information technologies throughout the industrialized world. Recent estimates suggest that in some high-technology industries, the equilibration period between the implementation of a new productive technology in an originating country and its adoption by foreign competitors can be as little as six to eighteen months.[52] This rapid diffusion of innovations provides a plausible explanation for the relatively tight productivity grouping among advanced economies that we see today.[53]

Several countervailing studies have contended that much of the economic growth observed over the second half of the twentieth century should be attrib-

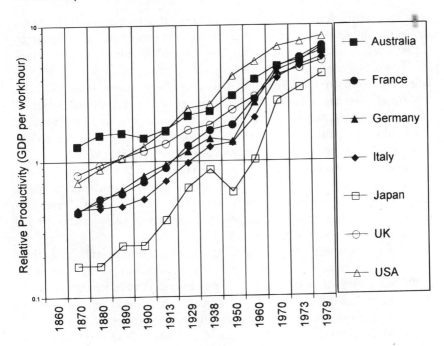

**Figure 1.4:** Labor productivity in GDP per workhour for seven leading industrial countries from 1870 to 1980. The convergence effect is clearly evident, with the ratio of relative productivities decreasing from 8:1 to 2:1 over this eleven-decade period. Source: Reprinted from *Phases of Capitalist Development* by Angus Maddison (1982) by permission of Oxford University Press.

uted to capital accumulation, rather than technical progress.[54] Surprisingly, this fundamental disagreement regarding the contribution of technology has persisted for more than four decades, another indication of the tremendous difficulty associated with isolating technology's role in economic growth.

While there is certainly no consensus, it appears from recent empirical analysis that capital investment and innovation are inexorably linked.[55] If a firm were to outfit a factory today with equipment from the 1950s, for example, it seems unlikely that it would be competitive, regardless of the capital-to-labor ratio. Likewise, access to new industrial technology is of little use without sufficient capital available to implement it. The investment of capital specifically to exploit a recent innovation is referred to as the *embodiment effect*. A symbiotic relationship between capital and innovation based on the embodiment effect is the most persuasive model for growth in rapidly changing high-technology industries.

Although the direction of causality is not clear, it seems reasonable that innovations are the driving force behind rapid productivity growth, particularly in high-technology industries. Capital accumulation plays a critical supporting role, enabling maximum benefits to be derived from productive technology. From a dynamic standpoint, one can imagine that capital expenditures are triggered by the introduction of recent innovations. The effects of diminishing returns to capital apply only if productivity remains stagnant. As new innovations become available, the potential for capital investment to achieve high returns is restored.[56]

Since Japan has been the focus of much paranoia with respect to lagging American competitiveness, it is worthwhile to consider its case in a bit more detail. Until 1990, it was widely assumed that Japan would continue its explosive economic growth, overtaking the United States by the end of the millennium. The melancholy performance of the Japanese economy in recent years has deflated that prospect considerably. Nonetheless, Japan's industries reached near parity with their American counterparts by the mid-1980s, and today hold a commanding productivity lead in critical industrial sectors such as heavy machinery and consumer electronics.

Japan's productivity growth enabled the manufacture of globally competitive products. Their share of OECD exports grew from 14 percent in 1970 to 23 percent in 1987.[57] Surprisingly, these gains were largely at the expense of Europe as opposed to the United States. Although the largest gains in export share were registered in capital-intensive heavy industries, Japan's relative productivity increased in all major sectors when compared to the United States, as shown in Figure 1.5.[58] The recent slowing of the Japanese economy has been paralleled by a sharp decline in that nation's rate of productivity growth. In fact, U.S. manufacturing has gained some respectable ground on their Japanese rivals over the last few years.

Will rapid productivity growth return to Japan? It now appears that its intimidating expansion during the postwar period was largely a result of fairly benign catch-up, both in technology and capital accumulation. In fact, Japan's postwar economic expansion provides an excellent example of the complementary nature of capital investment and imitation-based technical progress. Unfortunately for Japan, the imitation well has gone nearly dry and capital deepening has reached the threshold of diminishing returns. In short, their future growth prospects, as well as our own, depend on increasing total-factor productivity through innovation and invention.

### What Fate the Laggards?

The fortunate few countries that are members of the convergence club share a common trait: They are able to benefit from technological advances occurring in other nations. Although capital deepening may accelerate the process, it is clear that the criteria on which "membership" is based involve the ability to procure, imitate, and absorb advanced technologies. This capacity appears to be lacking in the least-developed countries, which have failed to take advantage of available technology despite their pressing need.

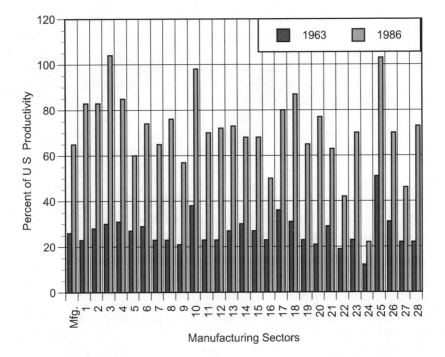

**Figure 1.5:** Japan's productivity in twenty-eight manufacturing sectors, and the average for all manufacturing, as compared to the United States, for the years 1963 and 1986. Source: Dollar, D. and E. N. Wolff (1993, pg. 12, Figure 1.3). Data used with permission.

In 1952, Alexander Gerschenkron noted that a nation that lacked basic technological capabilities enjoyed the greatest opportunities to borrow technology from developed countries.[59] He also observed, however, that the "advantages of being backward" applied only if a nation was not too far behind.

Recent studies offer two primary explanations for why some countries fail to hitch their wagons to the technological leaders. The first involves the degree to which a country is integrated into the world economy. An outward orientation with respect to both imports and foreign investment offers far greater opportunity to transfer technology, and helps develop a sophisticated domestic market to encourage local innovation.[60]

A second important factor is the level of education of the general citizenry. Studies have indicated that high levels of education in the United States have made a significant contribution to our strong economic growth, as has been the case for the other developed economies.[61] It is reasonable to presume that a certain amount of education (say at the secondary level) is required for a country to take full advantage of the flow of technical information available to it.

There are some notable examples of rapid productivity convergence being facilitated by a combination of an open economy and rising levels of education. The newly industrialized countries share a high degree of openness to trade and investment, and place tremendous emphasis on both general literacy and higher education. Even traditionally closed economies such as China's have achieved rapid gains through education and increased openness, as is demonstrated in Box 1.2; in this revealing example, a Chinese computer company rose from obscurity to become a threat to the likes of IBM and Compaq in less than two years.

The preceding discussion provides a reasonable explanation for why productivity convergence has been bimodal. Although the remedy for lagging productivity in developing countries is arduous, there are some shining examples of nations that have bridged the gap and become fledgling members of the convergence club.[62] But is it possible for a country to be ostracized from this elite group? What would be the fate of an advanced economy that consistently lagged in productivity growth?

An advanced economy with lagging productivity growth will not feel the effects overnight. There will be no sudden price shock for consumers, or even a noticeable jump in unemployment. In fact, it is the lack of obvious symptoms that makes long-term lags in productivity growth so insidious.

As I will demonstrate in Chapter 2, the primary effect of lagging relative productivity is reduced competitiveness in world markets. As a nation's industries slip farther behind those of its trading partners, its goods cease to be price competitive. If this happens in a single industry, wages must fall to compensate. When aggregate productivity across many sectors lags behind that of trading partners, the more likely result will be a currency devaluation.[63] In effect, this is the same as reducing the real wages of every worker in the country.[64]

A currency devaluation may damage national pride, but it can prove to be a sensible short-term remedy. Provided that wages are brought into line with world-market pressures, a nation's overall employment should not be affected by slow productivity growth. It is likely, however, that the nature of the work performed in that country will change.

If an advanced country's productivity growth were to consistently lag behind its trading partners, its industries would cease to be competitive in producing high-technology products. Market pressures would drive that country to export products that are competitive at lower levels of productivity. High-wage workers would be forced to take relatively unchallenging jobs at lower pay. As the following discussion will demonstrate, in an open world economy there is a direct linkage between the relative productivity of a nation's workers and the standard of living they can expect to enjoy.

## The Productivity Paradox

The patterns of trade in the global economy are determined by the varying ability of nations to provide goods at the lowest relative cost. Classical trade theory assumes that this cost advantage results from a relatively rich endowment of input factors such as labor and capital, or from the possession of unique natural resources. Each country will tend to specialize in what it can produce most cheaply relative to world prices, and will import those products which are costly to produce at home.

Classical theory, however, assumes that productive technologies are equal in all countries. While this assumption is generally true with respect to aggregate productivity among the world's advanced nations, there is growing evidence of specialization in high-technology sectors.[65] Under these conditions, classical thinking must be set aside if we are to understand how trade will progress among the advanced economies.[66]

In recent years, trade patterns among the industrialized nations have been increasingly characterized by intra-industry trade; countries with similar factor endowments are importing and exporting goods within the same industrial categories. This trading of "like goods" is an unexpected result from a classical standpoint, but can be explained by subtle differences in technological capability among nations within a given industry sector.

With wage levels being roughly equal throughout the developed world, goods that are desirable in one nation will likely be popular in other advanced economies as well. This is particularly true of technology-intensive products such as automobiles, telecommunications and medical equipment, entertainment products, personal computers, and office machines.

Once a global market is established for these high-tech goods, firms from various countries vie for competitive advantage. As the market matures, these firms will tend to specialize in a particular niche within their industry. Japanese automakers, for example, penetrated the world market through the introduction of low-cost, efficient transportation vehicles, and gained market share by delivering exceptional quality and reliability. German companies, on the other hand, have focused on the high-end automotive market, targeting enthusiasts who are willing to pay top dollar for highly engineered, sophisticated machines.

Each nation's industries emphasize inherent strengths in their industrial core to achieve world leadership in a specialized niche market. Today, imported cars from many countries are available to the world's consumers. Low-wage countries such as South Korea still serve the low-cost, utilitarian segment of the market. Italian automakers hold a commanding lead in unaffordable exotic cars. Germany, Japan, and the United States compete in various subsegments of the middle and high-end market.

Although aggregate trade flows may appear on the surface to follow classical theories, closer inspection reveals an emerging structure to international trade between technological equals. Industries within the advanced nations are specializing in market segments in which they have a subtle, but potentially sustainable, technological advantage.

---

## Box 1.2: The Legend Group - A Study in Convergence[67]

In just a few short years, China's domestic personal computer industry has grown from producing obsolete, low-quality clunkers to besting America's giant multinationals. Legend Holdings Ltd. of Beijing delivered more PCs into the Chinese market in the first quarter of 1997 than any other manufacturer, surpassing the likes of IBM, Compaq, and AST. This represents a startling rise for a company that was ranked sixth in the domestic market just two years earlier.

Exploiting its inherent labor-cost advantage, Legend has been able to undercut price leaders such as Compaq by more than 20 percent. With a network of more than one thousand retail outlets nationwide, it has developed the most effective distribution channel in the industry. In addition, it is currently exporting intermediate PC components to twenty-six countries, thereby benefiting from significant economies of scale.

All of this would mean little, however, if its products were not state of the art. Rather than depend on the usual channels for technology transfer, Legend established a design center in California's Silicon Valley in 1993, the first Chinese PC manufacturer to do so. In this way it was able to rapidly translate technological advances, and adapt them for its domestic customers.

The increasingly open Chinese market has been the catalyst for aggressive competition by domestic manufacturers. The market for PCs in China is expected to grow from $2.1 billion in 1996 to greater than $9 billion in 2001. For American companies to retain a significant share, they must carve out niches that are not heavily dependent on price competition. Hewlett-Packard's PC sales in China, for example, showed an 83 percent jump in the first quarter of 1997, due to its focus on high-end machines and its international reputation for quality and service.

Legend's ascendancy should be taken as a warning shot to America's technology industries. It appears that the rate of productivity convergence at the industry level can now be measured in months rather than years.

---

### Implications for American Competitiveness

What are the implications for American industry of the recent productivity convergence among industrial nations? Discussions of "national competitiveness" in recent days have focused primarily on increasing access to foreign markets and improving global market share of U.S. exports. Although these issues are critically important, this focus on "external factors" fails to acknowledge the responsibility of American firms to achieve world-leading productivities.

With respect to global competitiveness, the most fundamental issue may be simply defining the term. Does competitiveness mean increased exports, higher wages, improved bilateral trade deficits, enhanced corporate profits, or all of the above? The literature offers a virtual continuum of definitions, each reflecting the intellectual disposition of the author. The following statement captures the essence of these various perspectives:

> *An industry or firm is globally competitive if it can sustain a substantial international market share, while achieving high employee wages and attractive returns to capital.*

How do American industries fare against this definition of competitiveness? A good indicator is the share of world manufacturing exports held by American industries with respect to other OECD countries. It is evident from historical data that the United States has lost ground over the last several decades to Japan and several other industrial nations. Our export share declined from 18 percent to 13 percent over the period from 1962 to 1985, with losses in chemical products and steel leading the way. In fact, the United States had few export success stories during this period, except for war munitions.

The story is more encouraging in high-technology exports, as will be demonstrated in Chapter 3. It is also important to temper this evidence by noting that exports have, to a significant extent, been supplanted by the foreign direct investments of American firms. In fact, according to the *World Competitiveness Yearbook* for 1997, the United States ranked first among exporting nations in international competitiveness, far surpassing Japan, which plunged to eighteenth position.[68] Despite this recent resurgence, however, there is still substantial cause for concern, given the critical role that export-intensive industries play in achieving a high standard of living.[69]

In high-technology industries, competition occurs primarily between advanced nations with similar wages and levels of capital accumulation. Since all industrial economies have equal access to low-wage, semi-skilled labor in developing countries, optimizing labor costs alone cannot offer a sustainable edge. With all "generic" inputs being equal, competitive advantage in global markets will ultimately depend on achieving the highest levels of total-factor productivity.[70]

It is important for Americans to recognize that our mid-twentieth-century preeminence in productivity was a manifestation of the unique conditions following World War II. Today, we must accept that our world is populated by a growing cadre of technological equals. The United States is well positioned to benefit from the opportunities for innovation and specialization that these sophisticated global markets represent. To do so, we must learn to stake out our niches, take advantage of our strengths, and hold the productive high ground for all it's worth.

## Opportunities and Pitfalls on the Pathway to Growth

The proliferation of information and automation technologies that began in the early 1970s brought with it a great deal of anticipation among industrial stakeholders. Surely these amazing machines would usher in a new economic era in which productivities would skyrocket and profits would soar. Yet the years passed without a hint of these gains. In fact, the period during which information technology (IT) was gaining broad acceptance closely paralleled America's notorious productivity slump. Could there be a connection?

This possibility highlights an uncomfortable fact about technical progress as it relates to growth in productivity: There are at least as many ways to reduce productivity through implementation of new technologies as there are ways to gain. Even innovations that have proven to be highly beneficial carry with them hidden costs and risks.

Beginning in the early 1970s, American firms began to invest heavily in computers and automation. Unfortunately, these new technologies were immature, and unfamiliar both to decision-makers and to workers. Hardware was notoriously unreliable, and software was often buggy and poorly integrated. Interconnectivity, now the watchword of productivity improvement, was virtually nonexistent. Employees required weeks of training to use systems that rarely performed up to expectations, and were obsolete in a few short years. Under these conditions, it is not hard to imagine that measurable productivity gains from information technology and automation were rare indeed.

Unfortunately, the cost of adoption of new technologies is typically not calculated in terms of opportunity costs. If the dollars, and more importantly the time, required to implement these ineffectual improvements had been spent on R&D, for example, it is likely that America's productivity and innovation slump would have been less severe. It has been estimated that the cost of adopting information technology in the United States during the 1970s and 1980s may have run as high as 10 percent of GDP, if capital investment, implementation costs, and lost worker time for skills upgrade are considered.[71]

Most professionals are intuitively aware of the potential productivity benefits of information technology. Yet, despite these seemingly obvious boons, it is difficult to formulate a return on investment (ROI) calculation that accurately captures both the benefits and the costs of these new technologies.[72]

Basically, productivity can be improved in only two ways: either a new technology enables higher value to be provided to customers (presumably with an associated increase in revenue) or it allows a reduction in the costs of production. Cost reductions are not all equal, however. Reducing the recurring cost of manufacturing a product, for example, has a far greater impact on productivity than saving labor in production-support operations does. Likewise, accelerating the decision process within a firm can dramatically improve both productivity and profitability, but the ability to generate a flood of customized reports and display superfluous graphics is often more a liability than an asset.

A recent survey conducted by the Gartner Group estimates that 70 percent of enterprise-management software packages (often referred to as enterprise resource planning [ERP] systems) fail to be fully implemented, even after three years of effort. An increasing level of disillusionment on the part of executives has resulted from a combination of poor implementation planning, overselling of features by suppliers, complicated and misunderstood products, and mistakes made by users in attempting to cut costs. One management analyst stated tersely that the number of enterprise-management implementations that succeed "closely approximates zero."[73]

The days of "gut-feel" justifications for multimillion-dollar IT implementations are over. For years, executives and consultants have argued that the benefits of information technology are so obvious that there is little need to measure ROI. This attitude has changed dramatically in recent years, as those responsible for the corporate bottom line are becoming increasingly skeptical of the productivity gains derived from automation and computerization. In a recent survey of IT executives within U.S. companies, 80 percent said that their organizations now require them to demonstrate the potential revenue, payback, and budget impacts of proposed IT projects. This reflects a dramatic shift from just one year earlier, when only 45 percent of respondents performed ROI calculations as part of their project justifications.[74]

So significant is the problem of unproductive IT and automation implementations that many firms are *de-engineering* and *de-automating* their production operations. A recent survey of 360 companies performed by Standish Group International Inc. found that 42 percent of corporate IT projects were abandoned before completion. Given the fact that U.S. companies currently spend about $250 billion annually on computer technology, these failures represent a tremendous drain on national productivity.[75]

The hidden costs of IT implementation can surface even in the most innocuous of software projects. A recent upgrade of Aeroquip Corporation's fifty-person research laboratory from Windows 3.1 to Windows 95 cost the company $20,000 per person, when all hidden costs were totaled. The president of Aeroquip, Howard Selland, noted that most office workers at his company used only a small fraction of the computer power currently on their desktops.

A similar story unfolded at Chrysler Financial, which invested several years wrestling with a highly sophisticated NeXT Software computer system. In this case the problem was interconnectivity; Chrysler's sales force was still using Windows, and communication between the two platforms was so poor that the organizations could not even send e-mail to each other. After tearing out the NeXT system,

Chrysler Financial chose a far simpler Windows-based network, and by adding a few more workers, was able to get back on-line in less than ninety days.[76]

These examples highlight what may be the underlying cause of disappointing IT productivity: diminishing returns to scale. It is definitely not the case that "more is better" when it comes to information technology. As has been eloquently demonstrated by the Japanese *kanban* and just-in-time production systems, what is most important in productive technologies is the *right* information, not just more information. Furthermore, complex solutions require sophisticated users, aggravating the "skills gap" already being experienced by many U.S. firms. The enthusiasm of some information technology firms may have carried the development of many software and hardware products well beyond an optimal productivity-to-cost ratio.

The difficulties associated with gaining productivity from information technology have been mirrored in the automation of manufacturing processes. The most famous example involved the multibillion-dollar robotic automation project implemented by General Motors in 1985. In this case, assembly errors, such as robots erroneously putting Buick fenders on Cadillacs, caused a severe decline in productivity. Only after the robot population was reduced did productivity rise to acceptable levels.

The high cost and limited flexibility of automation equipment can prove to be a severe handicap in a world hungry for customization. Recently, Sun Microsystems decided to re-engineer its production lines after determining that its highly automated factory was not sufficiently flexible to meet growing customer demands for customized workstations. The result was a simplified cellular process that utilizes a balanced mix of conveyorized, mechanized, and manual operations, and can achieve a 30 percent faster cycle-time in less floor space.[77]

In summary, productivity leadership is not determined by the number of computers on employees' desks, nor by the complexity of the software system used to manage the shop floor. Only when new technologies result in high-value products, improved customer loyalty, or reduced production costs can they play a role in a firm's search for competitive advantage.

# 2
# The Global
# Technology Marketplace

## Globalization in Brief[1]

Anyone who has not spent the last decade in a sensory-deprivation chamber must be aware that the economies of the world are becoming increasingly integrated. Unfortunately, this awareness has likely come via a flood of oversimplified media hype, fear-driven rhetoric, and biased misinformation. The term *globalization* itself implies a monolithic process that could not be farther from the truth. Globalization is not a switch that is thrown: Yesterday we were in isolation and today we are one. Instead, the process occurs at many levels, in a sporadic and often tenuous way. The purpose of this chapter is to revisit this overworked topic from the perspective of the high-technology firm, and to establish an objective baseline from which global strategies can realistically be developed.

Throughout most of our lifetimes, we have witnessed the nations of the world become increasingly interdependent. Beginning shortly after World War II with the establishment of the Bretton Woods exchange-rate system and the signing of the Global Agreement on Tariffs and Trade (GATT), we have observed a steady increase in both global exports and foreign direct investment (FDI).[2] The multinational corporation has risen to prominence during this dynamic period as the lead actor on the world's financial stage.

It has been suggested that the rapid expansion of multinational corporations (MNCs) has been the driving force behind recent global integration. Motivated by the threat of aggressive global competition, firms are constantly seeking greater economies of scale, increased levels of specialization, and new sources of advanced technologies. Thus, MNCs are compelled to reach beyond their boundaries to seek new sources of competitive advantage.[3]

Globalization has become a catchword for a number of distinct trends, all of which tend to reduce the importance of national borders and increase the level of global economic interaction. Since much of the recent globalization process has been facilitated by advances in transportation and telecommunications, one might assume that global economic integration is a uniquely modern phenomenon. In reality, cross-country trade and investment are time-honored traditions. A look back into history will demonstrate that the levels of integration being experienced today had been achieved, and even surpassed, a century ago.

### It's Really Nothing New

Economic linkages between nations, such as those formed by the trade routes along the Mediterranean coasts or the Silk Road, have enabled the flow of goods and knowledge for much of recorded history. Despite the longevity of these early commercial channels, however, their economic importance was fundamentally limited. The relatively high risks and cost of early trade allowed for the exchange of only a trickle of merchandise. More important, economic dependency on foreign states was considered to be a risk to national security. With the prospect of war a constant threat, leaders were reluctant to establish permanent trade in essential agricultural and mineral products. Hence, increased cross-country integration was limited by the lack of a secure and lasting peace.

The first opportunity for a truly global economy did not occur until the late nineteenth century. The relative stability brought about by the hegemony of the British Empire made foreign direct investment an attractive prospect. Furthermore, the establishment of a world gold standard, in which the U.S. dollar and the British pound sterling were tied to fixed exchange rates, reduced the risks of capital mobility. As a result, the world's economies may have been more tightly integrated during the first two decades of the twentieth century than they are today.[4]

Global integration, as measured by the intensity of both exports and FDI, was on the rise in the early 1900s. Growth in exports averaged 3.5 percent per year, far outpacing the growth of real output. By 1914, the share of world production exchanged in global markets reached a level not surpassed until 1970.[5] Although much of the capital flows to foreign countries occurred through portfolio investment, some estimates indicate that direct investments accounted for as much as 35 percent of foreign assets held by the United Kingdom (then the world's largest international investor) in 1914.[6]

Prior to World War I, nearly 25 percent of the companies listed on the London Stock Exchange operated mainly or exclusively abroad. These "free-standing companies" were typically run by a resident team in the foreign country. The investors back home had little control over operations, and were often given inadequate or even biased information. Despite these limitations, enormous fortunes were amassed through foreign investments in mining, quarrying, and real estate. Such famous extraction giants as De Beers and Shell Oil were formed in this way.[7]

During this period, American manufacturing capability, transformed by the technological progress of the second industrial revolution, was beginning to display world-leading levels of productivity. American manufacturing technology and management expertise had evolved to support the large, geographically diverse continental market. Networks of regional plants, linked together by a strong central executive function, provided an excellent foundation for global expansion.

Pioneering multinational companies based in America included Singer Manufacturing Company, Kodak, United Shoe Machinery, and Ford Motor Company. Singer, for example, had operations in Europe dating to the 1860s, with facilities in four countries and sales offices in dozens by the 1880s. Particularly in the areas of manufacturing and marketing, the U.S. multinationals began to dominate their European rivals in numbers by the eve of World War I.[8]

The era of economic stability that heralded the rise of the first American multinational corporations came to a violent end with the onset of the war. Foreign investments in factories and equipment languished in enemy territory, and much of the industrial base of Europe was either damaged or destroyed. After the war, investor confidence was shattered, and stirrings of discontent from underprivileged groups were growing louder. With the onset of the Great Depression, world trade volume collapsed. Governments imposed high tariffs and capital controls in attempts to insulate their economies from the spreading global slump.

Nationalism became increasingly evident in the politics of the 1930s. *Laissez-faire*[9] was out, and the rise in popularity of central planning philosophies (particularly communism) was a portent of future world events.[10] The overall result of these tumultuous times was that multinational corporations accounted for a substantially smaller part of the world economy in 1949 than they had in 1929.[11]

After World War II, the world found itself with a new de facto leader: the United States. With its military and industrial might intact, the U.S. served as fair-broker and primary beneficiary of a new era of economic integration and cooperation.

American corporations consolidated their position as world leaders across the entire range of industries during the 1950s, due primarily to a decisive lead in technology. Advances resulting from wartime research, including jet engines, aeronautics, radar, and communications, fueled a growing image of the United States as the source for the world's advanced technologies. During this period, the U.S. share of world exports reached a historic peak of roughly 27 percent. A summary of recent OECD export share data for the U.S. and other advanced nations is shown in Figure 2.1.

As was noted in the previous chapter, during the early postwar period America held a substantial lead in productivity over the rest of the industrialized

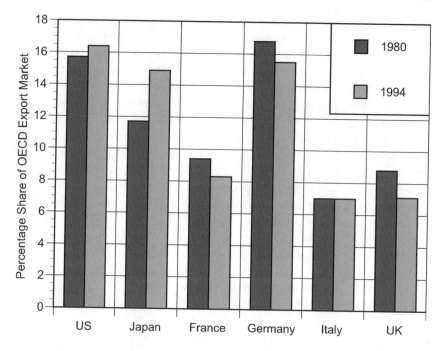

**Figure 2.1:** Share of OECD manufacturing exports, 1980 and 1994, for the six major industrialized countries. Source: OECD (1997, pg. 174, Table III.6.1).

world. This rather substantial gap brought to light the role of two important factors in global integration: *technology transfer* and *technology spillover*. Not only were U.S. manufacturers more efficient in their home country, but their affiliates in foreign countries were more productive as well. The ability of American firms to transfer technology from one global location to another represented an enormous strategic advantage.

A study by John Dunning during the early postwar period noted that this ability could yield a substantial benefit for the host country as well.[12] The direct transfer of technology through the employment and training of host-country nationals represented a known benefit. In addition, an indirect effect was observed, in which even unrelated industries in the host country displayed improved productivity as a result of the presence of American MNC affiliates. This boon to the local economy is attributable to technology spillover.[13]

The perceived benefits of the spillover effect stand in sharp contrast to the generally negative image of MNCs held by developing countries during the 1960s and 1970s. Seen as a last vestige of colonialism, investments by corporations from the advanced nations were subjected to severe government scrutiny and control. Often, several MNCs would be played against each other for a

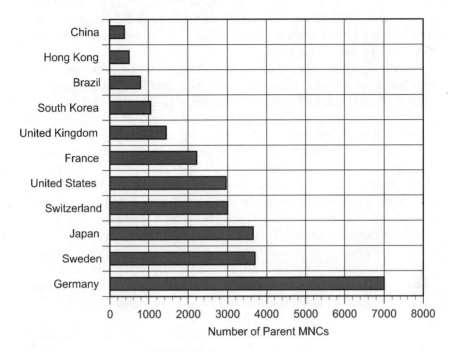

**Figure 2.2:** Number of multinational corporations based in several advanced and developing countries in the early 1990s. Source: UNCTAD (1995, Table 1.2, pp. 8 - 9).

scant number of "licenses" to operate in a developing market. Once a license was granted, the corporation could operate as a virtual monopoly, with some of the excess profits being diverted back to the host government. During this period, the expropriation of foreign subsidiaries by politically unstable nations was an all too common event.[14]

By the mid-1960s, three-fourths of the Fortune 500 industrial firms had manufacturing operations in at least one country other than the United States, and as of the early 1990s, thousands of firms worldwide could be called multinational. Today the multinational corporation is a global phenomenon, with American firms being joined by rivals from Europe, Japan, and even a substantial number of developing countries, including South Korea, China, and Brazil, as shown in Figure 2.2. Indeed, the egalitarian nature of foreign investment is a distinguishing feature of our current rendition of a globalized economy.

### The Current State of Global Integration

The extent of globalization can be estimated in several ways. The World Bank uses two aggregate metrics known as the *integration index* and the *speed of integration index* to gauge the level and rate of involvement in the global economy. By these measures, the United States is roughly average among advanced economies in level of integration, as shown in Figure 2.3.

Another useful metric is the ratio of product trade to manufacturing output. By this measure the world market for products has doubled its level of integration since 1950, with much of those gains occurring in the 1980s and 1990s. Furthermore, this estimate significantly understates the true level of integration, since it fails to capture recent growth in the exchange of services.

Over the last decade, there have been large increases in both trade volumes and capital flows between advanced economies. Although not as significant in absolute terms, trade volumes and investment flows to developing nations have also grown substantially. The traditional exchange of agriculture and energy products has remained important, but global trade is increasingly being dominated by technology-intensive products, as shown in Figure 2.4.

A large proportion of North-South commerce has been driven by an unprecedented demand for modern infrastructure, coupled with the privatization of government-owned monopolies in telecommunications, energy, transportation, and extraction. Major construction projects in developing countries have

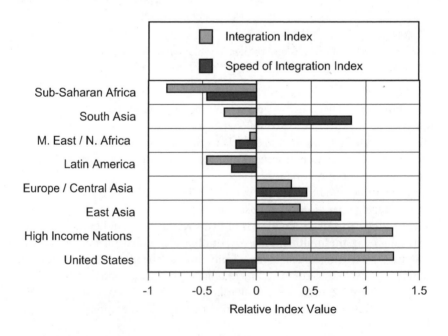

**Figure 2.3:** Values for The World Bank's integration index and speed of integration index for various regions of the world and the United States. Source: The World Bank (1996, Tables A2-1 and A2-2).

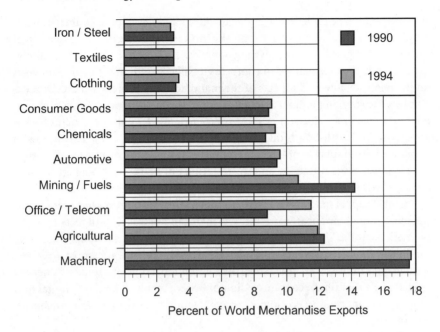

**Figure 2.4:** Percentage breakdown of world merchandise exports by product, 1990 and 1994. Source: World Trade Organization (1995, Chart IV.1, pg. 77).

brought together advanced engineering capabilities from the OECD economies, semi-skilled labor from the newly industrialized countries, and low-wage labor provided by the host nation. Recent trends toward liberalization of markets and privatization of state-controlled monopolies in Asia, Eastern Europe, and Latin America have opened up exciting trade and investment opportunities. Advances in telecommunications and energy technology have motivated the development of an entirely new generation of infrastructure throughout much of the world.[15]

A significant fraction of global commerce is characterized by intra-industry trade: the import and export of similar goods between countries. This growing trend, which has occurred both along North-South lines and among the industrialized countries, represents a significant factor in the recent globalization process.

Among the advanced nations, a reasonable explanation for intra-industry trade is provided by market demand and technological specialization. An increased desire for variety and customization in products tends to attract multiple entrants in a single industry, each of which specializes in a particular market subsegment. A good example of this behavior is the massive international trade in automobiles among industrialized countries, despite the presence of mature domestic industries.

Intra-industry trade between the advanced and developing countries, however, is primarily a result of a strong shift toward outsourcing the assembly of manufactured goods to lower-wage countries. This production strategy allows firms from high-wage nations to compete globally by optimizing their labor costs on the relatively low-skill aspects of the manufacturing process. An excellent example is the sourcing of final assembly operations for such products as automobiles and consumer electronics to Mexico's *maquiladora* plants. Much of what appears to be North-South intra-industry trade, therefore, is actually a flow of components and subassemblies being exported from parent firms to foreign affiliates in lower-wage countries, and the resulting finished goods being returned to the advanced nations.[16]

Global capital markets have also become increasingly integrated over the last decade, although not nearly as rapidly as merchandise trade. Until recently, capital markets in the United States, Europe, and Asia behaved as though they were in almost complete isolation.[17] Under the Bretton Woods system of fixed exchange rates in the 1950s and 1960s, the international flow of capital was heavily regulated. At that time, economists did not believe that high levels of capital mobility were important to economic growth. Stability was the paramount consideration.

After the fixed-exchange-rate system collapsed in the early 1970s, flows of capital dramatically increased. Currency trading expanded from $190 billion per day in 1985 to more than $1.2 trillion per day in 1995. Flows of capital investment into emerging markets increased from $50 billion in 1990 to $336 billion in 1996.[18]

Despite this impressive growth, global capital markets are far from fully integrated. There is still a significant risk premium associated with investments in developing countries, due to both currency exchange rate and political instabilities. If investors sense an overvaluation of a nation's currency, they will pull out en masse. Unless the country's reserve banks can sustain high interest rates to stem the flow, a devaluation is inevitable. This is the fate suffered by Mexico in 1994 and throughout much of Asia in 1997.

Unlike the flow of goods and services, the flow of global capital has historically been characterized by "panics and manias." Nowhere is this more evident than in the recent upheaval in Southeast Asia. In 1996, $93 billion in foreign investment capital flowed into Indonesia, Malaysia, South Korea, Thailand, and the Philippines. The following year, after fiscal mismanagement caused an investor panic, there was a net *outflow* of more than $12 billion in capital.[19] Although integration of capital markets is a desirable long-term goal, the benefits must be weighed against the risks of global financial crisis.

Progress in international logistics has been another force in global integration, particularly the recent dramatic improvements in transportation and communication technologies. The expanding use of *electronic commerce* to manage shipping documents and customs forms in many nations has greatly simplified the bureaucracy of trade. Modern container technology has fostered a revolution in the efficiency and productivity of the transportation industry. The availability of advanced intermodal shipping has significantly reduced the cost and risks of international freight forwarding. Finally, a new generation of transportation infrastructure, including advanced container-port facilities such as those in Singapore and Hong Kong, has slashed shipping costs and transport times throughout much of the world.

The very nature of modern products has facilitated the reduction of transportation and logistical barriers. Information-intensive products such as software and advanced integrated circuits impose minimal demands on international transport systems. Even those products that have traditionally been characterized by high mass and volume have slimmed down considerably. The use of lighter materials, the substitution of embedded microprocessors for bulky manual controls, and a shift to high-value, miniaturized components have improved the ratio of shipping cost to product value considerably. In some cases such as software and entertainment products, "transportation" can be provided through the telecommunications infrastructure rather than by physical movement. As a result, dramatic reductions in telecommunications and air-transport costs, as shown in Figure 2.5, have had a significant effect on trade flows in high-value, technology-intensive products.

Despite the impressive list of factors that are driving us toward economic integration, it should not be assumed that globalization is either complete or inevitable. There is still a strong tendency for countries to trade with their nearest neighbors, and even among the most prolific trading partners, domestic commerce still dominates over trade. The exchange of goods between provinces of Canada, for example, is roughly twenty times the volume of trade between a typical Canadian province and the United States.[20]

Developing countries have opened their markets considerably in the last decade, but there are still significant barriers to trade and investment in many countries. The stability of political systems in most of the developing world is questionable; corruption is rampant, particularly in countries that lack the resources to adequately compensate their officials. Furthermore, much of the world lacks adequate infrastructure, a fair and transparent judiciary, and stable financial institutions. All of these structures must be in place to support further economic integration.[21]

Although the evidence of globalization is all around us, it is important to recognize that the impact of this metamorphosis is certainly not all positive. Exposure to global trade and investment can offer hope and opportunity for those countries capable of "converging." Unfortunately, nations that have not developed the institutions and human capital necessary to fully benefit from global trade may find themselves falling ever farther behind.

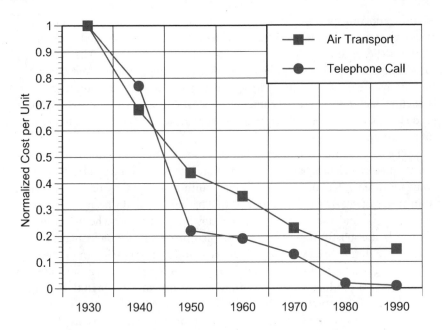

**Figure 2.5:** Reduction in the relative cost of air transportation and telephone communication from 1930 to 1990 (normalized to 1930). Source: International Monetary Fund (1997, Table 11, pg. 46).

Likewise, those individuals within our own country who suffer disbenefit due to import competition and shifts in industrial focus must be considered for appropriate redress. We are being drawn down the path toward globalization by powerful forces, but the outcome is by no means certain. History has shown that as our society evolves, there are always groups that must endure the negative ramifications of change. Without adequate concern for those who are disadvantaged in the globalization process, the political and social backlash may drive us backward into economic isolation.[22]

### The Role of Multinational Corporations

The world economy in the 1990s is still far from perfectly integrated, but the globalization of business activities has come a long way. Both trade and foreign direct investment are growing rapidly, with FDI recently surpassing trade in relative importance. United Nations estimates show that the total value of all exports of goods and non-factor services was about $4.8 trillion in 1993, whereas the total sales resulting from direct investments abroad were roughly $5.2 trillion.[23]

In the past, trade and foreign direct investment were assumed to be opposite sides of the same coin; a firm would either export goods to a foreign market or establish a manufacturing plant overseas to produce the products locally. In our current global economy, trade and FDI in the manufacturing sector tend to increase and decrease together, rather than in opposition. It can be said that trade and FDI are acting as complements rather than substitutes.[24] As with much of the globalization process, this trend can be directly attributed to the rise of the multinational corporation.

The standard model for MNC operation assumes that their investment decisions are based on optimizing both production costs and sales opportunities. Their physical presence transcends national boundaries; they are typically structured as a global network of facilities and capabilities, which can be exercised in concert to capture a particular market opportunity.[25] It is often said that MNCs are "footloose," tending to move their direct investments and resources to locations that provide the lowest labor cost or the best market access.

One result of the web-like structure of MNCs is that a significant portion of what appears in statistical data to be intra-industry trade is actually *intra-firm trade*. Components might be sourced by an affiliate in Southeast Asia and shipped to an assembly plant in Mexico. Finished goods could then be transported back to the domestic market, or routed to several distribution subsidiaries in key international locations. All of these transactions take place within the MNC, and outside the global marketplace. As of 1990, between one-fifth and one-third of all U.S. exports and imports could be attributed to intra-firm trade.[26]

The pattern of foreign direct investment among multinational corporations follows that of trade: a strong emphasis on the advanced nations, with a growing interest in developing countries. Roughly 85 percent of sales by American foreign affiliates occurred in OECD nations in 1990, with the outward stock of FDI by U.S. multinational corporations more than doubling in the period from 1985 to 1993. Despite leading in absolute terms, however, the substantial rate of growth in outward FDI by American MNCs is dwarfed by that of Japan's over the last decade, as shown in Figure 2.6.

One of the most intriguing aspects of the recent globalization phenomenon is that the very nature of enterprise is changing. In the 1970s and 1980s it was assumed that the vertical, command-and-control form of MNC would eventually dominate world trade. Powered by economies of scale, an oligopoly of corporate giants was expected to control the global marketplace, creating insurmountable entry barriers for smaller competing firms. This prediction has proved to be far from accurate, particularly with respect to high-technology industries.

The benefits of increased scale of operations in vertically organized firms were severely diminished by the inefficiencies associated with a rigid, hierarchical structure. The strategic decision-making process was weakened by distorted and outdated information filtering up through endless levels of middle managers. Inadequate local knowledge, particularly in foreign markets, limited responsiveness to customer needs. Perhaps the most crippling factor was the inability of

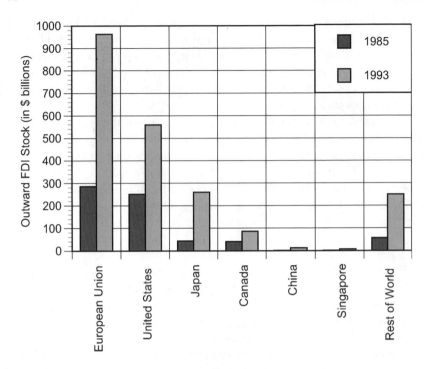

**Figure 2.6:** Outward stock of foreign direct investment by American multinational corporations in 1985 and 1993, as compared to other important regions. Source: UNCTAD (1995, Annex Tables 3 & 4).

these vertical corporations to optimize their mix of internal specialization and outside expertise.

The initial motivation for vertical organizational structures within MNCs was an almost paranoid fixation on controlling the availability, quality, and cost of inputs to production. At the core of these concerns was a lack of information; without reliable methods for identifying, monitoring, and managing suppliers, a horizontal structure was simply not a viable option.

In a sense, the recent revolution in information technology has changed the rules of industrial organization. Now companies can achieve a global reach through horizontal structures, and still effectively coordinate activities of these dispersed, specialized operations. Within the last decade, the vertical organization has largely been replaced by a far more efficient network structure. In some technology-intensive industries, complex webs of co-producers and suppliers have developed, which allow all participating firms to rationalize their production operations and to focus on their areas of specialization.[27] An excellent example of such a horizontal network has formed in the PC industry around the "Wintel" standard for personal computers.[28]

Companies are exploring a broad spectrum of enterprise linkages to extend their reach.[29] The use of supplier partnering, strategic alliances, joint ventures, industry networks, and research consortia is growing rapidly, as shown in Figure 2.7. Each of these relationships offers the opportunity to leverage specialized talent across national boundaries.

This discussion might leave one with the impression that MNCs have relatively little connection with their home countries. As I will demonstrate in later chapters, the core strategic behavior of MNCs is more a reflection of their national environment than one might expect. This attachment of MNCs to their home soil is revealed, for example, by a strong tendency to retain critical core operations such as research and development within their domestic facilities.[30] There is strong evidence suggesting that the economic, political, social, and technical environment of the home nation plays a pivotal role in the global competitiveness of multinational corporations.[31]

## The World Trading System

Throughout most of recorded history, trade and barter among regions has been the basis for both social and economic development. Some of the first examples of written script were cuneiform tablets etched with detailed lists of inventories and barter ratios. The exploration of the New World was motivated by the search for exotic trade

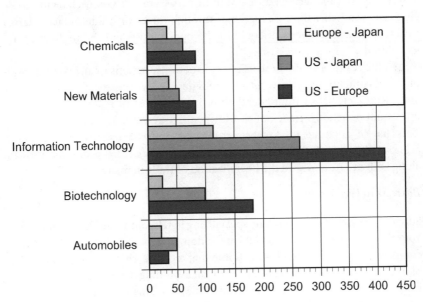

**Figure 2.7:** Number of strategic alliance formations during the decade from 1980 through 1989. Source: UNCTAD (1994, Table III.12, pg. 139).

goods from faraway lands. It could be said that much of early Western history was driven by a hunger for rare commodities that could not be produced at home.

The best way to understand why trade has been such a compelling force throughout history is through a simple example. Suppose that you live in a region which has as its sole asset an enormous herd of wildebeests. Your people have become proficient at herding wildebeests, but your cuisine has a definite lack of variety. One day you come into contact with people from a different region who are endowed with land and technology ideally suited to the production of rice. They have become quite efficient at rice cultivation but are also growing weary of the local fare. From a culinary standpoint, both groups would be made better off by exchanging wildebeests for rice.

You could try to cultivate rice in your region's arid soil, while the other group struggled in their wet marshes with an unruly herd of wildebeests. It is likely, however, that you would both be inefficient at these unsuitable tasks. Much more labor would be required if your people harvested rice locally than if you added a few more wildebeests to your huge herd. Hence, the effects of differing conditions, specialized technologies, and perhaps some economies of scale, combine to make this exchange highly advantageous.

Equalizing an unequal distribution of resources and technology is the driving force behind international trade; regions would not engage in commerce if they did not believe that they would be made better off. If the two peoples in this example were to come across a third group with a wealth of cocoa, each region could enjoy an even richer dining experience through three-way trade. Hence, a larger variety of available goods offers both the opportunity and the motivation for global trade.

In the real world, the process of trade between nations is surprisingly complex and contentious. The *world trading system* today consists of a cumbersome and incomplete set of multilateral agreements, a host of regional pacts, and a crazy quilt of national trade policies, tariff structures, and non-tariff restrictions. This chaotic tangle has recently been mitigated somewhat by the almost universal acknowledgment of capitalism as the preferred system for economic growth, along with progress toward the liberalization of trade in many nations.

### Barriers to Free Trade

Suppose that you own a company specializing in telecommunications equipment. Your firm has been growing rapidly in the domestic market and is ready to flex its competitive muscles abroad. After some market research, you decide that Brazil represents an excellent opportunity. With an exploding demand and the impending privatization of its national telecom monopolies, the future looks bright.

As a newcomer to international trade, you will discover that there are numerous obstacles between you and your foreign customers. National governments frequently impose *tariffs* on imported goods, the levels of which can be oppressive.[32] Although import duties worldwide have fallen from a postwar average of roughly 40 percent to a current average of about 5 percent, they are typically much higher in developing countries. In China, for example, there was a 15 percent duty on steel products, a 25 percent duty on plastics, and a 110 percent duty on automobiles in 1994.[33] In general, tariffs act to increase the price of imported goods in a foreign market, but do not directly limit the quantity that can enter that country.

Although tariffs are the most common form of trade barrier, there is a pantheon of more subtle trade-restricting measures, collectively referred to as *non-tariff barriers*. Quotas are applied to imports in many countries, limiting the quantity of a product that can cross their borders. Licenses are used to control the number of foreign companies that can operate in a given industry. In some cases, governments may pressure foreign countries to voluntarily restrict the quantities of specific categories of imports.[34] Each of these measures has the effect of protecting the competitive position of domestic companies at the expense of foreign interlopers. The list is long and onerous.

Returning to our example, even after you have solved Brazil's labyrinth of taxes and duties, you will still be subject to some tricky restrictions. The Brazilian government requires that many technology-intensive products, including telecommunications equipment, incorporate *local content*. This means that a specified percentage of the value-added operations in the manufacture of these products must be performed on Brazilian soil, using domestic workers. A policy of this type is designed to capture revenue for the domestic economy, and to promote technology transfer and spillovers.[35] Other technical restrictions, such as safety regulations, design standards, and certifications, round out the list of obstacles importers may encounter.

In most respects, trade among the OECD nations is a reasonably civil affair. Tariff barriers have been reduced on most imports to inconsequential levels through successive rounds of GATT negotiations. Recent multilateral trade discussions have focused on the harmonization and eventual elimination of non-tariff barriers as well. Among the advanced economies, overt obstacles have been replaced by a highly politicized game of strategic maneuvering and brinkmanship.

In the developing world, however, the trading environment is often much more challenging. Underlying the generally higher tariff structure and inconsistent regulations are even more insidious obstacles. Corruption is rampant in many parts of the world, both at the local and national levels. Inadequate protection of physical and intellectual property is common, and crime in some of the least-developed regions can represent a threat to both property and personal safety. The institutional protection of a fair judiciary is often lacking, with arbitrary decisions and inconsistent verdicts a common complaint.

These three factors of corruption, crime, and an unfair judiciary are collectively referred to by The World Bank as the *lawlessness syndrome*. A recent survey of businesspeople active in international enterprise revealed that twenty-seven of the forty-eight countries in the survey displayed unacceptable levels of all three lawlessness factors. The nations of sub-Saharan Africa and Latin America were well represented, but there were members of this dubious group from most geographic regions.[36]

Is there a way for multinational business to overcome the lawlessness syndrome? The United States adopted the Foreign Corrupt Practices Act in 1977 as a preemptive strike against American corporations participating in the bribery of foreign officials. Although this measure was intended to serve as a "righteous shield" to protect U.S. businesspeople from solicitations and scandal, it has limited the competitiveness of our firms in some developing markets. Unfortunately, there are other advanced nations that do not share our high ethical standards, and even consider bribes paid by firms to be a legitimate, tax-deductible expense.[37]

The ability to maintain the rule of law is closely related to a nation's level of development. The cost of effective law enforcement is considerable, and inadequately paid police are prime candidates for corruption. A transparent and impartial judiciary is even more problematic, requiring both a sophisticated institutional structure and a culture of ethicality.

First attempts at an effective legal system are often mired in bureaucracy. The judicial system in Brazil, for example, can be exceedingly cumbersome. In 1981, procuring a business license required 1,470 separate legal actions, involving thirteen government ministries and fifty agencies. Despite its inefficiencies, however, the Brazilian judiciary has a reputation for being relatively fair and predictable, as compared to other Latin American nations.[38]

Although the environment for business in developing countries is at times daunting, the opportunities for profit often outweigh the risks. More important, it is on these new battlefields that the competitions of the twenty-first century will be waged. American firms must learn to excel in this challenging environment if they are to participate in the modernization of the developing world.

### The World Trade Organization

Given the gauntlet of obstacles that firms face, it is surprising that international trade occurs at all. By comparison, however, the global trading environment today is far more liberal than at any time in recent history.

In the 1930s, for example, high tariff barriers nearly shut down global trade.[39] Motivated in part by the onset of the Great Depression, protectionistic sentiment grew in both the United States and Europe. The U.S. Congress enacted the Tariff Act of 1930, which created the highest general tariff and rate structure in American history.[40] Many European governments followed suit with the imposition of stiff import duties and quantitative restrictions. The result was disastrous for global trade. The value of imports to the United States dropped from $4.4 billion in 1929 to $1.5 billion in 1933, with exports declining from $5.2 billion to $1.7 billion over the same period.

Following World War II, the governments of the United States and Europe were anxious to avoid the ruinous economic conditions experienced after World War I. Geneva-based negotiations, convened in 1947, established the Global Agreement on Tariffs and Trade. The GATT laid out an agenda for the harmonization and reduction of trade barriers throughout most of the industrialized world.

Since that time, eight rounds of multilateral negotiations have taken place, addressing the reduction of both tariff and non-tariff barriers. The result has been a decrease in the average tariff rate for the OECD countries, from a range of 30 to 50 percent shortly after World War II to just over 3 percent today.[41] During this same period, the number of signatories grew dramatically. From the original twenty-three negotiating countries in 1947, the GATT had grown to 123 signatories by 1993. While the original group consisted primarily of industrialized countries, the new GATT represents a broad cross-section of the world's nations, from the very rich to the very poor.

The most recent GATT sessions, known as the Uruguay Round, have had a profound effect on global trade. Upon their conclusion in 1993, the Uruguay Round negotiations had successfully addressed a number of issues that are critical to future economic integration. The inclusion of trade in services, the establishment of international guidelines for intellectual property protection, and an improved dispute-settlement mechanism have expanded the scope of this multilateral agreement considerably.

It was evident during the Uruguay Round that the task of administering this unwieldy framework of tariff reduction schedules and non-tariff barrier disciplines would require a new and more powerful institutional structure. The World Trade Organization (WTO) was formed in 1995 to provide the structural support needed to keep the GATT agenda on track.

This has proved to be no small challenge, as there have been new issues raised since the Uruguay Round that threaten to derail global trade liberalization. Moreover, the much-touted dispute-settlement mechanism within the WTO has recently come under fire as being incapable of handling the complex decisions it will face. A recent dispute between Eastman Kodak and Japan's Fuji Photo Film Company was rejected by the three-person WTO panel after months of effort by the U.S. government, which had filed the case on behalf of Kodak. Although the suit ostensibly dealt with unequal access to Japan's huge domestic film market, the United States had hoped to use a favorable decision as a lever against the broader issue of Japanese protectionism.[42] The negative outcome of this case has prompted claims by U.S. policymakers that our agreement to abide by the rulings of the WTO places us at a competitive disadvantage in world markets.

In addition to its primary role of orchestrating multilateral trade-liberalization initiatives, the WTO will expand its agenda in the future to include several problematic issues. A critical debate is raging over the incorporation of human rights and environmental regulations into multilateral trade negotiations. These topics were first raised during the North American Free Trade Agreement (NAFTA) discussions in the early 1990s, and have recently been a political stumbling block to President Clinton's request for "fast-track" trade negotiating authority.[43] Additional issues for future WTO action include harmonizing international rules for foreign direct investment and telecommunications infrastructure,[44] and establishing global standards for fair competition.[45]

There is a final troublesome note regarding the WTO. The most populous and potentially the most economically important country in the world is not currently a member. China has engaged in talks to bring about its accession to the WTO, but as yet has not met the requirements for opening domestic markets, implementing financial controls, and ensuring intellectual property protection. Although this problem is likely to resolve itself in the near term, exclusion of China due to newly enacted human rights or environmental regulations could be a crippling blow to the cause of free trade.

### Regional Agreements

Simultaneous with the expansion of multilateral agreements under the auspices of the GATT, a number of regional trading blocs have formed. The GATT allows for the formation of preferential trading arrangements such as free-trade areas under certain conditions, provided that other members are not disadvantaged.[46] Although these groups have formed largely to encourage trade between geographic neighbors, their existence represents a challenge, and possibly a threat, to the WTO.

Most Americans are aware of the North American Free Trade Agreement, which provides for tariff elimination and other trade-enhancing measures among Canada, the United States, and Mexico. Similar regional groupings, with varying levels of economic integration, seem to be springing up everywhere, as shown in Table 2.1. As of 1994, there were sixty-seven such agreements worldwide, covering all five developed continents.[47]

Another familiar example of a regional economic bloc is the European Union. Since World War II, the countries of Europe (and with a bit less enthusiasm, the United Kingdom) have been moving steadily toward full economic integration. The latest agreements, signed in Maastricht, Netherlands, in 1992, will complete this evolutionary process. The Maastricht Treaty holds each signatory government to strict financial disciplines, culminating in the creation of a single common currency, the "euro." The EU represents an economy that is larger than that of the United States, and is significantly more trade-integrated with the rest of the world.[48]

In South America, several industrializing countries, led by Brazil, have formed Mercosur, a customs union with rapidly expanding cross-border trade. On the other side of the Pacific, the Association of Southeast Asian Nations (ASEAN) garnered publicity in 1997 for its admission of Laos and Myanmar, and exclusion of Cambodia. It too plans to form a free-trade area, promising significant expansion of internal commerce. As a single entity, ASEAN represents a powerful economic bloc, with perhaps the highest aggregate growth rate in the world.

Not to be outdone, the United States has aggressively pursued the formation of both regional and supra-regional free-trade areas. In 1994 the members of the Asia-Pacific Economic Cooperation (APEC) signed the Bogor Declaration, committing the U.S., Japan, and most of the Pacific Rim countries to forming a free-trade area before 2020. In a separate initiative, many countries in the Western Hemisphere have committed to forming a Free Trade Area of the Americas (FTAA) by 2005.

Does all this activity mean the WTO is in trouble? Are member countries shifting their allegiances in favor of an isolated world consisting of regional blocs, as some authors have suggested?[49] There are several possible scenarios.[50]

First, it is possible that the regional trade agreements are simply a response to the rather unwieldy framework of the GATT negotiations. Countries anxious for free trade may be frustrated by the numerous distractions that are sapping WTO

| Preferential Trade Areas | Type of Grouping | Percentage Share of World Exports |
|---|---|---|
| European Union | Common market (economic union pending) | 22.8 |
| North American Free Trade Agreement (NAFTA) | Free-trade area | 7.9 |
| Mercosur | Customs union | 0.3 |
| Free Trade Area of the Americas (FTAA) | Free-trade area (by 2005) | 2.6 |
| ASEAN | Free-trade area (pending) | 1.3 |
| Asia-Pacific Economic Cooperation (APEC) | Consultative group (Free-trade area by 2020) | 23.7 |

**Table 2.1:** Several of the major regional trading groups, and their percentage share of world exports as of 1994. Sources: Schott, J. J. (1996, pg. 266) and *The World Factbook-1995*, Central Intelligence Agency.

energies. A more focused regional agreement can give the local economies a needed push while they wait for the rest of the world to catch up.

A second possibility is that the trend toward regional blocs is simply a matter of geographic proximity. The potential to save on the costs of transportation and communication by trading locally is a solid economic incentive. Again, this scenario would not represent a threat to the WTO, just an expeditious "side agreement" to move things along economically.

Finally, the growth of regional accords may be due to a desire for countries with a related past and strong geographical ties to "test the waters" of liberalized trade. By serving as a pilot project for solving complex issues, these regional agreements may prove to be a catalyst for expediting multilateral negotiations among the larger constituency of the WTO. The environmental and labor agreements incorporated into NAFTA, which are now serving as pathfinders for global negotiations on similar topics, are an excellent example.

These rather benign scenarios are strongly disputed by several prominent economists, who believe that a proliferation of preferential trade areas is damaging to both trade expansion and economic welfare. One important argument centers on the complications associated with *rules of origin*.[51] These often arbitrary regulations attempt to establish tariffs on products based on their country of origin. Since virtually everything produced today consists of an amalgam of components and labor from multiple countries, establishing product origins can prove to be an immensely complex and costly undertaking.

A famous example of how rules of origin can result in absurd contradictions involves the production of Japanese automobiles, by American workers, in American facilities, owned by Japanese manufacturers. In the 1980s, the U.S. Trade Representative, Carla Hills, made a strong case to Japan that these cars were of Japanese origin, and therefore should be counted as part of their voluntary export restraint (VER, essentially a voluntary quota). At the same time, she made an equally emphatic pronouncement that these cars were of American origin, when confronted with European import quotas on Japanese automobiles.

The most compelling argument against preferential trade agreements, in my opinion, is the most obvious one. Regional trade agreements will represent an obstacle to further economic integration, if politically motivated "regionalism" begins to dominate the sensibilities of policymakers. As long as these accords are based on a sincere desire to encourage and expedite trade, it seems likely that the cause of liberalization is helped, rather than harmed, by the formation of regional groupings.

## Free Trade and Protectionism

In its purest form, *free trade* can be defined as the free exchange of goods between nations, in the complete absence of government interference. If this ideal were the case, both consumers and producers in all trading countries would experience the same world market prices.[52] Although this may seem to be an innocuous concept, there are some important reasons why the free exchange of goods across borders has been the subject of heated debate for the last several centuries.

The source of this conflict can best be understood by considering two polarized (and admittedly oversimplistic) perspectives. Supporters of free trade, including many economists and industry leaders, would argue that free markets allow for the maximum economic gains from trade, and that interference by national governments would result in a loss of wealth. The opposing view would contend that without some form of government regulation or control, the negative impact of free trade on import-competing domestic industries and dislocated workers would be unacceptable. As one might expect, this opinion is typically held by those who stand to lose the most from increased global competition.[53] The stakes in this discussion are high: International trade and investment have been growing at unprecedented rates over the last decade.

Economic theory suggests that the free exchange of goods among nations offers the greatest opportunities for economic growth. Little theoretical effort has been expended, however, on solving the human issues of domestic job loss and industry collapse. In a sense, the debate over free trade comes down to a classic battle between the narrow interests of the disadvantaged and the broad interests of society in general. There can be no winners or losers in such a conflict; the only equitable solution is one based on conciliation and compromise.[54]

### The Free-Trade Debate

The desirability of free trade was noted as far back as the Golden Age of Greece. Plato argued that the benefits of specialization were a strong motivation for trade between regions. In *The Republic*, he observed that "More things are produced, and better and more easily, when one man performs one task according to his nature, at the right moment, and at leisure from other occupations."[55] This idea that division of labor might achieve greater efficiencies represents a fundamental source of increased wealth through trade.

The sea played an early role in enabling a second economic boon: gains from exchange. In the first several centuries A.D., theologians began to note that not all regions were equal in their ability to produce goods. Some territories were ideally suited to the cultivation of wheat, while others had the perfect climate for grapes or olives. In these superstitious times, the connection to Providence seemed obvious: God had intentionally scattered the bounty of the world unevenly about the planet, and created the sea to enable people of different regions to come together in trade. This so-called Doctrine of Universal Economy was based on the idea that each nation benefited from the exchange of a variety of goods.[56]

The rise of powerful nation-states in the sixteenth century brought about the first serious discussion of maximizing national wealth vis-à-vis trade. During this era, a number of benefit-enhancing strategies were proposed, under the collective label of *mercantilism*. The common element here was a desire to maximize the wealth and power of the home country. Since the world's economy was thought to be a zero-sum game, for the home country to gain wealth, it must do so at the expense of other nations. In the mercantilist system the welfare of individuals was almost always sacrificed to the advantage of the producer and trader. Production, rather than consumption, was considered to be of paramount importance. Much of this philosophy was focused on increasing a nation's hoard of "treasure," in the form of precious metal specie.[57]

After several centuries of nations squabbling over bullion, a more enlightened view began to emerge. The economists of the eighteenth century recognized that the world's economy was not based on a fixed amount of gold or silver, but rather was capable of growth. It soon became accepted, largely through the monumental work of Adam Smith,[58] that the gains from exchange and specialization held the potential for increasing aggregate wealth.[59]

Furthermore, Smith proposed that the best way to achieve these gains was to allow the market forces of supply and demand to act freely.[60] Smith acknowledged that some forms of government intervention might be necessary, but felt that these intercessions could be justified only under conditions of obvious market failure. As a result, the free-market philosophy espoused in *The Wealth of Nations* has become strongly associated with laissez-faire capitalism.

The potential for gains from trade was now evident, but under what conditions was trade beneficial? Suppose, for example, that a rich nation was considering trade with a poorer neighbor. If the rich nation was more efficient than its neighbor at producing *everything*, how could it possibly benefit from trade? In the early nineteenth century, the famed British economist David Ricardo proposed that gains from trade between nations were based on the relative costs of production.[61] Ricardo and others demonstrated that it is not necessary for a nation to have an *absolute advantage* in a specific product for trade with another country to be beneficial; it is only necessary that the nation have a *comparative advantage*.

The distinction between absolute and comparative advantage is a subtle one, but is critical to understanding the forces that drive international trade. I will therefore return to the two mythical trading regions introduced earlier, which will henceforth be referred to as Country W (for the land of wildebeests) and Country R (for the land of rice).

One day representatives from Country W and Country R met to discuss the prospects for trade between their regions. Both countries were currently producing the other's specialty domestically and were finding it to be very labor-intensive.[62] Hence, trade seemed like an excellent alternative.

The first step for the negotiators was to establish the relative value of wildebeests in terms of rice. Both groups agreed that the specialized technology and natural resources that enabled Country W to be efficient at wildebeest production were roughly equivalent to those that Country R employed to produce rice. They therefore decided that equal labor should represent equal value. Since Country W required one unit of labor to produce a single strapping wildebeest, while Country R needed a unit of its labor to yield one ton of rice, the nations established 'one wildebeest per ton of rice' as their *terms of trade.*

Under these trading conditions, would both countries be better off? Due to the dry climate and lack of specialized technology, Country W had previously required five units of labor to produce a ton of rice. Since it could now import the same ton of rice for a single wildebeest, it would save four units of labor, as shown graphically in Figure 2.8. By simply moving its labor from rice production to wildebeest herding, Country W would be enriched by the equivalent of four extra wildebeests.[63]

Similar gains could be enjoyed by Country R. By redirecting its labor to the industry in which it had an absolute advantage (rice cultivation), its people could feast on the same number of wildebeest steaks while producing four extra tons of rice. A combination of unequal factor endowments (climate, soil, etc.) and increased specialization could yield an economic gain for both countries.

This example assumes that the level of technological sophistication is roughly equal between trading partners. Now suppose that Country W possessed labor-saving technology that allowed it to produce both more wildebeests *and* more rice than Country R with a single unit of labor. In this situation, Country W would have an *absolute advantage* in the production of both products. Could these countries still benefit from trade?

Actually, not only would it still be advantageous for Country W to trade for rice, but Country R could afford to raise its prices. Suppose that technically advanced Country W produces either five wildebeests or two tons of rice with a single unit of labor. If Country R were to trade with Country W at its original terms, by exchanging two wildebeests for two tons of rice, Country W would be better off by three wildebeests. In fact, even if Country R were to improve its terms of trade to two wildebeests per ton of rice, Country W would still remain better off by one wildebeest.

Does it make sense that Country W would gain from trade, even though it is more efficient at both wildebeest and rice production? Although Country W has an absolute advantage in both products, Country R has a *comparative advantage* in rice. As long as the terms of trade allow Country W to buy the same basket of goods for a lower labor cost, it is better off trading with Country R. Country W derives these gains by reallocating its labor to the industry in which it is the most productive.

This is a general result. As long as each country exports the products that it produces relatively cheaply, and imports those products that are relatively costly

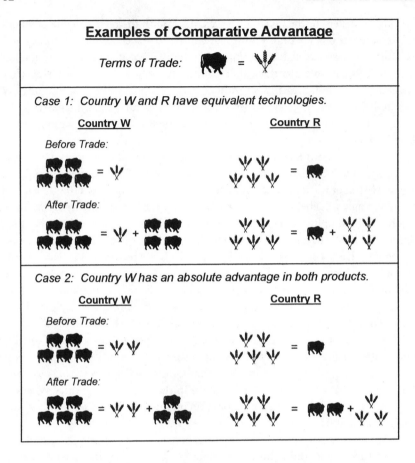

**Figure 2.8:** Example of comparative advantage between nations. In Case 1, the trading nations have equivalent technologies. In Case 2, Country W has an absolute advantage in the production of both products. In each case, however, the two trading partners are made considerably better off by trade.

to produce at home, both countries will gain from trade.[64] This does not mean that both countries have the same standard of living, however. Far from it. In this example, workers in Country W could buy ten wildebeests and five tons of rice with four units of labor. The same basket of goods in Country R would cost the local workers ten units of labor. Even though some of the technological differences between partners benefit the less-productive country, most of the gains from increased productivity go to workers in the advanced nation.

In a similar way, countries with more advanced technologies delegate activities that are lower in relative productivity to countries that can perform the tasks more cheaply. The advanced country benefits by freeing up labor to perform more valuable activities; the less-developed trading partner benefits from the advanced nation's productivity through improved terms of trade. At least according to theory, everyone wins.

Unfortunately, the liberalization of global markets can cause severe hardship to some groups within trading nations. In the more advanced economies, wages for unskilled workers can be driven downward by competition from lower-cost foreign labor. Capital owners in import-competing sectors will find their investments at risk. Regional economies can be devastated by the loss of traditional industries.

Among economists there is a general consensus that, under ideal conditions, free trade represents a "first best" strategy for global economic growth. However, the theories upon which this conclusion is based depend on some unrealistic assumptions about the nature of global markets, and tend to ignore the political and social costs of jarring change.[65]

The classical derivation of comparative advantage assumes, for example, that tastes and preferences are identical in each country. If consumer preferences vary a great deal within a country, however, some people may be made worse off by free trade.[66] A recent example occurred after the United Kingdom entered the European Economic Community. British consumers were forced to pay significantly higher European prices for agricultural and meat products, in return for access to much cheaper manufactured goods. Unfortunately, lower-income families, which spent a relatively large proportion of their earnings on food, were made worse off as a result of liberalized trade with Europe.[67]

Another reason why some groups may fail to benefit from free trade is a lack of *mobility*. Private capital owners in import-competing sectors may suffer losses from liberalized trade, since their facilities and equipment are often industry-specific. Although there may be excellent returns for capital in the semiconductor industry these days, for example, this is small consolation to the private owner of a failing textile mill. Equity owners in publicly traded companies, however, have significantly greater mobility, and are in a much better position to profit from new global markets.

Surprisingly, while highly skilled technology workers tend to be quite mobile, unskilled and semi-skilled labor is becoming increasingly *immobile*. Employment opportunities for factory workers are often tied to fixed capital assets such as production machinery. Rapidly increasing specialization at all levels of production limits worker mobility in the domestic market, while language, cultural, and economic barriers prevent most factory workers from relocating abroad.

Within the advanced nations, the negative impact of globalization has been greatest on less-skilled workers. Whereas in the 1960s and 1970s semi-skilled workers could exert some degree of monopoly power through collective bargaining, a recent surge in global outsourcing has enabled employers to substitute the services of low-wage foreign workers. Hence, business owners can shift production to take advantage of cheap labor from developing countries, while semi-skilled domestic workers are subjected to factory closures and layoffs.[68]

When considering the benefits of free trade, economists typically focus on the long-term implications, rather than acknowledging transient short-term effects.[69] Unfortunately, these short-term effects involve real people with homes, families, and dreams for the future. It may be true that over an extended period the negative impact of free trade will dissipate. Workers in the garment industry, for example, will eventually find other employment, and capital owners in "obsolete" industries will find ways to reinvest. There is no question, however, that some segments of the population will endure a disproportionate share of hardship from the dislocations caused by increased global competition.

The success of free trade critically depends on government acknowledgment and action on these issues. Some form of compensation must flow from those who benefit from increased global commerce to those who do not. Indeed, the reallocation of wealth represents a primary function of national governments during times of peace. Constructive programs of employee retraining, factory conversion, and technology transfer can achieve high benefit-to-cost ratios. Without effective policies to address these important social issues, however, any gains from free trade will be threatened by a groundswell of protectionism.[70]

### Trade Policy and Protectionism

The desire for economic growth has become almost universal in the modern world. It appears, however, that the pursuit of wealth is the only goal upon which all nations agree.[71] Differences in cultures, social norms, and political philosophies represent a continuing source of friction between even the most prolific trading partners. This paradox of common economic ambitions and conflicting social values has brought about a number of sticky international debates. Issues of human rights, environmental protection, corruption, and fair competition have all found their way into multilateral trade negotiations. The common element among these topics is that they straddle the line between economic growth and societal belief.[72]

The debate over cultural norms and accepted practices comes home to roost in the *fair-trade* rhetoric of special-interest groups. Fair-trade advocates assert that relaxed labor standards and poor environmental protection measures in developing countries give these competitors an unfair advantage. The underlying goal of these lobbying efforts, however, is not to protect the global environment or improve conditions for foreign labor, but rather to secure protection for American industries from low-cost foreign imports. If we strip away the euphemism, fair trade is simply another term for protectionism.[73]

There are two fundamental problems with the fair-trade proposition. First, can we be sure that our somewhat arbitrary standards for labor and environmental protection can be legitimately applied to foreign nations? Practices such as child labor, for example, may seem unthinkable to relatively affluent Americans, but might be a justifiable alternative to starvation in Bangladesh. Similarly, it is hard to envision the United States taking a strong position on global environmental regulations while we still choose to put the interests of business ahead of the well-being of the planet.[74]

Second, does it make sense to use global trade as a weapon of social change (assuming that is the real intent)? We no longer live in a world dominated by the U.S. economy. By holding trade hostage to unrealistic labor and environmental standards, we may simply be impoverishing ourselves. A far better way to use our considerable economic clout is by voicing our preferences in the global marketplace. Holding out the carrot of increased market demand for "labor-safe" and "green" products seems a far better solution than wielding the ineffectual stick of trade sanctions.[75]

Many developing countries, inspired by the "Asian miracle," are seeking economic growth at virtually any cost.[76] Exploding world demand for low-cost exports can represent an irresistible temptation for business leaders in these countries to exploit natural and human resources, as discussed in Box 2.1. For America to influence such important social issues as labor standards and environmental protection, it must modulate demand for foreign imports through informed consumers and natural market forces, rather than by imposing protectionistic trade barriers.

The most troubling aspect of protectionism is that *it fails to solve the problem.* Increased foreign competition cannot be blamed for much of the wage gap we are now experiencing. The jarring dislocations felt by American workers in traditional industries are caused primarily by rapid technological progress and fundamental shifts in the nature of work within our own country. Even if we were completely isolated from the rest of the world, the wage gap between skilled and unskilled labor would continue to grow, and rapid technological change would continue to cause labor turmoil.[77]

Humanitarian and environmental concerns provide a cloak of legitimacy to protectionistic trade policy, but special-interest groups represent the most common instigators of government action against imports. As with most aspects of our political system, a loud outcry from a highly visible (and financially generous) industry or labor union can distort the sensibilities of policymakers, often to the detriment of the broader constituency.

Several examples will serve to demonstrate the tremendous inefficiency associated with the protection of import-competing industries. In each case, pressure from private-sector labor and industry groups resulted in the application of either special tariffs or quotas on low-cost imports, with the goal of saving domestic jobs and industries. As can be seen in Figure 2.9, the cost of these measures to the American consumer has been unacceptably high.[79]

## Box 2.1: Growth at Any Cost?

I have one more story to tell involving our two trading partners, Country W and Country R. Unfortunately, this tale does not have a happy ending.

As was noted earlier, Country W was particularly well suited to herding large grazing animals. Its prairies were flat and covered with fodder, and the climate was ideal for maintaining a herd. This grazing land represented a natural factor endowment, which provided Country W with a comparative advantage in the production of wildebeests. In fact, it was differences in the endowment of natural factors that brought our two trading partners together in the first place.

The government of Country W recognized that this natural resource could enhance the nation's wealth, and decided to allow open grazing on all public lands. Naturally, the herding industry was elated. After all, export demand for meat was rapidly increasing, and the only limiting input factor was food for the animals. Each herding company began adding wildebeests as fast as it could, recognizing that it could profit from free fodder.

Before trade with Country R had begun, the vast grazing fields of Country W had been more than sufficient to feed its herds. By shepherding the animals from place to place, the land was given ample time to renew itself. Hence, from an environmental standpoint, the production of wildebeests had been sustainable.[78]

As the herding companies expanded to take advantage of free food, however, the tradition of rotating grazing fields was abandoned. Each herder recognized that he received all the benefit from exploiting the land, while suffering only a small disbenefit from the depletion of this vital natural resource.

This, of course, is a rendition of the well-known sociological metaphor "The Tragedy of the Commons," in which the actions of individuals, incentivized by personal gain, result in the destruction of a shared resource upon which they critically depend. In the end, Country W becomes a dust bowl, and is forced to request foreign aid from Country R to feed its people.

What is important to recognize here is that many of the environmental tragedies that we face, both in the developing world and in our own backyards, are a direct result of this benefit-seeking behavior. For too long we have treated the planet as an inexhaustible source of wealth. The time has come to include natural resources on the balance sheet of enterprise.

The most protected industrial sector in America over the last several decades has been apparel and textiles. This is not surprising, since this sector is estimated to have suffered nearly 90 percent of the job losses occurring in American import-competing industries.[80] In 1973, Congress enacted the Multifiber Arrangement (MFA), which was intended to protect these jobs from low-wage overseas competition. The MFA imposed significant trade barriers, in the form of high tariffs and quotas, on both raw fabrics and finished apparel.

There is no question that textile and apparel workers have been disadvantaged by increasing import competition from low-wage countries, and that this industry deserves some form of compensation. But at what cost? It was recently estimated that the MFA cost the American economy roughly $144,000 per job saved per year. This is more than six times the average wages typically received in this sector. In one-quarter of the job classifications covered by this measure, the cost was more than $500,000 per job.[81]

It is important to recognize that this expense was not borne by tax revenues; it came directly from the pockets of U.S. consumers, in the form of higher cloth-

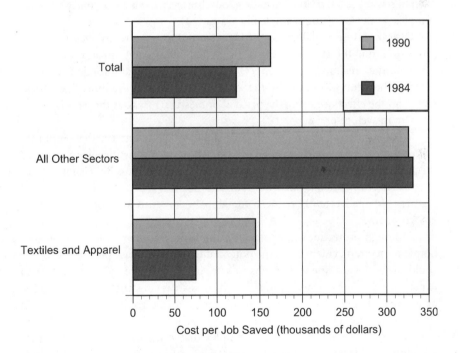

**Figure 2.9:** Cost to American consumers per job saved due to the application of special protection, for the years 1984 and 1990. Source: Hufbauer, G. C. and K. A. Elliott (1994, Table 1.4). Data used with permission.

ing prices. Furthermore, since the proportion of wages spent on clothing is substantially higher for low-income Americans, the financial impact of the Multifiber Arrangement has largely been felt by the relatively poor.

Another example that has received considerable media attention is the ongoing battle with Japan's auto industry. In the early 1980s, the U.S. government convinced Japan to impose a voluntary export restraint on its cars. This measure had the effect of increasing the average cost of automobiles (both Japanese and American) in the U.S. market by more than $1,000, with an estimated negative impact on American consumers of roughly $4 billion per year.[82] Much of the excess profits resulting from these price increases was paid directly to Japanese automobile manufacturers. In fact, the ability to maintain artificially high prices was so advantageous to the Japanese that they "voluntarily" retained these quotas even after the United States ceased to demand them.

If the costs of protectionism were borne exclusively by the consumer, the decision to erect trade barriers would represent a simple welfare tradeoff: Are the jobs that could be saved worth the exorbitant cost? When trade protection is applied to raw materials or intermediate goods, however, the issue becomes far more complex, as the following example illustrates.[83]

Over the last several decades, the U.S. steel industry has received considerable protection from foreign competition. In 1984, the Reagan administration negotiated voluntary restraint agreements (VRAs) with several major foreign suppliers of steel. The result was similar to the preceding automotive example: These measures had the effect of raising the price of domestic steel above the world market price by an average of 7 percent.[84]

What is different about this example is that steel is an intermediate input for a huge number of American products.[85] Import protection artificially maintained the price of steel above the world market price, thereby effectively raising the cost of *every American product made from steel*. In the process of protecting jobs in one segment, we inadvertently reduced the global competitiveness of thousands of U.S. companies.

Although protectionism appears on the surface to represent a simple economic tradeoff, it evidently has the potential to backfire catastrophically. No one could make the argument that jobs in steel production are more important to Americans than jobs in steel-using industries, such as construction, automobiles, or household appliances. Saving jobs in one sector by risking jobs in others is simply an unacceptable policy alternative.

Can we conclude from these examples that America should immediately abandon all trade barriers in favor of free trade? Unfortunately, such a sudden and dramatic liberalization of our markets would likely lead to economic disaster. Achieving gains from free trade depends on there being very few distortions to the world trading system. Given the current lack of global harmony in trade policy, it would not be advantageous for the United States to unilaterally drop all tariff and non-tariff barriers. The best pathway to global free trade is through a gradual, multilateral lowering of barriers, as is prescribed, for example, in the GATT.

Furthermore, the contentious realities of the current global marketplace provide several legitimate motivations for imposing tariff or non-tariff barriers. For example, there have been increasing numbers of assertions in recent years that some foreign markets are closed to U.S. imports. This has been a perennial problem with Japan, and has most recently become a point of contention with China.[86] If a foreign government restricts the quantities of imports or applies unreasonable regulations, the threat of a retaliatory tariff might provide the leverage needed to increase foreign market access.[87]

In recent years, the most common motivation for imposing trade sanctions has been retaliation for the *dumping* of imports into the American market. Dumping refers to a foreign competitor exporting products at a price that is below its production costs in an effort to penetrate the U.S. market or to increase market share.[88] Since dumping represents unfair competition, an injured party can petition the U.S. Commerce Department to undertake an investigation. If the claim is found to be legitimate, it is submitted to the U.S. International Trade Commission (USITC), which determines whether material injury has been sustained by the petitioner. A favorable judgment will result in an *antidumping tariff* or other remedy being imposed.[89] Export subsidies by foreign governments that artificially reduce the price of imports in the U.S. market can be addressed through the same antidumping measures. In this case, the unfair competitive practice might be offset with a *countervailing duty* to equalize the playing field.

If selectively applied, these retaliatory measures can actually move to restore perfect competition and increase overall economic welfare. Unfortunately, the criteria used to evaluate charges of dumping, and to determine the degree of material injury, are heavily biased in favor of the "injured" American petitioner. Antidumping cases have skyrocketed over the last decade, particularly in the United States, as shown in Figure 2.10. What was intended to be a trade-normalizing process has been corrupted by opportunistic petitioners into the latest protectionist method of choice.[90]

A final policy tool that has become important in recent bilateral trade negotiations is the *Section 301* provision of the Omnibus Trade and Competitiveness Act of 1988.[91] This provision mandates that the U.S. Trade Representative (USTR) take all appropriate actions, including the imposition of retaliatory tariffs or other remedies, to obtain the removal of any policy, act, or practice by a foreign government that violates an international agreement or in some other way is discriminatory to U.S. industry. This clause has been supplemented with the so-called *Special 301* provision, which specifically addresses countries that have a history of violating intellectual property laws or agreements.

Both of these provisions have been used extensively by the Executive Office of the President over the last decade to target countries which support "unfair" trade

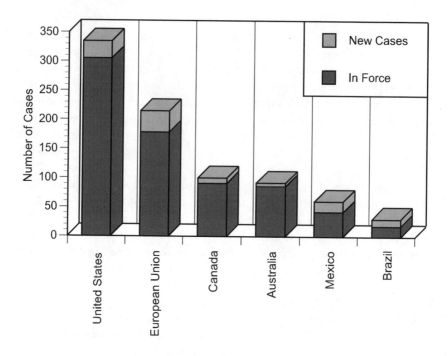

**Figure 2.10:** Number of new antidumping cases and measures in force as of mid-1995, for the top six initiating countries. Source: WTO Secretariat, 1995, *Focus*, No. 7, December, pg. 10.

practices. Thus, Brazilian computer and telecommunications import restrictions, Chinese market access, Canadian lumber subsidies, intellectual property rights in India, and a host of other trespasses have been addressed by the USTR through the Section 301 process. Similarly, the Clinton administration has recently made effective use of a "Super 301 executive order" to gain advantage in stalemated bilateral trade talks with Japan.[92]

Trade policy in America today consists of an arsenal of weapons intended to intimidate trading partners into accepting our judgment of fair competition and open markets. This rather one-sided posture has been effective in the past, but it may prove to be self-defeating as other economies begin to share center stage. Nonetheless, some level of activism may be necessary to secure fair opportunities for American firms in global competition.

The strategic importance (from an economic standpoint) of trade in technology-intensive products and services suggests that we have much to gain by enforcing a level global playing field. Moreover, some new thinking in international trade theory suggests that trade policy could become a weapon of competitive advantage. At least in theory, the application of selective tariffs and subsidies might tip the scales of the global economy just slightly in America's favor.

## Strategic Trade Policy

Beginning in the early 1980s, trade theorists began to recognize that some of the underlying assumptions of classical comparative advantage were not valid for modern international trade. It was observed, for example, that as a result of economies of scale and high entry barriers, global rivalries within an industry often took the form of a few strong firms engaged in *imperfect competition*.[93] In these oligopolies, pricing, investment, and output decisions were often dependent on the actions of competitors.[94]

A new body of work, referred to collectively as *strategic trade theory*, suggests that governments might exploit these market distortions to increase national wealth. In principle, governments could target "strategic" industries for selective subsidy and protection, in an effort to shift the competitive playing field to the advantage of domestic enterprise. Strategic industries would be selected based on their ability to provide spillover benefits to society. Technology-intensive industries are prime candidates, since evidence suggests that productive innovations "leak" into both related and unrelated sectors at a rapid pace.[95]

Unfortunately, there are several significant limitations to strategic trade policy.[96] First, at its current level of development there are few solid conclusions that could form a basis for national policy. For example, it is not clear which industries should be targeted, or what measures will garner the highest returns. Second, the potential benefits from these maneuvers are quite small, and are always derived at the expense of our trading partners. This leads us to a final negative aspect of strategic trade policy: It doesn't work if other nations retaliate in kind. Given the highly competitive nature of global markets, it seems unlikely that other nations would sit idly by as we drain profits from their industries.

Although the study of strategic trade policy has provided some welcome insight into the behavior of imperfect markets, the prospect of a costly trade war should temper our enthusiasm for implementing this modern-day form of mercantilism.

## Global Competition in High Technology

In the first chapter, I presented evidence that the advanced nations have become approximate equals in capital accumulation, labor skill levels, and general technical sophistication. The industrialized countries appear to be converging to common aggregate levels of total-factor productivity, presumably due to the rapid transfer, diffusion, and adoption of new technologies. This scenario led to the conclusion that trade in technology-intensive goods among the developed nations is increasingly driven by subtle differences in productivity within each industry segment. But how do we reconcile this picture of technology trade with the classical theory of comparative advantage?

Imagine that there are two distinct levels of trade in world markets. On the "lower" level, trade is driven by the forces of classical comparative advantage: unequal distribution of the factors of production. Countries specialize in products that they produce at a low cost relative to the world market. This classical tier would be characterized by trade in most raw materials, agricultural products, commodity manufactured goods, and generic services.

On the "upper" tier of the global marketplace, all the factor inputs that form the basis for classical trade are equalized.[97] Despite input costs being roughly equivalent, comparative advantage still applies; in this case, differences in total-factor productivity will determine the patterns of trade.

In the commodity-driven lower tier, comparative advantage is largely beyond the control of individuals or even governments; labor and capital prices are determined by market demand, and natural resources are inherently immobile. In the technology marketplace, however, firms can establish monopoly positions and value leadership through intentional acts of innovation and specialization. This process of *creating comparative advantage* determines the outcome of global competition in technology-intensive industries.

### A Few Misconceptions

In recent years, trade issues have received unprecedented levels of public scrutiny. Unfortunately, this democratization of the trade debate has spawned a good deal of misinformation, particularly with respect to the negative impacts of global competition on the American standard of living. It is worthwhile considering several of these controversial issues under the harsh light of objective observation.

The description of international trade that I have provided thus far has suggested a benign role for national governments. Furthermore, I propose that cooperation among trading partners is almost always preferable to conflict. Can this be right? Is America not at "war" with foreign governments, economically speaking?

It has been asserted that the military tensions of the Cold War have been replaced by an economic conflict among nations, in particular Japan, the European Union, and the United States.[98] While it is true that competition in the global marketplace is becoming increasingly contentious, it is important to recognize that these raging battles are being fought between *firms*, not governments. Does foreign competition represent a significant threat to our national welfare, or have we inappropriately transferred our political aggressions onto a new battlefield?

First, it should be noted that our exposure to "threats" from the global economy is still rather low. The total value of U.S. trade with foreign markets in 1995 represented only about one-tenth of our GDP, with exports totaling roughly 8 percent and imports amounting to nearly 11 percent.[99] Although this percentage is growing, the domestic market will continue to dominate our economic lives for the foreseeable future.

Much of the current debate over trade policy deals with threats to American employment by low-wage foreign workers. These concerns may be valid with respect to the dislocation of American workers discussed earlier, but increased global trade has little effect on overall levels of employment. The rate of unemployment is determined by macroeconomic factors; trade can impact the mix of employment and the wage rate, but does not determine the availability of jobs in general.[100]

The economic expansion of both advanced and developing countries is often viewed as a threat to America's well-being: if Third World countries are growing, it must be at the expense of domestic jobs and wages. Voices of reason among economists have repeatedly pointed out, however, that economic gains in other countries do not necessarily reduce our own welfare.[101] In fact, there is a high likelihood that we will be better off with wealthier trading partners. They are, after all, our customers.

Viewing nations as customers can be taken to excess, however. Recent promotions of exports by the U.S. Commerce Department have begun to stretch the limits of economic credibility. Although the expansion of foreign markets for American products is definitely a good thing, the gains from trade result from access to low-cost imports, not opportunities to export.[102] It is important to separate the "micro" from the "macro" arguments here; a $2 billion contract for 747s might allow the Boeing Company to hire more workers and pay extra dividends to shareholders, but it will have a negligible effect on the overall growth of the U.S. economy. The reason we sell products overseas, from a macroeconomic standpoint, is *to pay for low-cost imports*. The gains from trade come from the reduced cost of imported items, not from the increased profits of export-sector firms.[103]

There are, however, some legitimate grounds for concern with respect to our place in the global economy. The expansion of investment in the developing world may cause capital to become more costly for American companies. The lack of access to markets in advanced nations might limit the ability of our infant industries to become competitive. Predatory pricing by the newly industrialized countries can destroy the profitability of American high-technology products before producers can recapture their investments. All these issues require serious consideration, and could potentially warrant aggressive action. It is therefore critical that we strip away the misconceptions regarding global competition, and focus on the more subtle battles that must be waged.

### The Terms-of-Trade Effect

As I demonstrated in Chapter 1, the primary source of sustainable economic growth is increasing productivity. There has never been a country that has shown increased domestic productivity over a significant period without also showing substantial economic growth.[104] Does this mean that foreign competition has nothing to do with our standard of living? Well, not exactly. There is at least one way in which gains in productivity by foreign competitors can reduce our economic welfare.

The effect is known in trade theory as *factor-price equalization*.[105] Conceptually, it implies that there is a linkage of sorts between the production functions of trading partners. As long as the partners do not compete in the same markets, there is no linkage. Once they begin to produce the same commodity for sale on the global market, their input-factor prices must converge to the ratio of their total-factor productivities. If all other input costs are roughly equal, the trading partner with the higher relative productivity can support higher wages. If that lead erodes, however, the wage gap will narrow. If productivities become equal, then wages will tend to equalize as well.[106]

The general result from trade theory states that if two countries produce and export the same commodity, global market forces will equalize the price under free trade. This will have the ultimate effect of equalizing the price of the factors used to produce it. Hence, if two countries with differing labor costs produce the same commodity using similar technology, factor-price equalization will tend to pull downward the compensation of the higher-wage nation.

The message here is clear: Changes in productivity between trading partners can reduce the effective standard of living of the lagging country. The brunt of the impact will be felt in import-competing sectors, but the entire economy will ultimately suffer. The strength of the effect will be determined by the magnitude of changes in productivity rates and by the number of products in which the trading partners directly compete.

Until the late 1960s, the American automobile industry led the world in productivity. As a result, our cars commanded high relative prices on the global market. At about this time, Japanese automakers began to export cars of better quality which, due to high productivity and relatively low wages, could be sold at a lower price. Over the next decade, market forces drove down the demand for U.S. cars, creating enormous pressure to reduce production costs. Ultimately, American autoworkers were subjected to severe wage cuts and layoffs.

If productivity within a nation lags in only a few industries, the negative effects will be limited. Lagging productivity across many sectors, however, could force a currency devaluation. If this occurs, buying power in world markets is reduced, resulting in a decrease in the effective standard of living. This is known as the *terms-of-trade effect*.[107]

The only way for us to avoid the erosion of our terms of trade is to focus our attention on achieving the highest possible productivities in all exporting sectors. Since virtually all of America's technology-intensive industries will soon become competitors in global markets, why restrict ourselves to those that currently export? We should be grooming the next round of world leaders, while nurturing our current stable of technological giants.

Globalization offers opportunities for an improved division of labor in almost every industry. Labor-intensive products are inherently better suited to production in developing countries, while advanced economies will tend to export products that require high skill levels and the accumulation of capital and technology. The challenge for the United States is to accept this global division of labor, and accelerate the development of social programs to transition relatively low-productivity industries and workers away from wage-slashing foreign competition.

### The New Trade Environment

At the beginning of this section, I described a two-tiered conceptual model for the new global trading environment, which is summarized in Figure 2.11. The lower tier is driven by factor costs, with production technologies being roughly equal. The upper tier is technology-driven, with the cost of inputs being nearly equivalent. Although this model does not represent a rigorous description, it does provide some useful insights into the behavior of current world markets.[108]

The bifurcation of international markets did not occur suddenly. Shortly after World War II, economists began to note that new forces might be supplanting classical comparative advantage as the determinants of trade patterns. In the early 1960s, Staffan Linder suggested that the principles governing trade in manufactured goods were not the same as those determining the flows of primary products.[109] While he agreed that trade in raw materials and agricultural products was driven by unequal factor endowments, he argued that the response of entrepreneurs to perceived needs in the marketplace was a primary determinant of trade in manufactured goods. This concept was carried further by noting that tastes among the advanced economies were becoming increasingly similar. Hence, *overlapping demand* became an early explanation for intra-industry trade between developed nations.[110]

A dynamic model for trade in technology-intensive products was proposed by Raymond Vernon at about the same time, in which the idea of a *product life cycle* was first introduced.[111] Borrowing the idea of entrepreneurial innovation as the gestation for new products, Vernon constructed a time-based model, in which products evolved through phases from creation to standardization. For technology-intensive products, he proposed that the level of maturity of the product determined the location of production, and ultimately the flow of trade. The initial innovation, along with early production, would likely take place in an advanced country. As world demand grew, competitors from other developed countries would enter the market, and the increased competition would drive production toward low-wage countries. In the final stage, the product would become standardized, prices would drop to commodity levels, and the cost of inputs (in particular, labor) would become the primary driver for trade flows.

## A Two-Tier Model for Global Trade

**Upper Tier (Technology-driven):**
> Trade flows determined by market demand and created comparative advantage
> Leadership based on absolute advantage in productivity
> Imperfect (monopolistic) competition
> Technological specialization among nations
> Intra-industry trade in differentiated products among advanced nations

**Lower Tier (Input-driven):**
> Trade flows determined by classical comparative advantage
> Leadership based on unequal endowments of input factors
> Perfect (price-taking) competition
> Generic production technologies equal among nations
> Intra-firm North-South trade based on rationalizing production and global division of labor

**Figure 2.11:** Summary of a two-tiered conceptual model for global markets in technology-intensive products.

Today, this concept of a linear diffusion of new innovations throughout the globe has been replaced by more sophisticated thinking. Multinational corporations no longer wait for "maturity" to establish production of high-technology products in low-wage countries. Except for the most technically challenging cases, production costs are typically optimized from product launch. Each step in the value chain, including raw material procurement, manufacture, test, distribution, and service is planned from the onset to optimize global cost and efficiency.

This represents a new form of industrialization, in which the tasks involved in a product's creation are divided among nations of the North and South, based on the relative cost of inputs and patterns of market demand.[112] The final products, however, will flow among the advanced nations in patterns determined by a complex mix of scale economies, imperfect competition, and technical specialization.

Conceptually, MNCs utilize the input-driven (lower) tier of the two-level model to execute all the physical aspects of the production process. The relative skill levels of labor and productivity of capital are carefully weighed for each step in manufacture and distribution, and nations are selected to serve as a base for suppliers, or as sites for foreign affiliates. This rationalization process must also take into account tariffs, local content requirements, intellectual property protection, and other obstacles. If these issues are considered during the initial formation of production strategy, firms have the opportunity to maneuver around some of the more cumbersome trade barriers.

It is on the technology-driven (upper) tier, however, that the winners and losers in international competition are decided. Since firms from every advanced economy have equal access to the input-driven tier, in principle any cost advantage achieved there can be matched by all competitors.[113] Leadership in this conceptual marketplace, therefore, is determined not by the costs of inputs, but by how well firms make use of them. In other words, it is advantages in total-factor productivity, not input costs, that determine the victors in global markets for technology-intensive products.

This leads us to a strategic imperative. If created comparative advantage is defined by the singular dimension of total-factor productivity, there can be only one global leader. It is therefore necessary for firms to achieve an *absolute advantage in total-factor productivity* to hold the dominant position in world markets for technology goods.[114] Our discussion of competitiveness has come full circle. Rather than being bound by forces that are largely beyond the control of firms, comparative advantage can be created by achieving the highest global levels of total-factor productivity.

In recent years, it has been suggested that the rapid globalization of technology will render national borders irrelevant. This popular view of a world linked by the global research networks of MNCs suggests a diminishing (and possibly obstructionistic) role for governments.[115] In reality, there is mounting evidence that the environment for innovation within the home nation of MNCs is a primary determinant of their success.[116] In the new trading environment, competitive advantage stems from a resonance among national policies, regional environments, and corporate strategies. As the next chapter will demonstrate, the extent of *technoglobalism* is highly overstated. Nations still provide the catalysts for creating and sustaining global market advantage.

# 3

# The American
# Technology Enterprise

## Technology-Driven Productivity Growth

The broad discussions presented in the first two chapters followed different paths, but arrived at a common conclusion regarding technology's role in the global economy. Trade patterns in technology-intensive products are increasingly determined by specialization among economic and technological equals, and leadership in total-factor productivity is the key to global competitive advantage.

In this chapter, we will explore the current state of American technology enterprise, and in particular the role of industrial research and development (R&D) in achieving global market leadership. The underlying goal in the sections that follow is to solidify the nebulous concept of total-factor productivity, and to begin identifying strategies that maximize it.

### Productivity Through Innovation

Studies of economic growth in the early postwar period defined productivity as a residual: that portion of output that could not be attributed to growth in factor inputs. Although it was generally accepted that technology played a critical role in economic development, it was assumed that innovations that yielded gains in productivity were a random outgrowth of basic scientific research.

Entrepreneurs, on the other hand, had recognized for decades that new productive technologies could be a source of windfall profits. What was not clear to either economists or entrepreneurs was the process by which scientific knowledge could be effectively harnessed to achieve increased productivity.[1]

The origins of modern thought with respect to industrial innovation is often ascribed to Joseph Schumpeter, who proposed a basic process for the intentional creation of productive technology. In his classic mid-twentieth-century work on the driving forces of capitalism, he suggested that in the hunt for profits, the entrepreneur gathers general scientific knowledge and transforms it into productive technology through privately funded industrial research.[2]

Schumpeter proposed a three-step process that begins with a discovery or invention derived from basic scientific knowledge. The entrepreneur provides the creative link between the initial discovery and its first application in production, through the funding of industrial R&D. The final step in Schumpeter's process involves diffusion of the resulting innovation to other enterprises, primarily through imitation.

Unfortunately, this early framework for industrial innovation provides limited insight into what has become an increasingly complex and disorderly process. The view of innovations as isolated "linear" events has been supplanted by a recognition of the interdependent, evolutionary, and revolutionary behaviors of technological progress. Furthermore, the scope of industrial innovation is now seen as encompassing all aspects of modern enterprise, including product development, process engineering, marketing, distribution, service, and product-line extension and adaptation.[3]

Before we embark on a pragmatic assessment of the innovation process within American technology industries, it is worthwhile taking a final look at the theoretical basis for economic growth and productivity. Recent work has extended the classical understanding of technical progress presented in Chapter 1 to include the realities of endogenous technological change. The incorporation of intentional and controllable innovation into growth models has yielded a clearer picture of the nature of industrial innovation, and the methods by which both firms and nations can profit from its pursuit.

### New Theories of Technology-Driven Growth

One of the first empirical tests of the idea that industrial innovation resulted from intentional, profit-seeking action was undertaken by Jacob Schmookler shortly after World War II. In a study of more than one thousand inventions in four different industries, he demonstrated that in essentially every case, the stimulus for innovation was profit. He concluded that his study "controverts the time-honored belief in a continuous increase in rate of technical progress."[4] In other words, productive innovations are driven by market demand and the profit-seeking actions of entrepreneurs.

Based on this influential work, economists began considering models for growth that included assumptions of endogenous technological change. Initially, gains in productivity were attributed to "learning by doing," primarily through the expanded industrial use of capital equipment.[5] The resulting innovations were assumed to "leak" into the public domain after a time to become universally available. Hence, in these early models, the technical knowledge developed by firms was treated as a *public good.*

A public good is assumed to be created for non-economic reasons, and, as such, is free to all users. If industrial innovation were truly a public good, firms could not exclude competitors from taking full advantage of their hard-earned technology. With no opportunity to gain a competitive advantage, there would be little incentive for firms to invest in research and development, and economic growth would grind to a halt.

One of the first researchers to propose formal models of economic growth that identified investment in R&D as the source of industrial innovation was Paul M. Romer.[6] The so-called "new growth theories" developed by Romer and others suggest the possibility of sustainable long-run growth, based on the unique economic properties of technical knowledge.[7] Early models had predicted that growth would eventually peter out, primarily due to diminishing returns to capital investment. Romer pointed out that productive knowledge, created through firms' investment in R&D, could overcome this slowing effect by exhibiting *increasing returns to scale.*

When knowledge is treated as an economic commodity, it displays some striking properties as compared to physical artifacts. A piece of machinery, for example, can perform its function either in Pittsburgh or Los Angeles, but not both. Physical objects are distinct; each unit requires inputs of materials and labor to produce it, and can be bought and used by only a single individual. Economists refer to such "hardware" products as *rival goods,* meaning that use in one application precludes simultaneous use in other applications, as discussed in Box 3.1.

Technology, on the other hand, is a *non-rival good.* The use of a piece of productive knowledge by one individual does not preclude its use by others, even simultaneously. The innovations incorporated into a new automobile design, for example, can be embodied in every unit produced. Whereas the sale of a piece of hardware is a one-time event, the technology embodied in products can be sold and resold indefinitely. Furthermore, since the knowledge content of a product has a zero marginal cost, producing additional units does not require a proportional increase in the cost of knowledge, and therefore results in increasing returns to scale.

There is only one problem with this argument from an economic standpoint. Since technical knowledge is non-rival, it can easily be expropriated by other firms. The rival nature of physical products makes them almost perfectly *excludable* from use by others. Laws have been on the books since ancient times to protect individuals' rights to ownership of real and personal property. Knowledge, on the other hand, can be excluded from use by others only through the legal protection afforded by intellectual property rights. Since these laws often prove to be an ineffective barrier to imitation, some privately funded knowledge almost always spills over into the public domain. Hence, technology is considered to be a *partially excludable* good.[8]

This is not just a semantic distinction. If we accept that productive innovations are created by intentional investment in industrial research, these non-recurring expenditures must be recaptured in some way by firms. The ability to exclude the use of a new innovation from competitors allows firms the opportunity to gain monopoly profits from their investment. This is essential to the reimbursement of R&D investment, and provides the incentive for firms to continue the innovation process.

The degree to which firms can profit from R&D investment is determined by two factors: the opportunities available in a given industry to gain competitive advantage through improved technology, and the *appropriability of returns* from innovation. With this conceptual framework in mind, we will examine the methods by which firms profit from investments in research and development.

### R&D Takes Center Stage

In the early postwar years, the growth of total-factor productivity in American industry was tightly correlated to levels of investment in research and development. Not surprisingly, this linkage was found to be the strongest in the most technology-intensive industries.[10] Beginning in the late 1960s and continuing throughout the 1970s, however, industrial R&D spending steadily declined, following a pattern that closely correlates with the two-decade slump in U.S. productivity described in Chapter 1. From a peak of 3 percent of gross national product (GNP)[11] in 1964, R&D investment slid to a low of 2.3 percent of GNP by 1975.[12] Did reduced R&D spending contribute significantly to the now-infamous productivity slowdown?

It is easy to accept that there is a positive correlation between total-factor productivity growth and investment in research and development. What is still not clear, however, is the process by which firms go about creating productive innovations. There is an obvious connection between advances in basic science and advances in commercial technology, yet many profitable new innovations are based on science that is decades old.[13] How do firms translate available scientific and technical knowledge into innovations that improve production efficiency or create customer value?

For us to gain some insight into the innovation process, we must consider the factors that motivate investment in industrial research at the level of the firm. As noted previously, the production function used in aggregate growth models becomes product-specific when considered at the firm level. Every output of industrial enterprise requires a unique combination of raw materials, skilled and unskilled labor, capital, and manufacturing technologies. Hence, the degree to which productivity can be enhanced in a given firm or industry is dependent on its unique technological and market environments.

## Box 3.1 - A Taxonomy of Knowledge[9]

The task of describing the nature of human thought seems better suited to poetry than to economics. Yet when knowledge is treated as a commodity, economists feel obliged to subdivide and classify it. One approach to this categorization process involves the determination of the *type of good* a piece of knowledge represents.

At the two extremes are *public goods* and *private goods*. A public good is available for use by any individual at no cost, and hence has no direct economic significance. Basic scientific research published in academic journals is a commonly cited example of knowledge as a public good. A more familiar example would be the television and radio signals broadcast over the airwaves.

A private good, on the other hand, is not available for use by individuals without some compensation flowing to the owner. While enforcement of the private-goods aspect of a physical object is straightforward, it is considerably more difficult to ensure the private use of knowledge.

Another approach to classifying the economic potential of knowledge involves determining its degree of *rivalry* and *excludability*. A rival good can be used by only one individual at a time, whereas a non-rival good can be used by many individuals simultaneously. A good that is excludable can be withheld for private use, while a non-excludable good can be accessed by all who desire it.

Knowledge as a public good is both non-rival and non-excludable, whereas patented knowledge is non-rival and at least partially excludable. The degree to which privately funded R&D is excludable determines the potential for firms to appropriate returns from their investment. That portion of R&D knowledge that cannot be excluded diffuses into the public domain to form a technological foundation upon which other firms can launch their innovative efforts.

A final thought is worth considering. There is an important class of public goods that is rival but non-excludable; examples include a common pasture, the air we breathe, and the water we drink. These goods are available for public use, but due to their rival nature, they can be damaged by unrestrained demand and opportunistic behavior. This is yet another manifestation of "The Tragedy of the Commons," in which free choice in the presence of rival, non-excludable goods can lead to waste and potential devastation.

In technology-intensive industries it is common to consider the alternative forces of *market-pull* and *technology-push*. In the former case, a well-defined need identified in the marketplace is satisfied through innovative product development. In the latter, newly available technologies provide the impetus for firms to create new markets or to revolutionize outdated products. The development of cellular phone technology in response to market demand for increased mobility is representative of market-pull, whereas the initial introduction of personal digital assistants (PDAs) in the early 1990s is characteristic of technology-push. Without the presence of both a technological opportunity and a clear market demand for a new product, however, the risk of investment in R&D is unacceptably high.[14]

Different industry sectors vary considerably in their opportunities for technological advancement, based in large part on their proximity to the frontiers of science.[15] It is essential for firms in the computer or semiconductor industries, for example, to expend large sums on R&D, while a more modest research investment is warranted in the shoe business. Likewise, the technical sophistication of suppliers, distributors, competitors, and most importantly customers can be a significant determinant of R&D efficiency.

Generally, the opportunities for productive innovation fall into two categories: those related to products and those related to processes. This is made clear by noting that total-factor productivity can yield economic gains in only two ways: either from an increase in the value of an output (i.e., product innovations) or by more efficient use of inputs (i.e., process improvements).

Enhancing the value of existing products can be accomplished through innovations that provide higher quality or increased variety. Customers will pay a premium for a product that delivers improved performance, displays superior reliability, or offers unique and desirable features. In an immature market, competing firms will typically pursue performance advantages, focusing their R&D efforts on reaching the next rung on the "quality ladder."[16] As the underlying technology for the product line matures, this process will reach a point of diminishing returns. Further quality improvements will fall below the consumer's threshold of "least discernible difference," and will fail to capture additional revenue.

Once the market for a product has reached its optimal level of quality and performance, it will begin to broaden. Firms will invest in differentiated products that cater to increasingly narrow subsegments of the customer base. The initial returns on this type of innovation can be high, particularly if a significant degree of customization can be achieved. However, as the number of varieties proliferates, products will become increasingly better substitutes for each other, and the marginal returns to R&D investment due to product differentiation will decline. Ultimately, such a market will evolve into a commodity structure, in which low recurring production cost becomes the only source of competitive advantage.[17]

In this mature market scenario, process innovations rise in importance. Historically, this arena has received scant attention by American industry. This is unfortunate, since the correlation between innovation and gains in total-factor productivity is significantly greater for investments in process improvement than for product-related R&D.[18] Innovations of this type can range from revolutionary implementations of automation or computer control to a continuous stream of minor improvements bubbling up from the factory floor.

Process innovations almost always require embodiment in capital equipment. In the United States, firms typically display a high ratio of total-factor productivity to labor productivity, meaning that we tend to depend more on clever product innovations than on capital deepening. In this environment, relatively low-cost solutions such as demand-pull production systems and just-in-time supply-chain management may be more acceptable to U.S. managers than capital-intensive automation and robotics. It is important to note that, all other factors being equal, a low-cost process innovation will offer higher returns than a high-cost improvement.

The opportunity for a firm to innovate new products or processes is a necessary condition for profiting from investment in R&D, but it is not sufficient, however. Unlike commodity production, which derives its profits from a small margin between cost and price, a technology-intensive firm must secure some level of monopoly profits from its investment. Since these excess profits are based on the unique knowledge incorporated into the product, the ability to appropriate a significant fraction of the total value of that knowledge is crucial to a firm's survival.

A number of studies have been performed on the relative rates of return for investments in R&D. In each case, it was found that firms could appropriate only a fraction of the returns from their innovations; some portion of the benefits from a new product or process technology always spilled over into the public sector. Despite this partial loss of monopoly profits, investments in industrial research and development have been shown to offer average rates of return of 30 to 50 percent.[19]

From a policy standpoint, the appropriability of returns from research and development offers an interesting paradox. On the one hand, firms require adequate incentives to continue investing in industrial research. Since the cost of innovation in technology-intensive sectors is especially high, a significant fraction of all returns must flow back to the firm for R&D investment to continue. On the other hand, the spillover of technology into the public domain has been shown to offer significant benefits both to other industries in the same sector and to society in general.[20]

The primary method by which firms can appropriate returns from R&D investment is through intellectual property rights (IPR) policies, including patent, trademark, copyright, and trade-secret laws. Unfortunately, patents and other IPR measures are fundamentally limited by the nature of the property they are intended to protect. One of the early researchers in this field, Kenneth Arrow, pointed out that "no amount of legal protection can make a thoroughly appropriable commodity of something so intangible as information."[21]

In a survey of 650 R&D managers spanning 130 industries, it was found that patents were considered to be generally more effective at protecting innovations in processes rather than in products, and were more suitable to excluding duplication of ideas rather than protecting royalty profits. A number of reasons were cited for the limited effectiveness of patents, including weak enforcement by firms, disclosure through patent documentation, and the potential for "inventing around" a protected innovation. A sampling of the perceived effectiveness of patent protection in several technology-intensive industries is given in Figure 3.1.[22]

Although most technology-intensive firms consider patent and copyright protection to be their first line of defense, these legal means may prove to be of limited value in future markets.[23] Our technology base is expanding at a geometrically

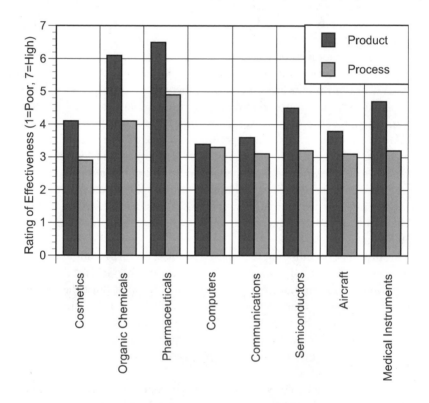

**Figure 3.1:** Perceived effectiveness of patents for products and processes in several industries. Results based on a seven-point scale of perceived effectiveness, with seven implying the highest. From a survey of 650 R&D managers reported in Levin, R. C. et al. (1987, pg. 797, Table 2).

increasing rate. In many cases, the patenting process is far too slow to provide any useful protection. With an application period of eighteen to thirty-six months, it is likely that the profitable life of a product in these industries will be exhausted before patent protection is in place. Likewise, in high-innovation-rate sectors such as semiconductors and software, there is considerable potential to invent around a patent, or to introduce an aggressive new technology that supplants existing innovations.

In this dynamic environment, firms have several alternative means of appropriating returns from R&D investment. Often the so-called "first-mover" advantage is sufficient to secure a dominant market position in high-technology industries. In scale-intensive sectors such as aircraft and semiconductors, being the first to expand production volume offers both market-share and marginal-cost advantages. Furthermore, the effects of the learning curve on production costs may allow the first mover to always stay slightly ahead of the competition.

In industries with strong consumer awareness, a first-mover advantage can be extended significantly through investment in brand identity, another form of intellectual property. This strategy has been used effectively for high-technology products ranging from pharmaceuticals to microprocessors.

Finally, and perhaps most important, firms that enjoy either a first-mover advantage or a market-leadership position have an "invisible" edge in maximizing the appropriability of returns. Considered from a purely dynamic standpoint, the ability to build on top of an existing technology base can enable firms to achieve a sustainable productivity lead. This capacity for *technological absorption and accumulation* is a critical factor in the long-run competitiveness of firms.

On the other side of the appropriability issue are the benefits that society derives from technology spillovers. While the private returns to R&D investment are estimated to be roughly 30 percent, several studies have shown that social returns can be significantly higher.[24] These spillovers occur in a number of ways, ranging from reverse engineering of new products by competitors to careful observation of the sales and marketing strategies of innovating firms. Some estimates of the private and social rates of return for technology-intensive products are shown in Figure 3.2.

Although spillovers typically cost the innovating firm monopoly profits, their impact on society is unambiguously positive. Many of our most competitive industries would not exist without the benefits of access to "public domain" commercial technologies. If Bell Telephone Laboratories, for example, had been given ironclad patent coverage on the transistor and the laser, hundreds of subsequent innovations in both related and unrelated industries might never have occurred.

Policymakers are therefore confronted with a tradeoff between the rights of firms to appropriate returns from R&D, and the potential benefits to society of encouraging some level of technology spillover. Should intellectual property laws be strengthened to provide additional incentive for firms to innovate, or should

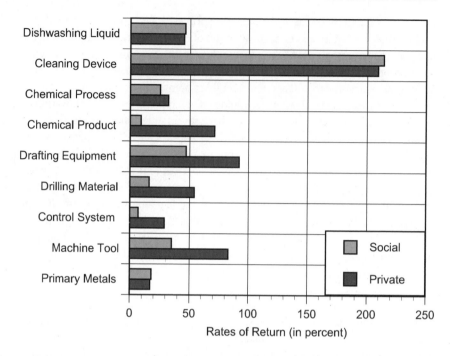

**Figure 3.2:** Social and private rates of return for several technology-intensive (as well as not so high-tech) innovations. Source: Mansfield, E. R. (1996, pg. 117).

they be weakened to increase the outflow of spillover benefits? One could argue that the opportunities for innovation in high-technology industries are so great that some intentional spillovers, perhaps through cooperative ventures and increased publication in the open literature, might be a responsible gesture.

Although it seems unlikely that this altruistic argument will inspire generosity on the part of technology firms, an appeal to self-interest might be more convincing. There is growing evidence that in our current technology-based economy, spillovers are the engine that drives economic growth. Without the public-goods nature of spillover knowledge, new competitors could not overcome the financial barriers to market entry, and new industries would not be as likely to form. In a sense, spillover knowledge reduces the marginal cost of innovation for all enterprise.[25] This synergistic effect may prove to be essential to sustaining both technological and economic growth.

## The Global Impact of American R&D

The apparent recovery of American productivity in the 1990s offers a ray of hope in an otherwise dismal last two decades. But is this turnaround sustainable? Has the source of these gains been an upsurge in industrial innovation, or has our recent obsession with corporate downsizing and restructuring created an artificial, and very temporary, rise in an otherwise downward slide?

Since the mid-1980s, much of corporate America has been distracted from the business of R&D. Our reactionary response to threats from foreign competition has been to squeeze every drop of waste and inefficiency from our organizations. Unfortunately, this process rapidly reaches a point of diminishing returns. While we were obsessing (with some justification) on business processes and quality improvements, levels of R&D spending in the United States suffered a significant decline.[26] Even the recent upsurge in R&D investment may be insufficient to ensure America's continued leadership in innovation.[27]

Although the hard lessons of the last two decades have been essential to our future survival, it is time to turn our attention outward again. The history of American research and development is characterized by vision, leadership, and genius. The shocks of increased global competition may have diverted our attention, but we still possess these vital elements in surprising abundance. Through exposure to global competition, our firms have acquired a deeper understanding of operational efficiency, and have gained a renewed sense of pride in the products they produce. It is now time to shift the engine of American innovation into high gear.

### The Merging of Science and Technology

There is an obvious logic to the idea that research in basic science forms the foundation for technological advance. Despite the reasonableness of this view, however, the exact opposite is more often the case. Throughout much of human history, practical discoveries have served as a catalyst for scientific curiosity, rather than vice versa. Although it may seem perverse, large bodies of technological know-how have been assembled well before a foundation of science was available to support them.

The advances of the second industrial revolution, for example, were made by practical individuals seeking useful solutions to well-understood problems. The methods used in Thomas Edison's Menlo Park facility were characterized by brute-force experimentation. Often hundreds of permutations of a design would be tested in the search for performance improvements. In 1883, during one of Edison's endless attempts to improve the electric light bulb, he accidentally discovered evidence of the existence of electrons. Although the so-called "Edison Effect" had no practical value at the time, it eventually motivated basic research which today forms the foundation for atomic physics, electronics, and much more.

Likewise, the Wright brothers (who incidentally were the owners of a bicycle repair shop rather than a research lab) had no understanding of aerodynamics or aeronautics when they achieved the first man-powered flight in 1903. These fields of study did not even exist until their brief jaunts raised compelling scientific questions.

So the story goes for many of the seminal inventions of the last two centuries. The development of the Bessemer process for steelmaking in 1856 resulted in a revolution in the cost and quality of this important industrial material. This innovation was discovered by an untrained experimenter, however, without the benefit of a foundational understanding of chemistry. The subsequent problems associated with implementing his discovery prompted several steel companies to invest in some of the earliest examples of research and development. It could accurately be said that the growth of steelmaking and the expansion of the petroleum industry in the late nineteenth century provided the impetus for the development of modern organic chemistry, materials science, and chemical engineering.[28]

Today America leads the world in basic research, but this preeminence is a relatively recent phenomenon. The rise of American ingenuity significantly predated the rise of our scientific prowess. By the mid-nineteenth century, a host of homegrown inventions were making their way across the Atlantic. Breakthroughs in industrial processes, such as the gun-manufacturing techniques developed at the American Springfield and Harpers Ferry armories received considerable attention from the British and the Germans. In fact, the adoption of the method of interchangeable parts by the British in 1854 represents one of the first examples of technology transfer flowing eastward from the United States to Europe.[29]

Perhaps the greatest strength of American innovation during the late nineteenth century was a growing network of mechanical and technical specialists, particularly in the machine-tool industry. New ideas diffused rapidly through this community, with many important breakthroughs resulting from an early form of R&D collaboration. Firms from various fields worked together to meet a growing demand for productive innovations in manufacturing and agriculture. In fact, it has been said that the most important innovation of the late nineteenth century was the method of invention itself.[30]

### The Rise of the R&D Laboratory

American industry had become a hotbed of innovation in the late nineteenth century. The Germans, however, receive credit for establishing the first in-house industrial R&D departments. Beginning in the 1860s and 1870s, the German dyestuff industry began developing specialized chemistry labs with the specific goal of product improvement and innovation. Companies such as Hoechst, Bayer, and BASF have carried this tradition into modern times by maintaining world-class chemical R&D capabilities.

In America, the earliest R&D organizations were formed, surprisingly enough, by the railroad industry. In 1875, the Pennsylvania Railroad hired a Yale Ph.D. chemist to establish a laboratory for the testing of rail steel and other supplier goods. Similar labs were formed at Carnegie Steel and other growing industrial giants, primarily for the purpose of performing mundane testing activities rather than exploring the frontiers of science.[31] The seeds of the future had been sown even in these humble beginnings, however. Breakthroughs in the identification of high-quality iron ores and other valuable innovations served to demonstrate the economic potential of industrial R&D.

By the early twentieth century, several European writers began to warn of the rise of American technological dominance. Through significant advances in materials and processes, the U.S. steel industry had become the world leader, enabling the export of a broad range of industrial and transportation products. In the period before World War I, formal research labs were established by Kodak, General Electric, Du Pont, and other industry pioneers, as shown in Figure 3.3. Although these early R&D departments were focused primarily on practical problems, the unprecedented rise in the importance of industrial research elevated popular interest in the basic sciences as well.

Throughout the inter-war period, American R&D drew steadily closer to the scientific frontier. In budding industries such as plastics and electrical machinery, breakthroughs in basic science ran in lock step with the evolution of new product

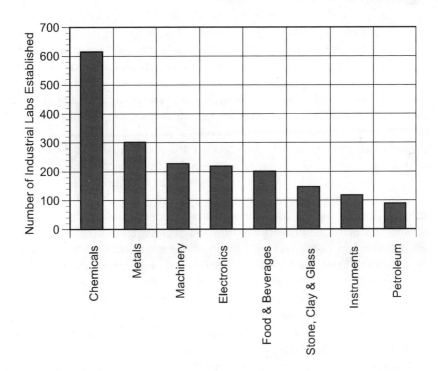

**Figure 3.3:** Formation of industrial laboratories during the period from 1899 to 1946 for several American industries. Source: Mowery, D. C. and N. Rosenberg (1989, pg. 62, Table 4.1). Data used with permission.

innovations. As the technological content of products increased, a new respect was given to academic education in the sciences. Until shortly after World War I, American industry had turned a blind eye toward formal training, preferring the practical, hands-on experience gained through apprenticeships in leading companies. This attitude rapidly changed as the need for a deeper understanding of electronic and chemical phenomena became crucial to commercial success. Thus, a bond began to form between the American educational system and technology-intensive industries, a bond that became the source of our enormous technological capacity in the postwar era.

### A Uniquely American Formula

As I have indicated, the growth of American technological capability in the early twentieth century was not based on our prowess in the basic sciences. Our early success in industrial innovation can be traced to two other important factors: our capabilities in mass production, and the development of an unmatched system of technical (as opposed to scientific) education.

Beginning in the late nineteenth century, the U.S. Congress passed several measures to create land-grant colleges and agricultural-experiment stations across the country. Educational programs in scientific agriculture and various engineering disciplines grew rapidly due to the stimulus provided by these acts; the number of American engineering schools, for example, increased from six in 1862 to 126 in 1917.[32] The dissemination of technology was also facilitated by government sponsorship, resulting in American agriculture becoming the most productive in the world during the inter-war period.

Although levels of secondary education at this time were similar throughout most of the advanced economies, America was distinct among nations in the percentage of its population having access to a college education. By early in the twentieth century, the United States led the world in the number of university students per one thousand primary students by more than a factor of two. More important, the enrollment in practical fields such as science and engineering steadily grew throughout the early part of the century. This trend was reinforced by heavy industry involvement in the development of engineering curricula at several major schools.[33]

In parallel with the rising levels of human capital in the United States, a unique organizational structure for exploiting technological advances was evolving. Since the turn of the twentieth century, America had enjoyed a unique advantage in the development of its industrial capability: a large and increasingly affluent domestic market. Our leadership in mass production was a natural outgrowth of the demand for affordable products in vast quantities. The assembly-line processes developed by Ford in 1908 were coupled with organizational structures that exploited scale economies and logistical efficiencies to yield an unmatched production capability. American consumers, unlike their European counterparts, readily accepted the standardized items produced by these methods, leading to the dawn of American mass production.

### Postwar Technological Dominance

Shortly after World War II, the OECD performed a famous series of "technology-gap" studies, in which it sought to identify the sources of what appeared to be an insurmountable American technological advantage. They concluded that it was not the level of scientific or technical knowledge available to U.S. industry, but rather the ability of American organizations to exploit it, that was the distinguishing feature. From its roots in the scientific management theories of Frederick W. Taylor, American industry had evolved a formula for practical exploitation of technology that proved to be highly successful.[34]

Whatever the cause, the United States was firmly entrenched at the scientific frontier of almost every field of study by the early 1950s. Much of this lead was an outgrowth of huge government investments in aviation, electronics, and communications during the war years. These investment levels were augmented by aggressive R&D programs within private industry during the 1950s and 1960s, as shown in Figure 3.4.[35] Increased government sponsorship of basic scientific

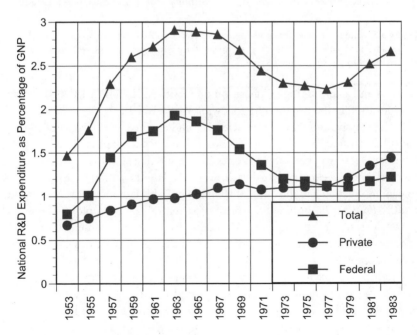

**Figure 3.4:** U.S. expenditures on R&D by government and business enterprise as a percentage of gross national product, 1953 through 1983. Source: Mowery, D. C. and N. Rosenberg (1989, pg. 127, Table 6.2). Data used with permission.

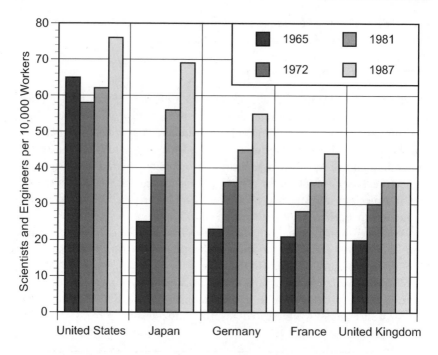

**Figure 3.5:** Number of scientists and engineers engaged in R&D per ten thousand workers in several advanced nations, for various years during the postwar period. Source: Nelson, R. R. and G. Wright (1992, pg. 1952, Figure 8).

research and a continuing emphasis on university education fueled a golden age of innovation in the two decades following World War II.

The massive mobilization of scientists and engineers in support of the war effort had created an unexpected economic bounty during peacetime.[36] In 1945, Vannevar Bush, who had served as director of the White House Office of Scientific Research and Development during the war, called on the nation to take a forceful role in the creation and diffusion of technology. In his famous report *Science: The Endless Frontier,* he outlined a strategy for centralized technology policy and proactive research sponsorship. Although few of its recommendations were adopted, this document had a profound influence on America's self-image as a world technology leader.[37]

Industrial capacity and rates of innovation continued to expand throughout the 1960s, causing Europeans to again sound a warning of America's "insurmountable" lead in science and technology. During this period, the number of scientists working in R&D in the United States per ten thousand employees was three times that of other advanced nations, as shown in Figure 3.5. The average number of years of post-secondary education among American workers doubled between 1950 and 1973, providing our industries with an enormous advantage in human capital. During this same period, gross expenditure on research and devel-opment reached an all-time high of 3.1 percent of American GDP. This was almost three times the investment levels of both the advanced European nations and Japan.[38]

The combination of enormous R&D investment, devotion to human-capital development, and the world's largest and richest domestic market seemed to create an unstoppable juggernaut. Yet within less than two decades America's global dominance in technology declined into a virtual dead heat among ad-vanced nations. Was this an unavoidable result of economic convergence, or did we fail to adapt to the changing nature of technology enterprise?

### Has America Been Asleep at the Switch?

Over the last three decades, America's success in the international trade of high-technology products has been something of a bright spot. Nonetheless, the U.S. share of the world market for technology-intensive products declined from 29.5 percent in 1970 to 22.3 percent in 1987. During this same period, Japan increased its share almost threefold, from 7.1 percent to 16 percent, as shown in Figure 3.6.[39]

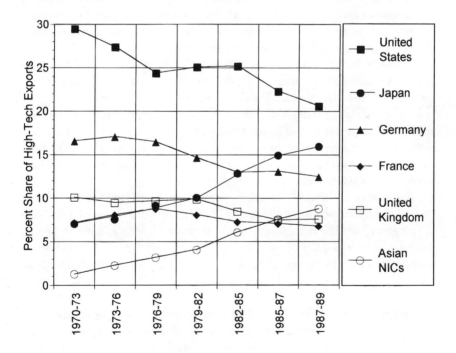

**Figure 3.6:** Share of world trade in technology-intensive products over the period from 1970 to 1989 for various advanced nations. Source: Tyson, L. D. (1992, pg. 23, Table 2.3). Data used with permission.

The U.S. trade balance in high-technology goods also suffered over the last three decades, particularly with respect to bilateral trade with Japan.[40]

These statistics support what has become intuitively obvious: We are still a power player, but America is no longer the dominant force in global technology. Has the erosion of our leadership been due to convergence factors that are largely out of our control, or have we brought this decline on ourselves? Although much of our fate has indeed been driven by inexorable changes in the global economy, we must accept responsibility for having reacted slowly and reluctantly to growing world competition.

As I have indicated, during the first half of the twentieth century America exploited a powerful combination of mass-production capacity, an educated workforce, massive R&D investment, and the world's largest domestic market to achieve technological hegemony. On close examination, it is evident that there has been a steady erosion of our position in each of these vital areas during the postwar decades, thereby diminishing our unique global advantage.

The recent liberalization of the world economy has allowed even tiny states such as Singapore and Hong Kong to exploit vast economies of scale. America's commitment to free trade, for example, has converted our affluent domestic market into an international "swap meet" for technology products.[41] In this environment, virtually any firm can access markets of sufficient volume to optimize production costs, regardless of the size of their home country.

Similarly, a well-educated and technically sophisticated workforce has become a recognized requirement for membership in the "convergence club."[42] Today, Japan leads the world in the density of scientists and engineers in its workforce, with the United States and Germany lagging behind in a virtual tie. Massive investment in research and development by Japan and the nations of the European Union over the last twenty years has erased our advantage in the R&D arena as well.[43] In short, the sources of our erstwhile technological leadership have become the common property of all advanced nations.

But what of the uniquely American organizational structure that allowed our firms to exploit innovation so effectively? Unfortunately, the nature of technological progress itself has changed during the postwar decades, and our industries have been slow to adapt. What was our crowning advantage in the 1950s became a substantial liability by the mid-1970s. Only in recent years has American industry belatedly accepted the organizational changes mandated by this new technological environment.

Perhaps the most fundamental change has been in the process of innovation itself. Throughout the early part of the twentieth century, technological advances were assumed to follow a simple linear progression, from basic science to large-scale industrial development, and ultimately to practical applications and innovations. This model neatly described some of the great achievements of the twentieth century, including lasers, nuclear energy, the transistor, and particularly the mission-oriented development efforts financed by government defense R&D. With the endorsement of Vannevar Bush, among others, this "pipeline" view of innovation became the model for industrial research and development in the early postwar years.

Major firms established corporate R&D centers far from their operating divisions and in complete isolation from both markets and competitors. It was presumed that ideas would "flow downhill" from these centralized laboratories to design and engineering departments, and ultimately into the marketplace. Since the linear model for innovation was a conceptual parallel to the vertical industrial organization so popular in the 1960s, these structures formed the backbone of virtually all major American companies.[44]

By the 1970s, the industrial organizations that had served us so well were becoming a significant handicap. As technological progress became increasingly international, the assumption of a one-way flow of innovation began to severely limit the ability of American corporations to absorb important breakthroughs from abroad. Little attention was given to training, production engineering, process improvement, quality, or reliability, since these factors did not fit into one-dimensional linear thinking.[45] More important, there was a definite arrogance associated with the central R&D lab, which later evolved into a pervasive resistance by American firms to exploiting ideas that originated outside their own organizations. As I will demonstrate in the next section, this so-called "not-invented-here" (NIH) syndrome cannot be tolerated in an increasingly competitive global environment.

## Technology Absorption: An American Weakness

The greatest danger with mental models such as the linear innovation process discussed earlier is that they can take on a life of their own. Although the linear model offered at best a limited perspective on the nature of technological progress, it became entrenched in postwar American organizational structures. Adopting a one-way flow of knowledge as an operational strategy resulted in increased technological arrogance and isolation. While firms from other advanced nations were learning as much as possible from America's leading industries, we consistently ignored the breakthroughs of our foreign rivals.

This failure to acknowledge the value of foreign technology harmed us in several ways. First, by clinging to obsolete organizational and R&D structures we allowed our leadership in vital industries to evaporate. By failing to learn lessons from German engineering and Japanese production, we repeatedly missed opportunities to remain competitive. Second, while American firms were struggling under the burden of the NIH syndrome, foreign firms were honing their ability to acquire, refine, and adapt our advanced technologies.[46] In the process of digesting the enormous backlog of American innovations, these firms developed a powerful capacity to absorb and accumulate knowledge. By comparison, American industries have been left ill-prepared to exploit the vast pool of global knowledge on which a modern R&D strategy must be based.

Our world is increasingly characterized by "poles" of technological specialization distributed among the advanced nations. In this section, I will demonstrate that in such an environment, the ability to absorb and accumulate technical knowledge represents a critical source of sustainable competitive advantage.

### The True Source of Japanese Competitiveness?

The rise of Japan's technology industries over the last thirty years has been the cause of considerable soul-searching among American policymakers and business leaders. As with most emotional topics, these discussions have been plagued by paranoia and misinformation. Amid the (somewhat justified) accusations of unfair trade practices, government subsidies, collusive behavior, and other conspiratorial trespasses, there has been only limited acknowledgment of the remarkable capabilities demonstrated by Japanese firms. Therefore, in the interest of fairness, I will highlight at least one important competitive advantage of Japanese firms that seems to have been lost in the general uproar.

First, we should recognize that there is no one source of Japanese competitiveness. The rise of Japan from humiliating military defeat to a position of global respect and economic influence must surely have resulted from a complex interplay of factors. Some frequently cited sources of Japanese competitive advantage include intensive private and government-sponsored R&D investment, considerable attention to technical education, a uniquely cooperative industrial environment, and the protection of a large and sophisticated domestic market from import competition.[47] Each of these facets represents a piece of the puzzle, but there is a more fundamental advantage that is often overlooked. Japanese industries have followed a very different model for research and development than that of American firms. By adopting a more balanced and outwardly oriented R&D strategy, Japanese firms have secured generally higher returns on investment than those achieved by U.S. firms over the last several decades.[48]

We can better understand the differences between U.S. and Japanese R&D strategies by breaking down research investment into several categories. In a study performed by Edwin Mansfield in the late 1980s, the composition of R&D investment for fifty matched pairs of U.S. and Japanese firms was compared, as shown in Figure 3.7. The most striking difference between the two investment profiles was the strong emphasis by American firms on product innovation, as opposed to process improvements. In particular, U.S. companies demonstrated a strong focus on developing entirely new products, whereas Japanese firms balanced their investment in new product development with an emphasis on adapting existing

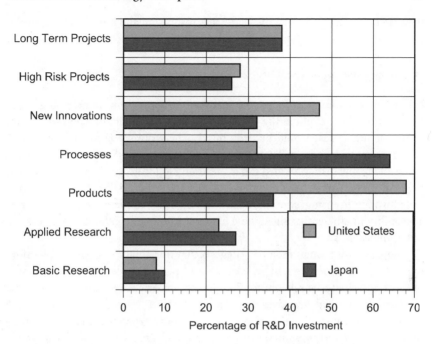

**Figure 3.7:** The composition of R&D expenditures for fifty matched pairs of firms from the United States and Japan. Note the dramatic difference in the levels of process-related investment by Japanese firms. (Entries do not add up to 100 percent, because several categories overlap.) Source: Mansfield, E. R. (1988, pg. 226, Table 2).

designs, along with innovations in advanced tooling, manufacturing methods, quality measures, and other downstream processes.[49]

In general, studies have shown that although Japanese firms are no better than American companies at creating new innovations, they are significantly faster and more effective at absorbing technology from outside sources. They have become adept at the rapid transfer and diffusion of technology through years of experience in scanning the scientific frontier. By maintaining close linkages between R&D laboratories, design departments, marketing organizations, and the factory floor, Japanese companies are better equipped to deploy this information across the entire product development process.[50]

Although American firms have recently made great strides in achieving a more balanced emphasis on products and processes, there is a conspicuous lack of attention paid to the absorption of knowledge from outside sources. In a highly competitive global market, the opportunity costs of squandering precious R&D funds on reinventing existing technology can be devastating. Enhancing a firm's ability to access and utilize outside knowledge may not yield immediate returns, but in the longer term the benefits can be substantial. American firms must revise their R&D strategies to make the aggressive absorption and long-term accumulation of advanced technologies an explicit priority.

### Developing the Capacity to Absorb

It is impossible for a firm today to research every aspect of new technologies the way IBM or AT&T could in the 1970s. Both the cost and the time required to accomplish such a feat would be prohibitive. Therefore, R&D must increasingly draw from outside sources. There is a vast body of technological knowledge that is accessible to any firm, provided it has the ability to locate, transfer, and absorb it. Beyond this pool of public goods lies a deeper layer of specialized expertise accessible through licensing, strategic alliances, and other cooperative arrangements. The ability to efficiently utilize these external resources is referred to as a firm's *absorptive capacity.*[51]

Research on human memory suggests that our ability to absorb new information is strongly dependent on accumulated prior knowledge. Memory development appears to be self-reinforcing, in that experience in learning a specific category of material improves one's ability to learn related knowledge in the future. In a recent study, a group of students with no experience in learning computer languages was compared to a group that had already learned Pascal (an introductory programming language).[52] When these two groups were exposed to a new language, LISP, the students with prior experience in programming learned much more readily. More important, the experienced students were better able to make use of the new knowledge to effectively solve problems.

This perspective on human learning suggests several properties of absorptive capacity. First, the ability to absorb knowledge is a strong function of the type and quantity of prior accumulated knowledge. Thus, it appears that effective learning is highly associative and contextual. Second, intensive study of related subjects imparts some form of tacit knowledge that allows individuals to both absorb and effectively utilize new information. It is not sufficient to "nominally" acquire explicit knowledge; a critical *tacit capability* is needed to make the explicit information fully intelligible and useful.

Since an industrial organization is a collection of individuals, we can assume that these features of absorptive capacity will carry over into the domain of the firm. It is the nature of an organization, however, that its capabilities are more than just the sum of its parts. Hence, a firm's absorptive capacity must be thought of as a mosaic of individual capabilities, linked through both formal and informal structures. Fortunately, the skills and human capital resources required to absorb knowledge are very similar to those needed for original innovation. Therefore, developing an absorptive capacity can be thought of as a complement to in-house innovation, rather than a substitute.

Before we consider some of the strategic implications of this concept, it is worthwhile to review the most important sources for acquiring outside technical knowledge. The most obvious way to add knowledge to a firm is through hiring personnel who embody the desired information and tacit skills. This can be accomplished through a selective recruitment plan, or by the acquisition of an entire company. The acquisition approach has come into favor with many large American multinational corporations, to the extent that it has become a latter-day substitute for in-house innovation in some cases.

There are several problems associated with this method of absorption, especially when taken to an extreme. First, there is a significant time lag between the addition of personnel and the point at which they become fully integrated into an organization. In the case of an acquisition, the "culture" of the absorbed company may not be compatible with its new parent, particularly if the scale of the firms is dramatically different. Even in the case of a new employee, it may take several years for that individual to learn the "system," gain the respect of his or her colleagues, and develop the working relationships needed to synergize with the broader organization. The strategy of absorbing technology by simply "buying it" must be undertaken with some trepidation.

The above discussion highlights an important aspect of absorptive capacity. For firms to be successful at exploiting technology drawn from outside sources, they must become adept at selecting promising and potentially profitable innovations. Often there are several parallel approaches to a given technological problem being pursued by various suppliers. A firm with little in-house expertise in these fields will have a difficult time understanding the subtle risk, cost, and performance tradeoffs, and could easily invest in a suboptimal solution.[53]

Studies have shown that for a firm to become skilled at the selection of new technologies, it must have internal experience in related research and development.[54] The intensive learning gained by a team of scientists and engineers struggling with similar problems is a powerful source of absorptive capacity. By bringing a firm's organizational experience in related technologies to bear on the selection process, there is a much higher likelihood of choosing the most promising approach.

The ability to absorb knowledge at the frontier of technology requires a significant storehouse of applicable experience. Semiconductor manufacturers from Taiwan and Korea, for example, fund "duplicate" projects, in which their R&D organizations reproduce the results of foreign firms, just to keep their teams' abilities honed.[55] Although the projects are typically not taken beyond the "proof-of-concept" stage, these learning opportunities pay important long-term dividends when the firms prepare to launch new product lines.

Beyond the direct purchase of a company or recruitment of new employees, there are a number of alternative sources of external technical knowledge. The survey of 650 R&D managers and executives cited earlier in this chapter included questions regarding the most important sources of knowledge about new product and process technologies. The results, shown in Figure 3.8, demonstrate that internal R&D is considered by this group to be an important complement to several other avenues of absorption.

Over the last decade, the production function of firms has evolved beyond a vertical structure into a network of forward and backward linkages, stretching from raw-materials suppliers to the final customer. This supply chain offers excellent opportunities to harvest technical knowledge, provided that a firm has the proper interfaces to these outside channels. Employees with recurring access to outside information, whether through suppliers, customer contact, or scientific affiliations, play the role of *gatekeepers* within an organization. It is essential that these individuals be selected carefully, and made aware of their responsibilities with respect to the gathering and dissemination of knowledge within the firm.

Although links to outside sources of knowledge are crucial, the internal lines of communication within a company are equally important. Firms that segregate personnel into functional departments, or divide participants in the product development process by geographic or organizational barriers, will be less effective at transferring vital technical information to the proper hosts. The rich literature available on the methods for developing *learning organizations* focuses primarily on individual and team processes.[56] While these techniques are useful, the most important first step is to break down arbitrary walls and avoid restrictive levels of specialization and division of labor. Each employee, from the receptionist to the director of R&D, should have a basic understanding of a firm's entire line of business, and should be connected to the firm's network as a potential host of absorbed external knowledge.

Ironically, the most valuable source of specific technological knowledge for firms is usually their competition. Absorption of knowledge from rivals can occur willingly, through cooperative agreements, or unwillingly, through information "leaks." The number of cooperative agreements between firms, particularly in the most R&D-intensive industries, has grown sharply in recent years. By avoiding duplication of background research, firms within an industry can significantly benefit from cooperation. This concept of industry-wide absorption and diffusion of foundational technology will be discussed in more detail in subsequent chapters.

Assuming that competitors refuse to share their proprietary information willingly, there is still a wealth of knowledge to be gained from spillovers. This leakage of vital information can result from movement of personnel between companies, informal communication networks, conversations at technical meetings, discussions with common suppliers or customers, patent applications, and even through mirror-technology assumptions.[57] The rate of leakage is surprisingly high, ranging from twelve to eighteen months for detailed information on planned new product introductions to reach competitors in many industries.[58]

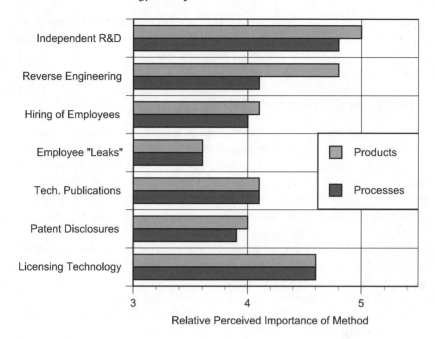

**Figure 3.8:** Perceived effectiveness of alternative methods for learning about new product and process technologies. Scoring based on a survey of 650 R&D managers and executives, with a score of seven implying that a method is very important, and a score of one indicating a relatively unimportant method. Source: Levin, R. C. et al. (1987, pg. 806, Table 6).

Since much of the leakage of proprietary knowledge occurs through informal and interpersonal channels, geographic proximity can increase a firm's access to these spillovers. This fact has prompted firms from both the advanced and newly industrialized countries to locate R&D facilities within technically active regions of the United States.[59] These foreign laboratories certainly benefit from absorbing spillovers, but this is typically not their primary goal. Rather than attempting to "steal secrets" from American firms, they derive benefits from accessing our pool of highly skilled scientists and engineers.[60] Although this justification for foreign R&D facilities seems harmless, if we recognize that our technical workforce embodies tacit (as well as explicit) knowledge of our industries, their strategy seems less benign. In fact, I will assert in Chapter 8 that establishing facilities to perform pre-adoptive research in regions of technical specialization is an excellent way to enhance a firm's absorptive capacity, while simultaneously harvesting a wealth of human capital.[61]

Finally, it should be recognized that the lack of an absorptive capacity can have severe consequences. Failing to invest early in a new technical specialty may preclude entry into that field at a later date. Absorptive capacity is highly path and history dependent. It is very difficult to jump into an already developed field without the benefit of preparatory learning. The prospect of being *locked out* of a new industry should be taken very seriously; with the rate of technological change accelerating, a running start may be essential to a firm's ability to "grab the baton" of a new and promising technology.[62]

### Gaining Advantage Through Accumulation

The capacity to absorb technical knowledge is becoming a necessary condition for firms to achieve global leadership in high-technology industries. It is not, however, sufficient to sustain a competitive advantage over extended periods. The explicit and tacit knowledge gathered by a firm over years of intensive learning must form a platform upon which all future innovations are based. Thus, absorption has a critical counterpart: the *accumulation* of specialized technical knowledge.

Knowledge can flow into an organization through several channels, as shown in Figure 3.9. Some fraction of this information is absorbed by appropriate hosts within the firm and becomes an integral part of ongoing innovation activities. As this knowledge accumulates in the collective minds and organizational structures of a company, the level of technological sophistication will continue to rise, eventually reaching and even extending beyond the scientific frontier.

Unfortunately, this process does not happen automatically. It is far more common for firms to allow critical accumulated knowledge and tacit learning to slip through their fingers. No amount of absorptive capacity can save such a company from "reinventing the wheel" at the beginning of every new project.

Technological accumulation is often misrepresented as the acquisition of embodied physical capital. A clear distinction must be made between a firm's production capacity and its level of accumulated technical knowledge. The ability to operate a sophisticated production system is not the same as the capacity to change or improve it.[63] Technological accumulation is achieved through enhancing the learning capabilities of employees and by creating structures within a company that "trap" vital knowledge for future use.

The patterns of accumulation tend to vary considerably between firms and industries. Companies that lag behind the state of the art may need to create learning opportunities that can rapidly build up individual and team capabilities. An excellent strategy for firms entering a new technical specialization is to identify an industrial or government sponsor who is willing to fund engineering development projects in a closely related field. Although a commercial product may not directly result from this engineering activity, the firm gains access to sophisticated customers, and can use outside funds to educate its design team in an entirely new discipline.

Firms that have achieved a higher level of technical accumulation can use their historical knowledge base as a focusing agent for future product develop-

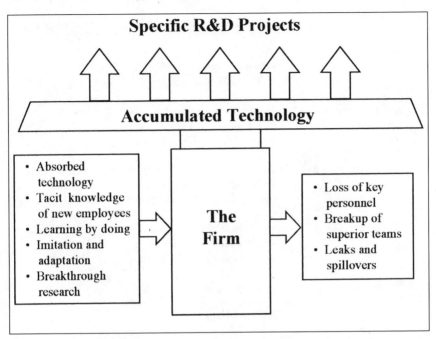

**Figure 3.9:** Conceptual diagram of the flows of technical knowledge into and out of a firm. Ideally, knowledge is both absorbed from outside sources and created internally through intensive learning activities (i.e., product or process research). As specialized knowledge accumulates, it forms a growing platform upon which the firm can launch commercial R&D projects. Significant attention must be paid to retaining this critical pool of explicit and tacit knowledge, however, since it can easily be drained from a company through several channels.

ment. These companies are in the best position to expand the technological envelope with breakthrough innovations. It is critical that such firms fully utilize their accumulated base of skills and competencies to extend their current lines of business. There are surprisingly few examples of companies that have successfully entered into unrelated fields and achieved a competitive advantage. Extensions of a firm's product line should make intensive use of its "platform" of specialized knowledge and tacit abilities (i.e., their core competencies).

There are three important components to the accumulation of technical knowledge within a firm: 1) the development of tacit knowledge, either through training or intensive learning-by-doing activities, 2) the ability to make important historical lessons "stick" through formal organizational structures, and 3) protection of the current stock of knowledge by guarding the channels through which it can escape.

As technology advances, higher levels of both formal education and tacit knowledge are needed to participate in innovative activities. While the university

| Method of Learning | Description | Knowledge Type |
| --- | --- | --- |
| Learning by Doing | Know-how and experience gained through repeated performance of an activity | Tacit |
| Learning by Using | Similar to learning by doing, but highly specific to a machine or process. Hands-on experience. | Tacit |
| Learning by Absorption | The gathering of information from outside sources, either public knowledge or industry spillovers. | Tacit & Explicit |
| Learning by Interacting | Similar to learning by absorption, but dependent on relationships and involvement in the community. | Tacit & Explicit |
| Learning by Creating | Learning through the intensive, trial-and-error process of creating new knowledge. | Tacit & Explicit |

**Table 3.1:** Several methods for accumulating technical knowledge within a firm. Adapted by the author from Malerba, F. (1992, pg. 848).

system can train the minds of graduates as preparation for more practical learning, it is the firm that plays the central role in creating human capital. Productive learning can be accomplished in a number of ways, as shown in Table 3.1. Whatever the method, every technology-intensive firm should dedicate a significant portion of its discretionary funds to employee training and development.

Multinational corporations in both Korea and Taiwan, for example, have built substantial training and education programs into their operations. The largest electronics firm in Taiwan, Tatung, has established the Tatung Institute of Technology. Similarly, Samsung, in South Korea, founded the Samsung Advanced Institute of Science and Technology.[64] There are domestic examples as well. Motorola, which also has established a corporate "university," invested more than $60 million in employee education in 1990. In each of these cases, highly competitive global firms have recognized the critical need to educate employees beyond the generic learning provided by universities.[65]

Investment in company-specific training and education imparts a higher level of specialized knowledge to employees. This background is only a starting point, however, for the detailed lessons learned during applied research and product development. Often this type of education comes in the form of "trial and error." Unfortunately, many firms fail to capture these hard-won lessons, thereby ensuring that the same failures will be repeated in the future. Similarly, the great achievements of a project team are rapidly swept off to production, while the tremendous knowledge base accumulated within the team is allowed to dissipate.

Without a doubt, the most challenging aspect of technological accumulation is establishing organizational structures that capture the lessons learned during research projects. There are a number of software products that purport to solve this problem, including data-warehousing, workgrouping, and knowledge-management packages.[66] While I recommend that firms avail themselves of these modern miracles, there are some fundamental organizational issues that cannot be solved by software.

First, the opportunities for knowledge accumulation during the design and development processes are severely curtailed if a product team is given unrealistic cost, schedule, or performance constraints and forced to produce a *point design*.[67] This often occurs when a firm believes that it is behind technologically and is attempting to catch up with its competitors. Unfortunately, this myopic approach often compels the design team to bypass the accumulation of knowledge for the future. For example, given a longer time frame and sufficient funding, a design team might develop a modular *product platform*, which can be easily scaled, customized, and adapted to global markets in the future. The strategic benefits of taking a broad and forward-looking approach to R&D activities cannot be overstated.

A second important issue in the retention of accumulated knowledge involves the collective learning of teams. The social and intellectual interactions of a technical team are enormously complex. It often takes years for a project team to begin to harmonize their personalities, egos, and competencies in an optimal way. Once a team jells, it can become a powerful creative force, achieving far greater creativity and productivity than a newly formed group.

Although this is common knowledge to those who work on such teams, many executives in technology-intensive firms are ignorant of the precious resource that an experienced team represents. Clearly, it is not possible (or even desirable) to keep the constituency of teams fixed for extended periods. However, firms should recognize the unique relationships among team members, and try to retain the essence of a successful team on future projects. Swapping out one or two members will likely not affect the team's dynamics. Transferring the key technical leaders to other divisions, as seems to happen frequently, is a tragic waste of collective learning.[68]

The most frustrating aspect of competition in high technology is that after training, cultivating, and honing key employees, they are free to migrate to other companies. The mobility of technically trained professionals is very high, particularly in rapidly growing industries with extensive tacit knowledge content, such as semiconductors and biotechnology. Although there is no way (short of chains and cuffs) to completely forestall employee defection, much can be done to stem the outward flow.

The first step is for firms to recognize the tremendous value of the tacit knowledge embodied in their key personnel. While investment in R&D projects is often risky, investment in the retention of top researchers and engineers is essentially a sure thing. A more enlightened compensation system for these irreplaceable human assets would go a long way toward ensuring a more stable workforce. Profit-sharing and stock-option plans represent a good starting point, but these could be supplemented with a results-based bonus system and even a reward structure for activities that promote technological absorption and accumulation. It is surprising how much effort is expended by firms to protect patents and copyrights, and how little attention is paid to retaining their vital stockpile of tacit knowledge.[69]

### Accumulation Within Industries and Nations

The concept of technological accumulation can easily be extended beyond individual firms to encompass entire nations. The aggregate technological expertise of a nation's industries represents a critical determinant of the current standard of living and prospects for future growth.[70] In this broader context, it is the responsibility of governments to foster the role of industry as enhancers of human capital, creators of tacit knowledge, and accumulators of technological advantage.

It is natural for firms within a given industry to cluster, both technically and geographically. Competitors, suppliers, distributors, and integrators each have a common stake in the development of domestic and global markets. If the level of proximity and cooperation is sufficiently high, a powerful feedback mechanism will tend to drive an industry toward a global leadership position. This positive feedback can be further augmented by the presence of complementary trade, competition, and technology policies. Ultimately, this resonance among firms, industries, and government can result in "poles" of technological specialization forming within a nation.[71]

As I noted earlier, the accumulated technical knowledge within an individual firm represents a platform upon which their competitive advantage can be based. This platform can be further elevated by the accumulated technological capacity of an industry, region, or nation. Since the requirement for global leadership in technology-intensive industries is the maximizing of total-factor productivity, it is easy to see that building a strategy on technological "high ground" provides a considerable competitive advantage.

## The Myth of Technoglobalism

Just how important is high technology to the American economy? According to U.S. labor statistics from 1996, 9.1 million jobs were directly assignable to technology-intensive industries. Even this impressive number is most likely an underestimate; government statistics fail to capture the numerous technical jobs that are embedded within "nontechnical" sectors. After accounting for these programmers, network technicians, and other technical support personnel, the total employment base for high technology could be 60 percent higher.[72]

From the standpoint of revenue, the high-tech sector contributed more than $420 billion to the U.S. GDP in 1996, and has been growing at an average annual rate of 7.9 percent over the last decade, as shown in Figure 3.10. The surge in revenues observed in 1995 and 1996 was matched by increases in aggregate R&D investment. The three hundred largest high-tech companies invested a total of $104.8 billion in 1996, representing 88.3 percent of the total industrial research conducted in the U.S. The leaders in research and development were automobile giants General Motors ($8.9 billion) and Ford ($6.8 billion), followed by IBM ($3.9 billion), Hewlett-Packard ($2.8 billion), and Motorola ($2.4 billion). Intel was ranked number nine in the nation with $1.8 billion in R&D, and Microsoft was ranked number twelve with $1.6 billion.[73]

The available evidence suggests that high technology will grow at a far faster rate than the American economy well into the twenty-first century, creating high-wage jobs and impressive returns on shareholder investments. But where will these jobs be created? Which nation's economy will benefit from the high-value-added R&D activities of American firms?[74]

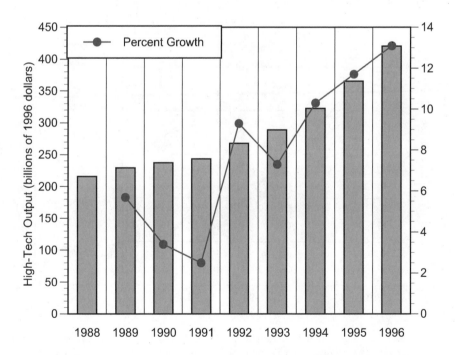

**Figure 3.10:** Total output of the U.S. high-technology sector in billions of chain-weighted 1996 dollars, and percentage change per year. Sources: Bureau of Labor Statistics, *Business Week.*

There is a growing concern that the global proliferation of advanced technologies will ultimately reduce the importance of national borders and government actions. The notion of *technoglobalism* presumes that multinational corporations are forming complex networks of technological resources, without regard for nationality or geography; in their search for the best available human capital and specialized knowledge, it is presumed that MNCs are globalizing their R&D activities, thereby rendering the arbitrary boundaries of the nation-state largely irrelevant.[75]

The evidence for technoglobalism at the aggregate level is circumstantial, but compelling. It is true, for example, that firms from all nations have significantly greater access to technology than ever before. Likewise, the levels of R&D investment among the advanced economies have been equalizing over the last twenty years, and their rates of growth are showing signs of converging, as shown in Figure 3.11. Although these trends imply a homogenizing of technological intensity, they offer no direct evidence that innovation itself is becoming a global activity.

To investigate this question further, we can consider three possible scenarios for technoglobalism, representing increasing degrees of global integration, and

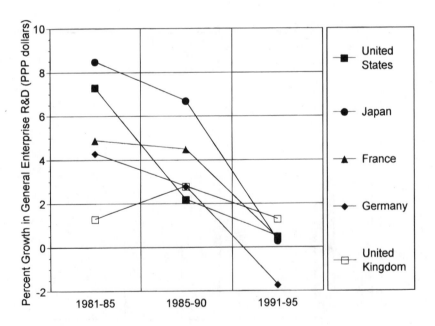

**Figure 3.11:** Rate of growth in business-enterprise R&D investment for various countries from 1981 through 1995. Source: OECD, 1997, *Science, Technology and Industry: Scoreboard of Indicators*, pg. 115, Table 1.1.1.

compare them to available empirical data.[76] The first model assumes that MNCs are exploiting their technological advantage in global markets, but are performing little overseas R&D. The second model includes both exploitation and collaboration, assuming that firms are accessing both markets and specialized technologies on a global scale. Finally, a borderless model is considered, in which the generation of industrial innovations by MNCs occurs without regard for nation or location.

At the outset, I will stipulate that multinational corporations are aggressively exploiting their technological advantages in global markets. Firms from smaller European countries, for example, have utilized both regional and global markets to gain economies of scale and to help offset the large non-recurring investments required to develop technology-intensive products. Today, even companies with access to large domestic demand, such as the United States and Japan, launch new products almost immediately into the global marketplace. Hence, it is reasonable to accept that the most limited model of technoglobalism is generally valid.

Likewise, there is some interesting indirect evidence that suggests that MNCs are availing themselves of specialized technology and human capital on a global basis. The percentage of technical papers that are internationally co-authored doubled between 1974 and 1984. There has also been a dramatic upsurge in the number of strategic alliances and joint ventures between firms of different nations. In this case, three rapidly growing technology sectors—biotechnology, advanced materials, and information technologies—accounted for 70 percent of all recent cross-country agreements. These data underline the desire of firms in high-technology industries to share the enormous cost and risks of front-end research.

While the exploitation of international markets and accessing of technological specialization represent strong evidence for increased global integration, this does not necessarily imply a diminishing role for nation and location. For national borders to become truly irrelevant, the generation of innovations by MNCs must transcend geographic, cultural, and political influences. It is at this critical point that the available evidence takes a decidedly different turn.

First, it is important to recognize that only a small fraction of the world's R&D activity is performed outside of the most advanced nations. There is virtually no evidence, for example, that MNCs are financing significant research activities in developing countries. With respect to the advanced nations, there have been several studies performed to determine the level of industrial innovation occurring outside the home countries of multinationals. Two general approaches have been used: studies of the geographic distribution of patenting activity, and analysis of R&D investment in home and host nations.[77]

International patent data strongly imply that firms perform the majority of their innovation on their home soil, as shown in Figure 3.12. Japanese and U.S. firms perform a negligible amount of R&D in foreign locations, while (not surprisingly) the degree of globalization is somewhat higher for the smaller European countries. Investment data from MNCs of various nations reflect an even stronger tendency toward *technonationalism*. When this data is broken down by sector, it appears that firms involved in more technology-intensive industries perform the lowest percentage of R&D outside their home country.[78]

The available evidence suggests that MNCs follow a "polyp model" with respect to their R&D activities: Tendrils consisting of foreign direct investments and strategic alliances are outstretched into a global network, which feeds a central "brain" residing in the home nation.[79] Although the actual process of innovation has become increasingly complex and highly nonlinear, firms still depend on their home countries to provide the human capital and R&D support structure that is vital to their success. Contrary to the view that the economic and innovative environment of nations has become insignificant, it appears that these factors may still be major determinants of the competitive advantage of firms in the global economy.[80]

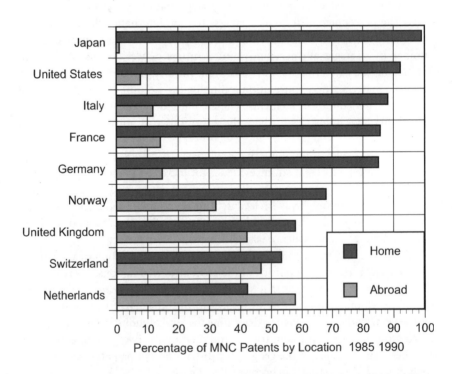

**Figure 3.12:** Geographic location (i.e., home or abroad) of U.S. patenting activities of large MNCs, according to nationality, 1985 through 1990. Source: Patel, P. (1997, pg. 207, Table 7.5).

One-dimensional terms such as technoglobalism tend to obscure more than they illuminate; the reality of multinational R&D strategy is both heterogeneous and complex. The leveling of general technological sophistication among advanced nations has not led to a convergence in either their methods of innovation or their profiles of specialization.

Firms in different industries are uniquely affected by market conditions and technology-specific factors. These differing environmental conditions will dominate their decisions regarding R&D location. The highest levels of foreign R&D, for example, are typically performed by pharmaceutical and chemical firms, which are driven by industry-specific concerns over transportation costs of raw chemicals, local regulatory and environmental restrictions, and the availability of skilled human capital in nations that specialize in these sectors, such as Germany or the United Kingdom.[81]

There are also distinct patterns in the strategic approach to research and development of firms from different nations. American MNCs have had a hard time coming to grips with the ramifications of globalization and national specialization. Our unique place in the history of global technology development tends to skew our perspective on the competitive benefits of overseas R&D activity. Nations such as Japan and Germany, which have learned to be comfortable with international commerce, are less threatened by the prospects of globalization and more objective about its benefits.[82] Concepts such as "free trade" and "deregulation," for example, have a very different connotative meaning among advanced economies outside of the United States.

Although the global diffusion of R&D activities is far from complete, I believe that it will increase dramatically in the next decade. As firms begin to recognize the strategic benefits of accessing specialized skills and technologies in foreign nations, an inevitable shift in R&D activity will occur, from the comfort of the home nation to the regional hotbeds of technological specialization around the globe. Ultimately, national governments must develop policies that attract high-value-added R&D activities, regardless of the home nation of firms. Those nations that cultivate their abilities as "good hosts" to technology enterprise will attract R&D investment from all countries, and reap the economic benefits of the innovation that takes place on their soil.

# PART 2

## Creating Advantage
## in Technology Markets

# 4

# A Framework for Sustainable Advantage

## The Impact of Knowledge-Based Products

Thus far, I have treated the output of industrial research and development as a series of commercially valuable, but only partially excludable, innovations to products and processes. Firms invest in developing new technologies in an effort to gain a market advantage, but then must vigorously protect those innovations from imitation by their rivals. This concern over imitation, however, assumes that the output of R&D is something tangible (e.g., products, patents, process methods, software code, etc.) that can easily be transferred to others either willingly or unwillingly.

This output-centered view of research and development fails to acknowledge the portion of a firm's R&D investment that becomes embodied in its team of scientists, engineers, and technicians through experience and learning. The tangible outputs of R&D investment tend to dominate the strategic actions of firms, while this human aspect of technology enterprise is often ignored. Yet among high-tech firms, the correlation between sustained market leadership and high concentrations of seasoned professional talent is well established.[1]

In this section, a three-dimensional model for the output of commercial R&D will be presented, in which the traditional "hardware" aspect of products represents only a starting point. The value created by a high-technology enterprise is far more than just the production of tangible goods. Global competition in technology-intensive industries is waged along two additional dimensions, which capture the explicit and tacit knowledge content of products.

### The Distinction Between Tacit and Explicit Knowledge

In the mid-nineteenth century, the United Kingdom was the preeminent source of the world's productive knowledge. This leadership position was the envy of other nations, and much effort was expended to imitate the "stream of gadgets" that Britain seemed to create so effortlessly.

During this period, economist Friedrich List began to formulate strategies that might allow his homeland of Germany to catch up with, and eventually surpass, its offshore rival.[2] List's policy recommendations centered on developing Germany's ability to transfer and adopt existing British technologies, with a focus on the rapidly expanding machine-tool industry. While a great deal of effort was expended on "industrial espionage" of various sorts, the German government also placed considerable emphasis on developing a national system of technical education and training.

German industrialists recognized that possession of a British tool, or even detailed plans for the device, was not sufficient to allow replication and improvement. Hence, a major effort was launched to attract British craftsmen and toolmaking entrepreneurs to Germany, to oversee apprenticeship programs for local industry. A combination of the knowledge transfer provided by these "experts" and the substantial base of skilled tradesmen resulting from an aggressive technical education system allowed Germany to surpass the United Kingdom in technological sophistication by the end of the nineteenth century.[3]

Blueprints can be transferred easily from one individual to another, but the ability to exploit this *explicit knowledge* depends on the capabilities of the receiver.[4] An advanced physics textbook in the hands of a novice does not yield a physicist; years of education and training are required to fully benefit from the book's contents. Those intangible abilities that are gained by individuals through experience, practice, and learning are known as *tacit knowledge.*[5]

The tacit knowledge contained in the heads of scientists and engineers may be even more important to sustaining competitive advantage than the transitory benefits of even the most commercially successful innovations.[6] In the current business environment, it is not sufficient for a firm to own a stable of valuable patents. The ability to regenerate innovations at a rate that keeps pace with the state of the art is essential to long-term market leadership.

Since explicit knowledge can be transferred from one location or person to another at essentially no cost, it approximates the perfect flows of information assumed in classical economic theory. Tacit knowledge, on the other hand, can be difficult and costly to transfer, making it considerably less mobile than its counterpart. Furthermore, the tacit knowledge held in the heads of individuals is only a starting point; organizations, relationships, teams, and other social groupings can possess unique capabilities that may be difficult to replicate.[7]

The intractable nature of tacit knowledge offers the seeds of a powerful strategic advantage. To be commercially useful, explicit knowledge must be in a form that is systematic, understandable, and, unfortunately, inherently easy to imitate. The appropriability of returns from the explicit portion of R&D is therefore limited by spillovers into the public domain. The long-term benefits derived from increasing a firm's base of tacit knowledge, however, are *almost perfectly excludable*. This fact has not been lost on Japan's Ministry of International Trade and Industry, for example, which for several decades has emphasized the funding of high-tacit-content R&D projects to protect against "leaks" to competitors in other nations.[8] Investment in R&D can be thought of as providing a "hidden" return to firms: the broadening and deepening of their employees' capacity to innovate.

Tacit knowledge takes on several forms, each of which offers its own unique advantages. At the most fundamental level is the knowledge gained through "learning by doing."[9] It is hands-on experience of this type that makes a journeyman machinist far more productive than a novice, despite identical tools and similar formal training. This kind of knowledge tends to be highly specific and specialized, and therefore cannot be easily extended to unrelated technologies. A broader form is represented by the engineer or technician who is capable of solving general problems based on years of trial-and-error experience.

The most powerful forms of tacit knowledge, from a competitive standpoint, are the multidisciplinary problem-solving skills of senior scientists, researchers, and system architects. These individuals have developed a skill set that transcends specifics and specialties, thereby allowing the integration of multiple facets of technology. Pure invention is most often the domain of such generalists, since they are best able to recognize the interconnections among technologies and to visualize complete solutions.

In its most refined form, tacit knowledge can only be described as talent or genius, unique capabilities that simply cannot be transferred to another individual. Although the famous seventeenth-century violinmaker Antonio Stradivari had many imitators, his instruments were distinctly superior to their best efforts. Nonetheless, much of the wisdom embodied in the minds of experts can be absorbed by others through observation, imitation, practice, and feedback. The time-honored social relationships of apprenticeship and collaboration provide the most effective mechanisms for the exchange of skill, experience, and insight.[10]

The potential to absorb the tacit knowledge of others is highly dependent on the intellectual preparation of the individual. Contrary to popular belief, the purpose of an advanced education is not to gather the critical information that one needs to sustain a professional career. At best, the university coursework required of scientists and engineers is a starting point for their future endeavors. The primary benefit of advanced education is that it *trains the mind*. Facts can be gleaned from any textbook, but the ability to conceptualize, analyze, solve, and absorb can be acquired only through years of rigorous mental exercise. It is this capacity to manipulate information and render unique and unprecedented results that is at the core of innovation.

*Tacit Knowledge and Strategic Advantage*

During the 1970s and 1980s, a growing body of empirical evidence began to support a commonsense conclusion: The education and training of a nation's workforce is highly correlated with increased productivity.[11] As a result, economists introduced a new term into growth accounting models to capture the additional productivity, above and beyond unskilled labor, which could be achieved by individuals with education and experience.

In the popular literature, this concept of *human capital* has come to be synonymous with the strategic value of a firm's labor force.[12] In formal economic terms, however, human capital more accurately reflects the generic education and training level of employees, while tacit knowledge describes capabilities that are far more specialized and unique. One could say that the level of human capital reflects the capacity of individuals to absorb and embody tacit knowledge.[13]

As the demand for technology-intensive products expanded in the twentieth century, so did the demand for human capital. Shortly after World War I, American industry employed only a few thousand scientists and engineers, but by the end of World War II that number had jumped to almost 46,000. By 1985, more than 600,000 engineers and scientists were needed to support an aggregate industrial R&D effort that had grown at a rate of roughly 6 percent per year for more than two decades.[14]

The continuing explosive demand for highly skilled technologists represents a potential bottleneck to rapid growth in total-factor productivity. Without ample supplies of human capital to perform research and development, rates of innovation will plummet. In terms of the growth of firms, this "input" to production is a likely cause of diminishing returns to scale; all the market demand in the world is of little value if a firm cannot muster the innovative talent needed to exploit new opportunities.

When tacit knowledge is embedded in the cultural and organizational context of a firm, it becomes nearly synonymous with *core competency*.[15] This ubiquitous term refers to the unique aspects of a firm that provide a competitive advantage.[16] The literature on core competencies emphasizes optimal division of labor, increased specialization, narrowed focus of in-house activities, and outsourcing of non-core functions. This process serves to winnow out all but the highest-value tacit knowledge, and if executed properly, it will tend to raise the average productivity of a firm. A business strategy that focuses on development and exploitation of core competencies takes maximal advantage of the tacit knowledge embodied in a firm's employees and social structures.

If tacit knowledge is firmly entrenched in the minds of individuals and in the collective wisdom of teams, how can a firm ever hope to replicate it? This question strikes fear in the hearts of executives faced with the prospect of geographic expansion. While generic functions can easily be relocated, the unique and subtle aspects of a firm's market advantage may be stewed in the juices of tacit knowledge. How can these capabilities be transferred to a foreign land?

The answer to this challenge for most companies is to not even try. Most multinational corporations choose to transfer the commercial exploitation of their innovations, but not the innovation process itself. A highly effective R&D organization represents a "golden goose" for any technology-intensive firm. Even a minor change in personnel, such as the expatriation of one or two senior researchers to a foreign subsidiary, might upset this delicate creative balance. Hence, relocating a firm's R&D capability is not something to be undertaken lightly.[17]

On a positive note, however, the intractability of tacit knowledge allows firms to hoard a treasure of innovative capability with little chance of public spillover. While the explicit portion of R&D must be waved about in the marketplace in the form of patent disclosures, sales material, and promotions, the finely honed team of technologists that spawned these breakthroughs can be kept far away from would-be imitators.

As the rate of technological change increases, the explicit portion of productive knowledge will develop a steadily diminishing shelf life.[18] It is the accumulated tacit knowledge within the research, development, and production organizations of firms that holds the promise of a sustainable long-term advantage.

## A Model for the Commercial Output of R&D

Suppose that you are an entrepreneur with a burning desire to enter the sporting-goods industry. Since you have little experience in this sector, you choose initially to manufacture wooden baseball bats, based on the assumption that they will be an easy commodity to master.

First, you must establish a production operation capable of fabricating bats. This requires the purchase of some specialized machinery, location of an appropriate source of hardwoods, and procurement of other generic facilities. After making capital investments and hiring some semi-skilled woodworkers, your factory is capable of producing a "plain vanilla" product. Unfortunately, you find that the discount stores are filled with cheap bats imported from low-wage countries. In this price-sensitive commodity market you cannot hope to achieve a comparative advantage. (Note the use of the term 'comparative advantage' to describe global competition in commodity products.)

After performing a bit of market research (which should have been completed before the factory was built), you discover that major retail sporting-goods stores stock baseball bats with recognizable brand names, which are sold at premium prices. Undaunted by your earlier failure, you open your checkbook and pay a hefty fee to license the brand name of a major bat manufacturer.

Despite the application of a recognizable trademark, however, your product is still inferior to that of the industry leaders. No respectable athlete would use your equipment, because a vital aspect of baseball-bat production is missing: the tacit knowledge associated with years of batmaking experience.

Companies such as Hillerich & Bradsby Company, makers of Louisville Slugger, produce hundreds of varieties of bats, some of which are tailored to the needs of individual professional athletes.[19] In fact, other than golf clubs, wooden baseball bats may be the most customized equipment in sports. Key employees of such a firm have decades of experience in selecting woods, tailoring profiles, and achieving the proper weight and balance. Short of stealing a significant number of employees from such an established company, your firm will have a difficult time penetrating the baseball bat market.

This intentionally low-tech example highlights three distinct attributes of a product that provide value to customers: the physical product itself, its brand identity, and the specialized skills required to produce it. By generalizing this example, the commercial value of virtually any product or service can be described using three orthogonal dimensions:[20] its *commodity content,* its *explicit content,* and its *tacit content,* as shown in Figure 4.1. With respect to our sports example, the generic machinery, labor, capital, and raw materials used to produce "plain vanilla" bats would fall along the commodity axis, the value of brand identity (a form of intellectual property) would be reflected on the explicit axis, and the skills and experience required to produce superior bats would be displayed along the tacit axis. The unique point in this "space" that is defined by these three coordinates graphically illustrates the total value being delivered to customers.

To demonstrate how this model can be used to describe the value of virtually any output of economic activity, let us consider a few examples. Products such as clothing, raw and processed materials, and simple manufactured goods fall almost entirely along the commodity axis. The tacit content of a commodity item is very low, and although the product might require a good deal of explicit knowledge (in the form of blueprints and the like) to manufacture, none of it is excludable from the competition, and so no monopoly profits can be gained.[21]

Services, on the other hand, fall primarily in the plane defined by the tacit and explicit axes. With little or no physical content, such items as entertainment products, professional services, health care, and insurance are heavily dependent on the knowledge and talent of individuals, along with the systems and processes that make up a firm's operational strategy. The only significant aspect of a service business that falls along the commodity axis is the capital required to finance its operations.

Since capital does not form the basis for competitive advantage between firms residing in advanced nations, companies that provide services have two options: Either they can exploit some excludable aspect of their services' explicit content (i.e., they establish an enforceable copyright, trademark, etc.), or they must

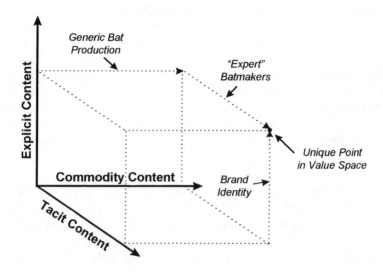

**Figure 4.1:** Diagram of a three-dimensional conceptual model that describes a given product or service in terms of its commodity, tacit, and explicit contents. This model provides insight into the sources of value for various product types, and can be used to determine optimal competitive strategies. The text example of a premium baseball bat is shown.

develop a valuable tacit capability that cannot easily be imitated, such as a highly trained team of applications engineers or field service technicians.

With respect to a specific innovation, the output from R&D must eventually be reduced to an explicit form for it to be useful in an industrial setting. Typically, the documentation describing an innovation might include patents, bills of material, drawings, schematics, CAD/CAM files, test specifications, and a host of other elements. Some of this explicit knowledge is often excludable through legal (intellectual property protection) or informal (secrecy and subterfuge) means, thereby ensuring a temporary flow of monopoly profits to the innovating firm.

From a static standpoint this is the entire story. Once a product is in production, the entire research team of a firm could quit (at least in principle) and profits would continue to roll in for years. If we consider a dynamically evolving market, however, the profits from individual products appear as momentary blips, while the continuous output of a superior research team is seen as generating a steady stream of revenue, capable of sustaining a company for decades. Even if the fabrication of an innovative product requires little tacit knowledge, *its initial creation critically depends on it.*[22] This reinforces my earlier statement that in a highly dynamic business environment, it is the tacit capability of a firm that holds the key to strategic advantage.

## Competitive Advantage in Evolving Markets

It is the nature of American enterprise to focus on near-term victories, rather than long-term prospects. This relatively short time horizon can lead to a narrow view of competitiveness. In high-technology industries, firms often achieve market leadership through "one-shot" breakthroughs. Even though the initial success of these companies is based to some extent on a lucky roll of the R&D dice, they may come to believe that they have discovered the pathway to long-term growth. What often follows is a series of high-risk projects, in which even a single failure can threaten the survival of the company.

Achieving market leadership is obviously an essential element in a successful long-term strategy. The methods a firm employs to gain early market share, however, may not include vital elements necessary to retain that advantage. In fact, the exact opposite may be the case: A firm that bases its competitive strategy on high-risk gambles cannot realistically hope to survive in the long run.

Likewise, firms whose advantage is based on methods that can be readily copied cannot sustain their leadership positions for extended periods. The most fundamental criterion for sustainable advantage is that those attributes of a firm that are at the core of its success must be resistant to imitation. Hence, traditional factors such as operational efficiency and high product quality must be discounted as potential sources of long-term dominance. Although these are clearly important attributes, they represent "table stakes" in modern global competition. Unless an advantage is sufficiently rare and excludable that it cannot be easily copied by competitors, the benefits it provides will be fleeting at best. Such generic strategies must be dismissed from our quest for the underlying sources of sustainable advantage.

### Dynamic Markets and Evolving Technology

In our search for sustainable advantage, we are looking for elements that underlie all strategic actions of technology enterprise, and provide a persistent edge to the firms and industries that possess them. Although the outcome of each market contest cannot be guaranteed, when averaged over an extended period these factors will become significant differentiators.

This discussion has a familiar parallel in the biological evolution of species. Indeed, the survival of firms in the global marketplace has many similarities to the processes of natural selection. The environment for technology-intensive enterprise has the same dynamic properties as the ecological environment, tending to shift and change in unpredictable ways. Those firms whose market advantage persists under rapidly changing conditions will tend to rise to the top, while competitors with static or transitory advantages will fall by the wayside.

This vision of enterprise as taking place in an amorphous and dynamic economic environment was first proposed by Joseph Schumpeter.[23] In his classic work

*Capitalism, Socialism and Democracy,* he observed, "The essential point to grasp is that in dealing with capitalism we are dealing with an evolutionary process. . . . Capitalism . . . is by nature a form or method of economic change. . . ."[24] Schumpeter goes on to describe the process of *creative destruction,* in which capitalism continuously discards unwanted structures in favor of more suitable ones.[25] Again, there is an obvious parallel with the behavior of ecological systems.

We can isolate the key elements of sustainable advantage by considering a simple thought experiment. Imagine a generic high-technology firm with a diversified product line and a burning desire to retain its current market leadership. Since this firm competes globally in many different markets, factors that are unique to a single technology or competitive environment cannot provide an aggregate long-term advantage. Furthermore, any strategies that can be easily copied by competitors must also be set aside. What actions can this firm take to ensure its long-term success in the face of rapidly changing technological and market conditions?

With no more information than is provided above, we can identify three elements that will offer such a firm a persistent edge over its rivals. First, we can increase its *rate of innovation.* A higher rate of innovation offers both the potential for gaining first-mover advantage and a greater degree of adaptability to changing market conditions.[26] Regardless of the type of technology product or the competitive structure of the market, rapid innovation is always desirable.

A second sustainable advantage can be derived from increasing the *appropriability of returns* from innovation. With all other factors being equal, a firm that can capture more profits from its innovations than its competitors can ultimately dominate its market. By plowing excess returns back into the R&D process, such a firm can compound the benefits of its increased appropriability. Much like the compounding effects of economic growth, a consistently higher rate of investment in research and development over a sustained period can produce an insurmountable long-term advantage.

Finally, the firm can improve its ability to *absorb and accumulate* technical knowledge. Whatever the conditions in the global marketplace, a substantial platform of accumulated knowledge is essential to sustained competitiveness. The benefits of rapid absorption and efficient accumulation of technology will compound in the same manner as an increased rate of R&D investment. The harnessing of accumulated knowledge within a firm is far less expensive than developing it from scratch, and represents in effect a subsidy to research and development. If technical knowledge can be acquired from outside sources or reused from previous projects, additional funds will be available to advance the firm's strategic agenda. An expanding pool of accumulated knowledge within a firm can create an intimidating barrier to new market entrants and can place competitors who lack the ability to exploit available knowledge at a significant disadvantage.

Together, these three elements represent a complete framework for sustainable advantage in high technology. Although they are distinct in their benefits, they are interrelated through their use of a common pool of tacit skills and complementary organizational structures. Enhancing any one element will almost certainly improve the value and effectiveness of the other two. In later chapters, I will demonstrate that this self-reinforcing triangle can be used to explain both the persistent success of individual firms and the technological specialization of nations.

## A Complementary Perspective

The primary motivation for creating a conceptual framework for sustainable advantage is that it can provide practical insights that can guide strategic actions. Often a simple model can be used to describe the organizing principles behind complex systems, with the goal of positively influencing the behavior of those systems.

There have been a number of efforts made in recent years to describe the nature of competitive advantage.[27] Recently, scholars have abandoned the unrealistic assumptions of "pipeline" innovation and perfect competition to embrace a web-like model reflecting the complex interactions among input factors, competition, customers, and suppliers.

Perhaps the most famous and potentially useful conceptual model for global competitive advantage was proposed by Michael Porter.[28] In his classic book *The Competitive Advantage of Nations,* Porter describes the determinants of national advantage for firms and industries, as shown in Figure 4.2. He identifies four interrelated elements, collectively referred to as the *national diamond,* which describe the competitive environment in a given industry and the potential advantages that can be derived from the creation of advanced and specialized factors.

The first point of the diamond represents the factor conditions within a given nation and industry. Porter discounts advantages in *basic factors* as offering little long-term benefit. He suggests that sustainable advantage can result only from the creation of unique and specialized factors, such as a highly skilled workforce or the presence of institutions that support specific industries. These created factors offer competitive advantage primarily because they reinforce a high degree of specialization and are difficult to duplicate.

The second and third points of Porter's diamond, demand conditions and related and supporting industries, reflect the forward and backward linkages of the value chain. The benefits of a well-developed and sophisticated supplier network include opportunities to collaborate on new product development and the potential to absorb technical knowledge from upstream specialists. Likewise, sophisticated and demanding customers can create a powerful incentive for firms to excel, and can reinforce the innovation process through high levels of intelligent feedback.

The structure and rivalry within an industry represent the final point in Porter's diamond. The presence of intense rivalry within an industry creates pres-

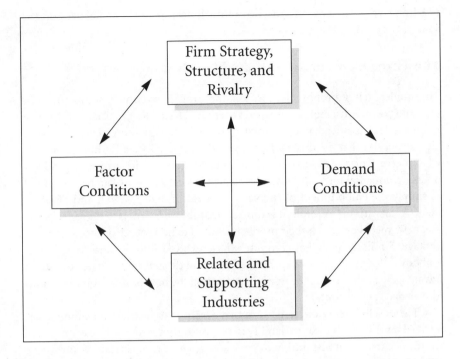

**Figure 4.2:** Porter's model of the determinants of national advantage, as represented by the "national diamond." Source: Porter, M. E., 1990, *The Competitive Advantage of Nations,* The Free Press, Figure 3-1, pg. 72. Figure used with permission.

sure for firms to innovate, both strategically and technically. Whether the source of rivalry is domestic or international, the structure and strategy of firms will be influenced most dramatically by the capabilities of their competitors.

The self-reinforcing nature of this model is immediately apparent. The presence of a strong supplier base, for example, will drive industries toward a more horizontal structure and higher degrees of specialization. A sophisticated customer base will motivate firms to provide increased differentiation and higher performance, which will potentially impact the strategies of both rivals and suppliers. Likewise, the creation of specialized factors may be either the cause or the result of evolutionary changes in customer demand, supplier networks, or inter-firm rivalry. Porter asserts that when all elements of the national diamond are strong, positive feedback will tend to drive firms within that nation toward global leadership.

While the reinforcing nature of Porter's model offers the potential for sustainable advantage, the broad determinants he has selected do not explicitly illuminate the evolving nature of these factors. As I will demonstrate in the next section, the four elements of Porter's model can easily be mapped into an alternative framework, thereby offering two mutually compatible perspectives on the sources of sustainable advantage.[29]

## The Three Elements of Sustainable Advantage

The model that I propose focuses on the process by which competition in technology-intensive industries evolves over time. Recall that in Chapters 2 and 3, it was asserted that leadership in total-factor productivity was essential to global advantage, and that the driving force behind productivity leadership among advanced economies is industrial innovation. The commodity aspect of products offers no potential for long-term advantage; physical hardware can be thought of as simply the embodiment of the value created through research and development. As an extreme example, it is unlikely that software firms would consider the magnetic media on which their product is stored to be a source of sustainable advantage. Similarly, personal computers have evolved into generic "boxes" that embody millions of hours of specialized tacit and explicit knowledge. Long-term advantage within the PC industry is based not on the box itself, but on the unique knowledge value it contains.

The development of a framework that describes industrial innovation is not a trivial task. There is great variability in the innovation process; one cannot plan for the occurrence of inspiration or genius. Furthermore, it can take years to reap the benefits of even the most valuable innovations. The development of electric-powered industrial equipment in the early twentieth century, for example, did not yield a substantial return until nearly two decades later, when the infrastructure necessary to provide adequate power to factories was finally in place. Yet underlying the chaos of creativity and innovation are fundamental driving forces that connect industrial research and development to market success. These vital elements essentially define the strategic environment for technology enterprise.

### Building a Framework

Although the determinants of success in industrial R&D are far from clear, the three elements that make up the *innovation cycle* shown in Figure 4.3 provide an unambiguous advantage to any high-technology company.[30] Moreover, the positive feedback between the elements in this model offers the potential to sustain that advantage for extended periods.[31]

Increasing the *rate of innovation* within a firm can be achieved in several ways. As was noted by Porter, an intensive rivalry within an industry can galvanize research teams and create an emotional focus for a firm's competitive strategy. Although rivalry across oceans is becoming increasingly heated, it is the competition among domestic firms that generates the greatest pressure to innovate. An advantage gained by a foreign competitor can easily be rationalized

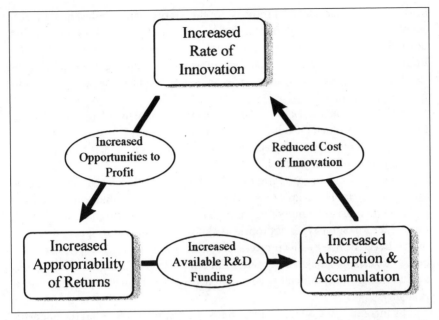

**Figure 4.3:** Model of the innovation cycle for technology-intensive firms. Note the self-reinforcing nature of this cycle; by increasing any one of the three elements of sustainable advantage, the other two elements are positively impacted.

as resulting from government subsidies or cheap labor. Within a firm's home nation, however, the playing field is decidedly level; domestic competition in high technology is almost entirely determined by the relative ability of firms to innovate.

It is worth reiterating an important point: Competitive advantage can be derived from innovations that are not directly related to a firm's products or services. Innovations in processes, organizations, training, marketing, sales, service, and strategy can all yield substantial long-term advantage. *Adaptability* in all aspects of enterprise is the key to success in a rapidly changing competitive environment.

A counterpoint to the pressure created by intense rivalry among competitors is provided by the pull of a demanding market. Rates of innovation are strongly influenced by access to large and affluent new markets, shifts in customer demand, or the introduction of revolutionary technologies. Other factors that impact rates of innovation include proximity to suppliers and customers, efficient infrastructure, and the establishment of industry-wide technical standards.

As the rate of innovation within a firm increases, its ability to appropriate returns from its inventions will also tend to rise. Rapid innovators are more likely to secure a first-mover advantage. By capturing the early market, a firm can extend the profitable life of its innovations, and by gaining a large initial market share, can potentially elevate its products to the status of an industry standard. Moreover, there are considerable cost advantages to be derived from being first down the learning curve of a new technology. As Intel Corporation co-founder Robert Noyce once noted, "A year's advantage in introducing a new product or new process can give a company a 25 percent cost advantage over competing companies. . . ."[32]

Similarly, rapid innovation offers increased potential to differentiate and customize a new product, introduce improved models, and create revolutionary substitutes. In all respects, an increased rate of innovation is the key to gaining and retaining technological leadership.

The *appropriability of returns* from R&D represents the second element of the innovation cycle. The degree to which a firm can keep vital innovations secret or preclude their use by competitors through legal means has a significant effect on rates of return. Some types of proprietary knowledge, such as recipes and process methods, can be more easily protected from imitation than more explicit and embodied forms. Advantages that result from unique tacit skills or other ambiguous sources are the most difficult to copy, and hence offer the highest levels of appropriability.

By increasing the rates of return to R&D investment, a firm has the opportunity to enhance its ongoing innovation efforts. In particular, these additional funds can be used to support a strategy of aggressive *absorption and accumulation* of knowledge, the third element of the innovation cycle. There is substantial evidence that the rates of diffusion, absorption, and accumulation of knowledge are a pivotal factor in the competitiveness of firms.[33] In particular, accumulated knowledge plays an important role in determining the *technological trajectory* of firms and industries. The self-reinforcing and path-dependent nature of accumulated knowledge serves to focus all three elements of the innovation cycle.

The accumulation of knowledge is becoming far less automatic as the complexity of products and processes increases. Furthermore, opportunities to develop a technological base within a firm are diminished when design teams are forced into a headlong dash to the marketplace. The temptation to shortchange the learning process can be irresistible, particularly when a firm has adopted a fast-second strategy.[34] It is critical, however, that companies look beyond these near-term constraints. By absorbing vital information from external sources and accumulating tacit learning and explicit knowledge through forward-looking internal design efforts, the long-term cost of innovation can be substantially reduced. Hence, the feedback loop is closed; accumulated technical knowledge shortens the time required for the next development effort, thereby increasing the rate of innovation.

Unfortunately, accumulated technical knowledge, along with all other facets of competitive advantage, is a rapidly depreciating asset. Human capital, infrastructure, organizations, and R&D strategies must all be continuously reevaluated and adapted to evolving market conditions. Without constantly tending to these sources of technological leadership, the positive feedback among the three elements of the innovation cycle will be damped out.

The most striking characteristic of this virtuous cycle is that many of the factors upon which it depends are beyond the control of individual firms. A high degree of cooperation is needed both along the upward and downward linkages of the supply chain and horizontally across related industries and sectors. Regional and national governments play a vital role as catalysts of cooperation, enforcers of competition and trade policies, and underwriters of basic research.

Evidently, at least for the case of high-technology enterprise, there can be no such thing as a "sustainable strategy of the firm." Sustainable market leadership is derived from the ability of companies to comprehend, anticipate, and adapt to the rapidly evolving external conditions that drive the innovation cycle. The development of competitive strategy, therefore, must transition from the domain of static choices and ossified five-year plans into a continuum of decisions and actions that align a firm's behavior with the changing global technology marketplace.

### External Forces and the Cycle of Innovation

Although the innovation cycle model that I have described appears straightforward, defining strategies to create and enhance it can be a daunting challenge. There is an endless list of factors that can impact the three elements of sustainable advantage, including firm-level capabilities, government policy, market forces, the state of basic scientific research, and the availability of human capital. The remainder of this book will be dedicated to reducing some of this intimidating complexity.

The first step toward comprehending this intricate milieu is to organize the factors that drive the innovation cycle into several categories, as shown in Table 4.1. The four groupings shown here are somewhat arbitrary, but offer at least some degree of consistency and insight. Each of these potential contributors to competitive advantage will be discussed in greater detail in Chapters 5 through 8.

The most direct contribution to the innovation process is made by the forces represented in the first category, *industry and market structure*. My description of this aspect of the competitive environment will focus on how individual firms can respond to (and possibly affect) these external conditions to their advantage. The opportunities in this category are endless; I will highlight several facets that have the highest potential impact on long-run market leadership.

The second potential source of reinforcement for the innovation cycle is enlightened *trade and competition policies*. This is perhaps the most contentious segment of the competitive environment, since it spans the domains of firms, industries, national governments, and international agreements. On the trade side, there is a strong temptation to exploit American political and economic dominance to gain a global advantage for our domestic industries. Both public sentiment and congressional debate have recently shifted away from the unqualified embrace of free-market capitalism toward a disturbing mix of protectionism and mercantilism.

Achieving sustainable advantage for our domestic industries will require that government place less emphasis on protectionism, and apply more energy to opening sophisticated foreign markets to U.S. technology products, ensuring fair competition, and enforcing global intellectual property rights. There is a growing chasm between our need for international cooperation in high technology and the increasing number of trade conflicts between U.S. and foreign firms.[35] It is the responsibility of government to *mediate* these disagreements, rather than aggravate them.

From a domestic standpoint, there are important opportunities to tailor competition policy to the needs of technology-intensive enterprise. Antitrust laws must be reevaluated in light of increased levels of industry specialization and the escalating cost of non-recurring R&D. Restrictions on the formation of alliances and consortia must be carefully considered, since these groupings are rapidly becoming essential to continued technological development.

In a similar vein, federal patent laws should be modified to achieve an optimal level of appropriability, while allowing the diffusion of new technologies to spur further innovation. In both of these cases, there is a difficult tradeoff to be made between the economic benefits to individual firms and the spillovers gained by industries, regions, and society in general.

The third segment of the competitive environment that can have a dramatic effect on technology-intensive industries is the external infrastructure provided by the U.S. government in support of industrial innovation. This collection of policies, institutions, and public goods has become known in scholarly literature as the *national system of innovation*. Unfortunately, early research in this area has done little to organize the complex maze of interacting factors, ranging from technology policy to human-capital development.

In the future, the national system of innovation will play an increasingly important role in promoting the long-term advantages of firms and industries. Certain critical activities that support the innovation cycle can be performed only at the level of regional and national governments. While firms are a source of specialized learning and tacit skills, the initial development of human capital will remain a public-sector responsibility. Likewise, the allocation of funding for basic scientific research is traditionally coordinated by national agencies. The U.S. government remains the largest single investor in R&D in the world, and must continue to absorb the risks of highly speculative research to facilitate the eventual commercialization of new technologies. In general, the national system of innovation provides a support structure that enhances the capability of American firms to innovate at the leading edge of technology.[36]

| Sources of Advantage | Potential Impact on Innovation Cycle | | |
|---|---|---|---|
| | Rate of Innovation | Appropriability of Returns | Absorption & Accumulation |
| 1) Industry and Market Structure (Chapter 5) | • R&D strategies of individual firms<br>• Collaboration and cooperation<br>• Domestic and foreign rivalries | • Degree of cross-licensing<br>• Mobility of skilled workers<br>• Profitable lifetime of new products | • R&D strategies of individual firms<br>• Level of spillover of "secrets"<br>• Willingness to form alliances |
| 2) Trade and Competition Policy (Chapter 6) | • Adapt antitrust laws to allow increased cooperation<br>• Subsidies for high-tech industries<br>• Ensure open markets | • Strengthening of foreign IPR law<br>• Antidumping actions<br>• Enforcement of trade agreements | • Tailoring of antitrust laws<br>• International treaties of technical cooperation |
| 3) National System of Innovation (Chapter 7) | • Basic research<br>• Pre-competitive R&D support<br>• Infrastructure and specialized factors<br>• Government procurement | • Tailoring of domestic IPR law<br>• Investment in human capital and tacit knowledge | • Support for pre-competitive R&D<br>• Human-capital development<br>• Diffusion of technology |
| 4) Regional Advantage (Chapter 8) | • Proximity of suppliers and customers<br>• Availability of human capital<br>• Access to outside expertise in non-core fields<br>• Intensified rivalry | • Local clustering of tacit skills<br>• Increased potential for specialization<br>• Higher potential for capture of spillovers in industrial clusters | • Enhanced absorption of spillovers<br>• Synergy among regional firms<br>• Enhanced tacit skill creation<br>• Proximity to basic research |

**Table 4.1:** Matrix showing four aspects of the competitive environment that can contribute to sustained competitive advantage, and their potential impact on the three elements of the innovation cycle.

Finally, the effects of *regional advantages* on the cycle of innovation will be considered. This last segment of the competitive environment has received a great deal of attention over the last few decades, given the obvious success of technology clusters such as California's Silicon Valley and the Route 128 region of Massachusetts. In a broader context, the advantages imparted by nations on their domestic industries were the subject of a ten-nation study performed under the leadership of Michael Porter.[37] His results confirm what has become intuitively obvious: There are substantial competitive advantages to be gained by harnessing national specialization, regional infrastructure, and local networks. In my opinion, regional advantage will have a greater impact on all three elements of the innovation cycle in the future than will any other aspect of the competitive environment.

It is worthwhile to pause for a moment to consider the approach that will be followed in Parts 2 and 3. In the latter portion of this chapter, I have identified three aspects of the innovation process that extend beyond individual market competitions to form a long-term foundation for sustainable advantage. Our goal will be to identify ways in which we can reinforce these elements through strategic action. In Part 2, I present a panoramic view of the competitive landscape for high-technology enterprise. Armed with this perspective, we can begin to explore how sustainable market leadership can be achieved. Part 3 begins with a survey of the tools available to develop and implement a dynamically adaptive competitive strategy. In the final chapter, I present several interesting models derived from high-tech industry case studies that are representative of what all future strategy must become: time-based, adaptive, and rooted in the tacit capabilities of the firm.

# 5

# Industry and Market Structure

## Rivalry as a Driving Force

The structure of markets and industries defines the envelope in which strategic actions must take place. Hence, this is a logical starting point for our survey of the global competitive environment. In this chapter, the effects of intensive rivalry on rates of innovation will be examined, followed by an overview of global R&D strategies in response to these market challenges. Finally, the potential benefits and risks of collaboration among firms, industries, and government will be discussed. Each of these subjects shares a common theme: They are instrumental to enhancing the innovation cycle and thereby securing long-term market success.

There is nothing like a heated rivalry to get the innovative juices flowing. Although the process of invention is often romanticized as a "flash of brilliance," the reality of research and development is more often characterized by endless hours of tedium, repeated failures, intense frustration, and enormous expense. No company invests in innovative activities unless compelled to do so by market forces.

Technology markets, by their very nature, are highly competitive. Whereas in commodity industries there are moments of rest between battles, in high technology the slightest pause can leave an opening for both domestic and foreign competitors. Even seemingly invincible Motorola, once the dominant force in the cellular-phone market, has lost much of its lead to rivals such as Nokia, Ericsson, and Qualcomm Inc. In 1994, Motorola claimed 60 percent of the U.S. wireless market; by 1998 its share had fallen to 34 percent, primarily due to the firm's inability to meet delivery schedules for the next generation in digital wireless technology.[1]

For American firms to survive in this dynamic arena, outmoded market strategies based on a mass-production mentality must be abandoned. Over the last two decades, American technology industries have lost virtually every global competition in high-volume, standardized products. On the occasions that we have achieved market leadership, it has been due to our exceptional ability to differentiate, customize, and invent. Out of the ashes of our postwar industrial model, an entirely new way of conquering markets has evolved. Rather than blunting our capabilities in futile head-to-head competition, many U.S. firms are successfully undermining low-cost foreign competitors by fragmenting commodity markets into high-value, specialized niches.

In the 1980s, America's giant semiconductor firms were under relentless attack by foreign rivals, first from Japan and more recently from Korea and Taiwan. U.S. industry leaders, including Intel, Advanced Micro Devices (AMD), and National Semiconductor, were nearly broken by rampant cost-cutting in high-volume products such as DRAMs and EPROMs. Despite unimaginable levels of capital investment in wafer fabrication facilities, U.S. companies could not match the yields and efficiency of these foreign invaders.

Just as the doors were closing on the domestic production of memory chips and other generic integrated circuits, a new breed of American start-up began to nibble at the edges of these huge commodity markets. Companies such as Cypress Semiconductor and Maxim Integrated Products began introducing dozens of specialized, high-value devices designed specifically for niche markets. Rather than producing large quantities of a few standard designs, these custom-chip houses were producing fifty or more different configurations on a single manufacturing line.[2] More important, these firms earned substantial monopoly profits by carving slices off the commodity sales of giant Asian chipmakers. Moreover, custom solutions proved to be more desirable and cost-effective for many customers than one-size-fits-all components, enabling these innovators to secure higher profit margins than their overseas rivals.

Several lessons can be learned from this experience. First, the American semiconductor industry provides an excellent example of a shift from the mass production of generic physical products to the marketing of embodied explicit and tacit knowledge. The competitive advantage of these new semiconductor entrepreneurs came from their ability to get close to their customers, understand specific applications, and design unique devices to meet these specialized needs. Low-cost wafer processing of standard components became secondary to achieving system-level cost savings through these tailored solutions. More important from the standpoint of sustainable advantage, the capabilities described above cannot be easily replicated by rivals and will evolve with the dynamics of the marketplace.

A second important observation is that even small segments can be highly profitable in global markets. The outdated truism that economies of scale are essential to competitive advantage has been mitigated by advances in flexible manufacturing and mass customization techniques. It is also incorrect to assume that competitiveness in niche-market products must be accompanied by high-volume production to defray capital investment and other fixed costs. This perception tends to drive companies toward vertical, product-driven strategies that limit adaptability and can lead to stagnation.

The need to derive sufficient returns from large capital investments can have a significant influence on a firm's strategic decisions. It is difficult to abandon "cash cows" in favor of promising new technologies when a firm is tied to a mountain of specialized equipment and tooling. This is why "outsiders" often have a competitive advantage in newly forming industries: They are not distracted by pressure to recover sunk costs. To remain the leaders in rapidly changing markets, firms must be ready to abandon mature products in favor of the next desirable embodiment of accumulated knowledge.

There are two general strategies for meeting the challenges of technology-intensive markets: the *first mover* and the *fast second*.[3] The first mover in a new market has several inherent advantages, as discussed previously. Along with the familiar learning-curve and economies-of-scale advantages, however, come some significant disbenefits as well. Trailblazing can be a risky business. Many small firms have entered the market with revolutionary products, only to find customers less than receptive. Likewise, transitioning an excellent technical design into rate production can be fraught with setbacks. Although the first-mover strategy appears to offer clear advantages, firms without sufficient experience, funds, or customer feedback can easily fail.

A fast-second strategy, on the other hand, offers the potential advantage of learning from the mistakes of the first mover. In particular, customer needs can be more successfully addressed, production problems may be avoided, and most important, the cost of development can be dramatically reduced. It is far less costly to improve on someone else's design than to break entirely new ground. The fast-second strategy is often employed by "industry giants" that can leverage their other competitive advantages, such as a strong supplier base and distribution channel, to capture market share.

The introduction of the first computed tomography (CT) X-ray scanners in the early 1970s offers an excellent example of the limitations of first-mover advantage and the effectiveness of a well-executed fast-second strategy.[4] The first mover in this case was a small development laboratory owned by EMI Ltd., a British electronics company (which incidentally provided recording and production facilities to the Beatles). In this case, the innovators could not adequately protect their new product from competitors through legal means. Despite filing several patents, the developers recognized that most of the core technologies involved in their product were in the public domain.

After achieving early penetration of the huge American market, they sought an alliance with General Electric, a world leader in medical-electronics equipment, to help work through production problems and gain access to the powerful GE distribution channel. GE declined the offer of a partnership, but soon launched a crash program to develop its own CT scanner. Despite EMI's substantial first-mover advantage, GE captured a majority market share by 1978. Ultimately, EMI was forced to sell its medical-electronics division to GE, making General Electric the undisputed world leader in CT scanning technology.[5] It is not uncommon for inventors to lose control of their innovations to highly competent and established competitors.

As the pace of technological evolution continues to accelerate, it will become virtually impossible for vertically structured companies to compete. The emergence of agile start-up companies in highly profitable niches has shifted the competitive playing field in many technology industries from cost competition among a few giant firms to increased levels of technical specialization within a network of firms.[6] This new horizontal industry structure mandates the need for increased levels of openness among suppliers and integrators, thereby driving firms toward higher degrees of collaboration along the supply chain.[7]

The presence of formidable rivals at multiple points in an industry network energizes the innovation process. Ideally, this heated competition can be supplied by domestic firms. A foreign threat, however, can galvanize entire industries into high rates of innovation.[8] History offers numerous examples of national pride providing the fuel for unprecedented technical progress. In the early 1900s, the French considered the surprising breakthroughs in aviation by American innovators to be a national insult. They had been the acknowledged leaders in lighter-than-air technology for a century, and considered their research into powered aircraft to be at the cutting edge. The success of upstart Americans such as Glenn Curtiss and the Wright brothers prompted a flurry of design improvements to attain longer flight times and greater air speeds. This early rivalry, combined with the heightened tensions in Europe prior to World War I, was largely responsible for the rapid development of the aviation industry in both countries.

A more familiar example is provided by the space race between the Soviet Union and the United States during the 1960s. Again, the combined forces of national pride and an underlying military agenda led to an amazing rate of progress. The U.S. space program matured from the embarrassing launch-pad explosions of the late 1950s to Americans riding dune buggies on the moon in little more than a decade.

By far the most effective rivalries are the personal contests between strong entrepreneurial leaders. More than in any other nation, America's technology industries have been shaped by powerful personalities. The explosive growth of the minicomputer industry during the 1970s, for example, was driven in part by a bitter personal rivalry between Digital Equipment Corporation (DEC) founder Ken Olsen and his former executive, Edison DeCastro, who had left DEC in 1968 to form Data General Corporation. In more recent times, the well-publicized feud between Microsoft's Bill Gates and Oracle Corporation chairman Larry

Ellison has been an important factor in the rapid evolution of the information-technology industry.[9]

Whether the source of competitive energy is a friendly contest or a bitter struggle, the rate of innovation is dramatically affected by rivalry among firms at the frontier of technology. The presence of demanding and sophisticated customers further reinforces a state of constant creative tension. Through the potent combination of opportunities for success and fear of failure, these warriors are driven to fight for technical leadership and customer allegiance in tomorrow's markets.

## Global R&D Strategy

American technology firms have steadily lost ground to foreign rivals over the last two decades in both the intensity and quality of R&D investment and the rates of return they have achieved. As noted previously, U.S. firms invest too little on scanning and assimilating external information, on developing process-related technologies, and on building internal storehouses of tacit and explicit knowledge.[10] An entirely new strategy for the allocation of R&D funding is needed to sustain long-term technological leadership.

The primary goal of R&D investment has traditionally been to create profitable new products. Unfortunately, this relatively narrow focus fails to capture the full scope of innovation in a technology-intensive firm. Much of what a firm does to achieve competitive advantage is not accounted for under the formal structure of research and development. Production engineering, testing and evaluation, process development, and industrial design functions, for example, are typically treated as being outside the boundaries of R&D. It is important that firms embrace a broader definition of innovation and establish a coordinated strategy encompassing all creative aspects of their enterprise.

The actual cost of innovation within most firms dwarfs their investment in formal R&D. For a new product to enter the market, a succession of high-skill activities must take place, beginning with prototype testing and ending with pilot and rate production. Just as core competencies can be distributed throughout a company, innovation can be widely dispersed. Although these creative activities can differ in form, the three elements of sustainable advantage directly apply. The rate of innovation on the production line or in the design department, for example, is just as critical to achieving market success as rapid progress in the research lab.

Likewise, appropriating returns from a proprietary process may be even more important than protecting the product itself from imitation. This is particularly true for products that have a unique formula or process recipe. Examples include the formula for Coca-Cola and the specialized techniques required for processing advanced semiconductors. Finally, the benefits of absorbing external "best practices" in all creative aspects of enterprise are clearly evident. The competitive advantage of accumulated knowledge is not restricted to the domain of research scientists.

To achieve the maximum benefit from investments in innovation, firms should reevaluate their strategies in light of the steps shown in Figure 5.1. The first task is to identify all innovative activities that directly contribute to the firm's competitive advantage. This audit must be done with an open mind, to avoid overlooking important hidden strengths. A sales force that is uniquely skilled at harvesting new customer requirements is as much a part of the innovation process as the "techies" in lab coats.

Once a firm's creative assets have been identified, their sources and levels of funding should be evaluated in the context of the strategic goals of the company. Ultimately, a coordinated plan should evolve that ensures optimal allocation of resources among all innovative activities. Most important, each creative area should set aside a proportion of its funding for enhancing the elements of the innovation cycle. Beyond project-specific tasks, a budget and plan should be established to: 1) promote the external absorption of knowledge, 2) implement a long-term tacit learning program, and 3) create an information archive of valuable explicit knowledge. Although it will take discipline to allocate precious discretionary funding to these long-term activities, a substantial investment in the future is essential to sustaining competitive advantage.

There are several recent trends in technology-intensive enterprise that must be considered when shaping an innovation strategy. An increasing number of products are incorporating aspects of several distinct disciplines through *technology fusion,* in some cases resulting in the formation of entirely new industries. Even the names of these hybrid fields often reflect the dual nature of the expertise they demand, including biomechanics, electro-optics, and mechatronics. The tremendous potential for breakthrough innovation in these multidisciplinary fields is still not fully recognized.[11]

Another relatively new aspect of technical progress is the potential for *network externalities.* A network externality is an economic benefit derived from systems that are interconnected and interdependent; as additional members join the network, each member is made better off. As new individuals are added to the telephone network, for example, all existing participants receive the benefit of access to a greater number of people. Similarly, as more people purchase cellular phones, subscriber rates will decline and service should improve.

Today, the effects of network externalities have infiltrated many aspects of technology markets. Industries such as telecommunications, information technology, and Internet software are driven by these external economies. In many cases, there is a significant pull-through effect, in which the proliferation of one network-based technology draws with it a number of supporting and

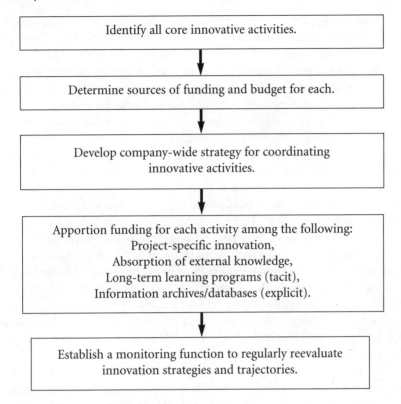

**Figure 5.1:** Process by which firms should reevaluate their investment in innovative activities. The goal is to apportion sufficient funding to all creative activities that enhance long-term competitive advantage.

related products. As Windows 98 spreads to PCs around the planet, for example, applications that support this operating system will tend to be pulled in its wake.

What this means to R&D strategy is that virtually every technology-intensive product developed today should be created in the context of external networks, ranging from meeting simple interface and communications standards to developing highly specialized components for a vast and complex metasystem. Modern electronic test equipment, for example, must communicate with an information network through a data interface, and interconnect with other instruments through protocols such as the VXI plug-and-play standard. At the other end of the spectrum, a piece of encryption software must be tailored to several layers of communication protocols and can function only within a narrowly defined external environment.

These two important trends, technology fusion and network externalities, create enormous pressure to look beyond the boundaries of the home nation for collaboration and expertise. The globalization of R&D activities has received considerable attention recently from both scholars and strategists, with very little consensus being drawn. On the one hand, the preponderance of data suggests that most large American firms tend to perform the majority of their core research on home turf.[12] Conversely, there are significant reasons why globalizing the R&D function offers unique competitive advantages.[13]

The most common motivation for extending research activities into foreign nations is to access technical specialization and unique human capital. In addition, there are strategic advantages, such as the ability to rapidly sense local market conditions and adapt products to regional tastes. Several authors have suggested that the location of all enterprise functions, including R&D, should be selected on the basis of optimal conditions for each activity.[14] This concept of global optimization, however, must be weighed against the benefits of economies of both scale and coordination derived from a more centralized innovation strategy. Clearly, the old IBM/AT&T concept of the insular corporate research center is no longer viable.[15] There are several reasonable alternatives, ranging from the *polyp model,* in which a coordinating center in the home country is fed by regional absorbers, to a completely decentralized *global network* of R&D centers.

As one might expect, the optimal configuration is most likely a hybrid of these two extremes. I call this hybrid structure the *neuron model,* as shown in Figure 5.2.[16] In this model, firms establish R&D centers in optimal locations for specific technologies or lines of business. Each centralized R&D organization is fed by a network of distributed absorption centers and is loosely connected to the other research centers of the firm. Each central R&D organization has primary responsibility for core research, strategic planning, and new-product development in a specific technical area. The absorption centers harvest specialized expertise, regional market data, and technical information while providing an in-country location for adapting product designs to local requirements. The central R&D locations serve as accumulators of knowledge for the firm and support open exchange of information and collaboration among their counterparts in other regions.[17]

While this model seems a bit complex, it is the most common structure observed among successful high-technology firms. One of the pioneers in this decentralized approach to research and development is Hewlett-Packard, which establishes relatively small autonomous divisions to pursue individual business segments. Each division is responsible for performing its own R&D and is free to locate in whichever global locations provide the greatest advantage. The capabilities and knowledge of these centers is shared throughout the H-P network, with corporate headquarters providing only high-level strategic planning and coordination functions.

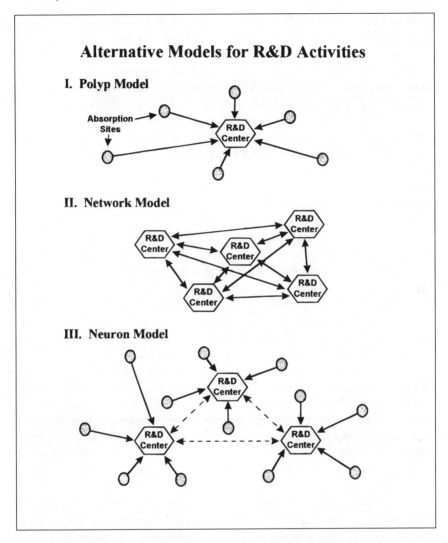

**Figure 5.2:** Conceptual drawings of three models for global R&D activities. The polyp and network models represent the extremes of reasonable innovation strategy. The neuron model is a hybrid, and has been successfully implemented by a number of high-technology firms.

One example of Hewlett-Packard's use of this approach is its establishment of central R&D organizations for both computer keyboards and specialized semiconductors in Southeast Asia. Technology-intensive firms that have adopted a similar model include Apple Computer, W. R. Grace, Motorola, Siemens, ABB Asea Brown Bovari, and many others. In each of these cases, firms have been able to balance the regional advantages of a global R&D presence with the increased security, focus, economies of scale, and interpersonal contact that a centralized innovation organization provides.[18]

While it is certainly true that many large firms continue to perform the majority of core R&D activities within their home nations, I believe this is primarily an issue of legacy rather than strategy. Almost by definition, the first "neuron" of any firm will locate its nucleus in its home country. As technological sophistication spreads to a greater number of countries and the levels of specialization increase, there is reason to believe that R&D activities from all nations will be drawn to these global centers of excellence.

## Alliances and Collaboration

Underlying the discussions of rivalry and strategy presented earlier are indications that cooperation among firms can represent a key factor in competitive advantage, both domestically and globally. The roles of strategic alliances, supplier partnerships, and various licensing agreements have become well established in technology-intensive industries. Leading firms are increasingly willing to look outside their boundaries to access critical expertise, gain a cost advantage, penetrate a vital new market, or diversify a product line.[19]

The most common forms of collaboration are based on the partners' substantial contributions of capital, technical expertise, and other assets. This commingling of valued property is in contrast to the "arm's-length" transactions of the traditional marketplace. Such agreements can be formed among component and systems firms, manufacturers and distribution channels, rivals in the same industry, companies in related industries, and myriad other combinations. The potential for collaboration can be further extended through cooperation among firms and universities, federal laboratories, and various levels of government.

These ventures can be loosely divided into two categories: competitive and pre-competitive. Competitive collaboration typically occurs when two or more firms recognize a market opportunity that requires cooperative action to capture. Since the topic of market-driven strategic alliances has received considerable attention in the popular business literature, I will focus instead on some less-trodden territory: the strategic benefits of pre-competitive collaboration.

There are compelling reasons for firms to establish cooperative agreements at the early stages of the innovation process. The most obvious motivation is financial: Collaboration provides an opportunity for firms to avoid duplicating costly exploratory R&D and to distribute the risks among several partners. By sharing in the uncertainties of front-end development, firms effectively increase the appropriability of returns on their research investment.[20]

Beyond the straightforward cost benefits of pre-competitive collaboration are more subtle rewards. It may be possible for firms to gain scale economies in their R&D activities, through the sharing of unique or expensive equipment, pooling of specialized expertise, and leveraging of shared databases. While scale economies may be a significant factor in some types of research, in general there is little evidence to support the idea that "bigger is better" with respect to industrial innovation.[21]

Perhaps the most frequently cited motivation for firms joining into pre-competitive R&D ventures is the gaining of exposure to new areas of expertise, with the prospect of anticipating the direction of technological advance.[22] The trajectory of a narrow industry segment is typically well known to competitors within that market. The most efficient way for firms to gather a "high-altitude" vantage on the evolution of entire industries is through collaboration. Microsoft, for example, forms strategic alliances as a preemptive strategy to anticipate new trends and ensure that it has a foothold in promising new technologies. Better still, this type of monitoring activity comes at a significantly reduced cost when compared to supporting a dedicated in-house "scanning" function.

One of the most important contributions that pre-competitive industry consortia can make to the innovation process is through the establishment of technical standards. As new technology-intensive industries emerge, it is essential that the resulting products fit within the existing base of hardware and infrastructure. The acceptance of new innovations is often predicated on their compatibility with telecommunications and information protocols, national safety, performance guidelines, and a host of industry-specific interface and data standards.

The formation of an industry consortium to serve as a sponsor for technical standards is a critical step toward achieving legitimacy for new technologies.[23] It can also provide a substantial competitive advantage to those firms that participate. Becoming the standard in an emerging market is the dream of every marketing organization, particularly if its product represents a "leverage point" for a broader industry sector.[24] Familiar examples include computer operating systems, microprocessors, cellular-phone modulation schemes, and data-formatting standards such as CD-ROM, digital videodisk (DVD), and VHS video.

A critical element in the success of Cisco Systems hardware and software products, for example, has been the ability to establish its operating system as the industry standard. In 1996, Cisco controlled more than 80 percent of the market for Internet data routers, largely due to global acceptance of its Cisco Internetwork Operating System (IOS). Cisco is a founding member of a sixty-company industry-standards body, the Gigabit Ethernet Alliance, which establishes standards for current and future high-speed data communications. By leading the effort toward rapid standardization, Cisco has enhanced its own rates of innovation and ensured that new industry protocols will conform to its product architecture.[25]

The potential to transfer explicit knowledge through collaborative ventures is fairly obvious. What may prove to be of even greater strategic importance, however, is the opportunity to harvest tacit skills. Consortia provide a proving ground for new technologies and a shared learning experience for all involved. Moreover, if the cooperative agreement is between firms and the university system, there is the potential to observe talented employment prospects in action. In this case, collaboration provides an opportunity to screen potential job candidates who have already gained a considerable base of industry-specific tacit knowledge.[26]

There has been an upsurge over the last decade in the number of international collaborative ventures. The reasons for this growth are much the same as those for the proliferation of domestic consortia, with the added incentives of increased foreign market access and the potential for collusive behavior in global competition.[27] The formation of international consortia makes sense from a long-term perspective as well. Typically, the challenges of innovation in a given industry are common across national borders. Technology is an equal-opportunity resource, and can represent an important common ground upon which business relationships can be built. In some cases, a domestic weakness can be bolstered by participation in global collaboration. More commonly, however, these cooperative activities provide a low-cost, high-efficiency method for absorbing knowledge from centers of specialization around the globe. I will return to the role of consortia in the rapid diffusion of technology in Chapter 7.

The possibilities for cooperative behavior in the pursuit of economic gains are endless, particularly when considering relationships that span firms, industries, and governments. A sampling of agreements that can provide significant long-term advantage is shown in Table 5.1.[28] While standard business arrangements are the most common, there is an increasing diversity of consortia, with charters that range from affecting government policy to promoting regional interests. One of the most interesting forms of cooperative behavior among competitors involves *tacit collusion,* in which firms informally agree to cooperative market strategies, as explained in Box 5.1.[29]

Implicitly, firms that enter into tacit collusion are extending good faith to other participants. Of course, since this kind of "agreement" cannot be explicitly documented (since it is often illegal), there is no recourse if cooperation breaks down. Often, tacit collusion takes the form of a moratorium on price competition among a few major competitors. Another possibility is the division of a market into informal "territories" that are serviced by only one firm. In either case, the result is reduced competition and higher profit potential for the colluding parties.

| Type of Collaboration | Potential Participants | Primary Impact on Innovation Cycle |
|---|---|---|
| Joint Venture | Firms | • Innovation Rates<br>• Appropriability of Returns<br>• Absorption & Accumulation |
| Cross-licensing Agreement | Firms / Government Labs | • Innovation Rates<br>• Appropriability of Returns<br>• Absorption & Accumulation |
| Tacit Collusion | Firms | • Appropriability of Returns |
| Standards Body | Firms / Government | • Innovation Rates<br>• Appropriability of Returns |
| Pre-competitive Research | Firms / Government Labs / Universities | • Innovation Rates<br>• Appropriability of Returns<br>• Absorption & Accumulation |
| Political Action Group | Firms | • Appropriability of Returns |
| Technology-sharing Agreement | Firms / Government | • Innovation Rates<br>• Absorption & Accumulation |
| Regional Promotion Body | Firms / Government | • Absorption & Accumulation |

**Table 5.1:** A sampling of various forms of collaborative agreements among firms, industries, and government. Note that the potential impact of these ventures on the innovation cycle is substantial, implying that cooperative behavior plays a central role in sustaining competitive advantage.

Although a positive outcome from any collaborative agreement is by no means certain, firms have a reasonable expectation of deriving strategic benefit from the arrangement. Increasingly, collaborative agreements are moving up the innovation food chain beyond strictly pre-competitive activities. The ability of such *proto-competitive* ventures to directly affect market structure and dynamics can generate both higher potential gains and additional friction between members.

The archetypical example of proto-competitive collaboration among firms is the Japanese *keiretsu*. These corporate groupings have come to symbolize both the benefits of collusion and the threat that such "anticompetitive" behavior represents to American industry. I will discuss these and other examples of market-shaping cooperation in Chapter 7. For now, I will simply note that the *keiretsu* are neither as anticompetitive nor as market-distorting as Americans typically believe.[30]

## Box 5.1: The Competitor's Dilemma

Gains from tacit collusion are based on an unspoken agreement among firms to execute strategies that increase the profits of all parties. This type of cooperation is illustrated in a classic thought experiment derived from game theory: "The Prisoner's Dilemma."

Two suspects are arrested for the same crime: a violent murder. The police separate the prisoners and each is offered the same deal: If each turns state's evidence on the other suspect, he will go free. If they both choose to squeal on the other, they will both hang, but if both prisoners refuse to implicate the other, they will both go free. Since neither prisoner knows that the other has been given the same offer, they will both be tempted to act in their own self-interest. However, they might reasonably assume that the offer will be given to both suspects, and through tacit collusion "agree" to stay silent.

The concept of tacit collusion can easily be extended to the real world. General Electric and Westinghouse, for example, were competitors in the steam-turbine industry during the early postwar period. Since a price war would hurt both parties, GE sent a "signal" to Westinghouse by widely advertising a price guarantee stating that it would reimburse new customers if prices fell in the future. This bold move communicated to Westinghouse that GE wished to enter into tacit collusion, with the result that both prices and margins in the steam-turbine industry remained at artificially high levels for more than a decade. Ultimately, the Department of Justice objected to what it called "conscious parallelism" and sued both firms in the late 1960s.[31]

A more recent example involves trading practices at the NASDAQ stock exchange. NASDAQ is structured to produce narrow bid-ask price spreads through aggressive competition among securities dealers. However, a study published in 1994 observed that a majority of the most frequently traded stocks were being bought and sold only in quarter-point steps, rather than the minimum eighth-point increments. This trading "anomaly" allowed NASDAQ dealers to appropriate extra profits for themselves. As a result, the authors of the study suggested that the dealers had entered into tacit collusion to garner higher profits from trades. Although the charges are still under investigation, there have been a number of class-action lawsuits filed against NASDAQ on behalf of investors who lost money in these transactions.[32]

The case of NASDAQ is unusual, in that tacit collusion is most effective among a small number of competitors. The potential to "cheat" on cooperative behavior rises dramatically as the number of participants in the arrangement increases.

As collaborative agreements become more competition-oriented, the potential for negative outcomes increases. Alliances that were initially formed to execute pre-competitive research may have their agendas corrupted by members wishing to derive some immediate benefits from their investment. This could tend to reduce the scope of projects toward narrow, mission-specific activities. Similarly, the more-advanced members of a consortium may feel that they are contributing disproportionately to the group. Their fear of "free riders" absorbing their proprietary knowledge may drive them to restrict access by certain members, or to even more-aggressive forms of uncooperative behavior.

It is important for firms to recognize that while collaboration can make a valuable contribution to their internal innovation process, it is at best a complement, not a substitute. Cooperative agreements are most effective when they serve to increase rates of diffusion and absorption of knowledge, rather than attempting to extend the scientific frontier. There is tremendous potential for American firms to benefit from international ties that enable the gathering of new knowledge. Beyond this limited scope, however, the opportunities for gains from collaboration must be carefully weighed against the cost and risk involved.

# 6

# Trade and Competition Policy

## Free Trade or Cautious Activism?

Collectively, trade and competition policies constitute a set of guidelines by which business is transacted on both a domestic and a global basis. There is a vital necessity for such rules, to protect intellectual property, ensure fair and open markets, and avoid anticompetitive behavior that might inhibit new entrants or restrict healthy rivalry. Within this context, trade and competition policies can have a significant impact on rates of innovation and the appropriability of returns from R&D investment.[1]

Unfortunately, the range of potential government action in this domain extends far beyond the benign "watchdog" role implied above. This is an inherently political arena, in which governments are heavily influenced by powerful and vocal constituencies. As with any political process, there is considerable risk that self-interested parties will distort the system, creating more losers than winners. Moreover, interference in markets is a notoriously risky business. Even if policies were devised that did not target special interests, the impact of such actions on global commerce would be far from certain.

With this warning in place, it should be clear that government intervention in free markets should be administered with a light hand. Enhancing the competitiveness of America's technology industries cannot be accomplished through market-distorting policies, nor will long-term economic growth be assured by such actions. As several scholars have observed, aggressive trade and competition policies are no substitute for an effective technology policy.

### Trade Activism as a Reasonable Response

The justification for the growing involvement of government in the international trade of technology goods is based on the disproportionate impact that high-technology exports have on the domestic economy. In particular, global leadership in these high-productivity industries offers a significant terms-of-trade advantage, resulting in reduced cost of imports and high wages in the export sector. The focus of American trade policy on technology-intensive industries is further supported by several studies indicating that a country's innovative capability is the main determinant of its overall export performance.[2]

There are other motivations for trade activism as well. Many of today's commercial technologies have their roots in the defense sector, raising issues of national security and "strategic importance" for products such as supercomputers and semiconductors. This thinking originated in the days of defense spinoff of dual-use technologies, and has carried over into today's commercially dominated technology base. The distinction between a strategically critical innovation and an economically valuable one is not obvious. Recent debate over the export of data encryption technology is an excellent example of the often conflicting demands of national security and commercial interests.[3]

Another area in which trade policy directly impacts the welfare of technology-intensive industries is in ensuring the consistent availability of the latest innovations from foreign sources. The ability of American firms to compete at the technological frontier depends on their having reliable global access to the best materials, components, and processing equipment. In many cases, however, these best-in-class products are manufactured by foreign firms. Hence, there is a growing concern that domestic companies may become dependent on overseas suppliers for critical inputs, leaving themselves vulnerable to shortages, price manipulation, and possibly even exclusion from obtaining these vital ingredients.[4]

The potential for industries to become dependent on *concentrated suppliers* may loom large in the future if the trend toward national specialization and polarization of technical activities continues. Unfortunately, the options for a policy response to this situation are somewhat limited. One possible remedy would be a government subsidy for domestic entrants into "critical" industries, which are perceived to be too highly concentrated from a global perspective.[5] Other options include aggressive antidumping actions to ensure a fair opportunity for American firms, and the filing of complaints against foreign subsidies and cartel-like behavior through the World Trade Organization dispute-settlement mechanism.[6]

Ultimately, however, I believe such actions are doomed to failure. Although trade remedies might prevent gross distortions of the market, the idea that there must be competitive domestic suppliers for all state-of-the-art inputs to production has already become untenable. Just as the vertical, command-and-control model for industrial organization has been discarded, so must our paranoia regarding foreign dependency be cast aside. Market forces cannot be resisted by domestic subsidies, as evidenced by unsuccessful attempts to bolster American semiconductor-equipment manufacturers, chipmakers, and liquid-crystal-display developers.[7] Industries will succeed or fail in global commerce because of their technical acumen and marketing savvy, not as a result of government intervention.

## Industrial Policy and R&D Subsidies

The issue of subsidies for commercial research and development has become central to trade-policy debate in recent years. Many American technology firms have rationalized their poor competitiveness in global markets as resulting from "unfair" subsidies by foreign governments. Japan's Ministry of International Trade and Industry (MITI) has been singled out as an example of blatant (and supposedly successful) industrial policy. Some objective light can be shed on this issue by considering two fundamental questions. First, how pervasive is the targeting of firms for government subsidy, both domestically and internationally? Second, have these subsidies actually produced "winners," or have they played an inconsequential role in global competitiveness?

On the first issue, it is well established that governments subsidize technology-intensive industries. In fact, the United States has led the world in such subsidies for the entire postwar period. This support has come in the form of defense R&D, directed military procurement, and "mission-oriented" activities such as the space program and the war on cancer. Under the guise of government-funded pre-competitive R&D, many of our leading export industries were born and raised. Examples include computers, telecommunications, satellite systems, semiconductors, lasers, aircraft, radar, pharmaceuticals, biotechnology, and a host of others.

While it is true that advanced economies such as Japan's have provided direct subsidies to commercial industries, the result has been neither important nor particularly successful. Targeting efforts by MITI, for example, have been much more modest than is popularly believed. The U.S. government provided ten times Japan's level of support for the semiconductor industry throughout the second half of the 1980s.[8] Likewise, it is a common misconception that Japan's industrial subsidies have had more successes than failures. Actually, the track record in this respect for both Japan and the newly industrialized countries has been disappointing. Only a handful of the more than sixty industries that MITI has targeted for subsidy have achieved global competitiveness, and there have been notable failures, including steel, aluminum, aircraft, and biotechnology.[9]

Unlike American government subsidies, however, which typically support pre-competitive technology development, industrial policy in many advanced economies has been focused on directly manipulating market outcomes. This distinction has been used to justify an unprecedented level of American government intervention in bilateral trade. The semiconductor industry in particular has been a poster child for *managed trade* over the last fifteen years. In this case, government involvement grew well beyond an occasional antidumping action to directly managing the prices and quantities of both imports and exports, as will be discussed in the next section. For now I will simply note that, as with many of our critical technology industries, flawed domestic choices in both strategy and policy have been the primary cause of past weaknesses in the merchant semiconductor sector, rather than unfair foreign-trade practices.[10]

Is it possible to reach any conclusions regarding the impact of trade and competition policy on sustainable advantage? Managed trade actions have been a mixed blessing at best, and a commercial disaster at worst. Although antidumping measures, for example, have been effective as an economic threat, the results have most often been higher domestic prices for critical inputs to production (such as DRAMs and EPROMs) and windfall profits for foreign producers who have essentially been given a license to charge monopoly prices for their products. These negative outcomes, combined with the arbitrary and inherently self-interested nature of antidumping actions, render them ineffective as a source of long-term advantage.[11]

This is not to say that industry-specific trade actions are never warranted. They may, however, be a bad investment. As with most business decisions, we must consider the opportunity cost of trade remedies as compared with other available options. Are these measures making any discernible contribution to American competitiveness? How, for example, does a countervailing duty create a sustainable advantage for American firms? The opportunity cost in this case is not just financial, although that is certainly a consideration. There is a political cost as well: By engaging in a steady stream of confrontational trade battles, we are poisoning the waters for far more profitable and beneficial cooperative actions. Are a few isolated "victories" worth destroying vital conduits for collaborative research and the absorption of technical knowledge?

A more promising application of trade and competition policy is the opening of foreign markets to American products. In this arena, all three elements of the innovation cycle can be significantly enhanced. Rates of innovation are driven upward through increased access to sophisticated markets. The appropriability of returns is enhanced due to economies of scale and improved opportunities to defray non-recurring investment. Finally, and perhaps most important, foreign markets provide a wealth of information on possible product differentiation and line-extension strategies. The importance of access to diverse, affluent, and technologically savvy markets by American innovating firms cannot be overstated.

## Market Access for the Right Reasons

There is little need to underscore the importance of increased access to global markets by American technology firms. Market access has become a primary goal of both the U.S. Commerce Department and the U.S. Trade Representative since the signing of the Omnibus Trade and Competitiveness Act of 1988. It is worthwhile, however, to reevaluate our motivations and methods.

## Better Markets, Not Just Bigger Ones

Economies of scale are not the only reason why access to large markets such as Japan and China is important, contrary to some trade-analysts' thinking. Their presumption is a relic of the mass-production era, and is supported more recently by experience in merchant semiconductors and other capital-intensive industries. For most high-technology products, however, economies of scale in production amount to a relatively minor factor. Software products, for example, have a near-zero marginal cost and are virtually unaffected by economies of scale. The deployment of mass-customization and flexible-manufacturing techniques further reduces the need for high volumes to achieve competitive marginal costs.

By comparison, the need to amortize fixed R&D costs over a sufficiently large sales base represents a vital necessity.[12] Although the United States still represents more than 30 percent of the world's market for technology products, the rapidly increasing cost of R&D and specialized production equipment is stretching appropriability to the limit.[13] Not surprisingly, an additional few hundred million sophisticated and well-heeled customers can make the difference between profit and loss.

In addition to the demands of the balance sheet, there are more far-reaching motivations for increasing access to global markets. If we look beyond an Ameri-centric view of the world, it is apparent that we are no longer the "first market" for many high-technology products. Until recently, innovations designed in America for domestic customers were adapted to foreign markets almost as an after-thought. Today, there is a high probability that the "next great idea" will evolve first in a foreign market. Without access to these trendsetting customers, our firms will lose first-mover advantage and potentially be locked out of new industries.[14]

In general, the greater the number and diversity of buyers, the greater the op-portunity to innovate. Even mature product lines can be extended through tailor-ing to niche foreign markets. This is not a manifestation of the outdated product life-cycle model; unsaturated foreign markets may offer entirely new applications for domestic technologies. Distributed power generation, for example, has a lim-ited market in the United States, but represents an essential lifeline in developing countries with low population densities. Likewise, environmental technologies developed for American water treatment and soil remediation are in high demand throughout the world. In these cases, the underlying science is mature, but meet-ing the needs of each new application requires considerable innovation.[15]

Achieving access to foreign markets is not simply a matter of overcoming tariff barriers; foreign nations may block imported products in a number of ways. Among the most common obstacles to market access, particularly in developing countries, are local-content requirements. In a sense, these are not barriers to market access for foreign *firms,* they are barriers to the import of foreign *products.* The European Union, for example, uses both import duties and quantitative restrictions to limit the import of products considered to be "high-value-added."

These actions have been taken to force foreign firms into manufacturing their products on EU soil, thereby creating high-wage employment opportunities. Thus, companies such as Motorola, National Semiconductor, and Hughes Electronics have established wafer-fabrication plants in the United Kingdom to avoid high tariff barriers on semiconductors, and several Japanese firms have located consumer-electronics factories on the European continent to circumvent quantitative ceilings.[16]

Although job creation is an important motivation for import restrictions, most often these measures are designed to protect domestic industries, particularly in high-growth "strategic" sectors. Many developing countries use import substitution and infant-industry protection to foster the development of "sunrise" industries, which offer long-term economic spillovers. Brazil, for example, has utilized a complex suite of local-content regulations in an attempt to orchestrate the modernization of their industrial sector. In some cases, such as personal computers, its attempts to protect infant industries have been abject failures. Efforts to attract foreign direct investment in aircraft and automotive industries, on the other hand, have had a generally positive impact on Brazil's technological capability.[17]

Local-content requirements represent a temporary obstacle, but the ability to bypass them through foreign direct investment softens their financial impact. Nations such as Japan and China, on the other hand, which place encumbrances on both imports *and* foreign investment, are more problematic. Japan is notorious for its apparently closed markets; among the advanced economies, the domestic Japanese market is uniquely resistant to both foreign competition and investment. In 1994, for example, the ratio of Japan's exports to imports in machinery and transport equipment was 5.4 to 1, meaning that Japanese firms exported more than five times as much merchandise in this category as was imported to its domestic market. Similarly, in the critical category of office machines and telecom equipment, the export-to-import ratio was 4.2 to 1. Similar ratios for the United States during this period were .81 to 1 and .70 to 1, respectively.[18] Based on this sampling of data, the U.S. market is more than five times more open than the Japanese market in these important high-technology sectors.[19]

### The Semiconductor Trade Agreement

There are as many anecdotes regarding the inaccessibility of markets in Japan as there are barriers to entry. Biased procurement of high-technology equipment by government-owned (or -influenced) agencies such as Nippon Telegraph and Telephone (NTT) is a frequently cited example. Foreign bidders on domestic projects have been rejected, despite proposing the lowest-cost solution, due to "type exceptions." Japanese procurements are often specified in both performance and design, meaning that they require a specific "type" of product. Hence, a superior product may be rejected due to its nonconformance to design requirements. Arbitrary technical standards of this kind have caused American fiber-optics and telecommunications firms to lose major contracts, and have forced U.S. medical and pharmaceutical firms to suffer endless delays due to Japan's labyrinthine product approval process.

One of the most famous market-access disputes between the United States and Japan was brought about by the decimation of America's commodity semiconductor industry in the mid-1980s. Complaints of unfair competition were lodged against the Japanese, both for below-cost dumping of semiconductors into the American market and for failure to grant equal access to American firms in the important Japanese domestic market.[20] For the last two decades, Japan has represented roughly 40 percent of the world's market for memory chips, microprocessors, and other silicon-based devices. U.S. firms claimed, with some justification, that lack of access to such a huge, leading-edge market was a substantial competitive disadvantage.

In the early postwar years, the American semiconductor industry was nurtured by a strong demand for miniaturization of military electronics. This sophisticated and captive market helped U.S. firms to dominate world production through the mid-1970s. In the late 1960s, however, Japan began developing a domestic semiconductor capability, expanding on its already mature technology base in transistors. By imposing tight border controls on imports and targeting high-volume processing technologies, the Japanese government aggressively supported domestic firms, including Toshiba, NEC, Fujitsu, Mitsubishi, and Hitachi. From the beginning, the focus of Japanese industrial policy in semiconductors was on capturing the high-volume commodity markets. With the benefit of a huge domestic demand for DRAMs, EPROMs, and other generic devices, Japan launched a highly effective global campaign in the late 1970s and early 1980s.[21]

Initially, U.S. firms countered this offensive with threats of GATT sanctions unless Japan abandoned its protectionist import policies. By the end of the 1970s Japan had responded by eliminating virtually all formal trade barriers with respect to semiconductors. Unfortunately, these measures failed to improve the rapidly eroding world market share of American merchant semiconductor producers.[22]

In June 1985, the U.S. Semiconductor Industry Association (SIA) filed a formal petition against unfair trading practices by Japanese companies, under Section 301 of the Trade Act of 1974.[23] Several antidumping suits were initiated over the next several months, including an unprecedented "self-initiated" filing by the U.S. Commerce Department, citing Japanese firms with dumping 256 Kbyte and one-megabyte DRAMs on the U.S. market. These aggressive actions prompted a flurry of negotiations between the United States and Japan, seeking mutually acceptable price controls and increased access to Japanese markets.[24]

The signing of the Semiconductor Trade Agreement (STA) in September 1986 was a landmark of sorts in global high-technology trade. Charges of unfair competition in U.S. markets were suspended in return for an agreement by Japanese chip manufacturers to voluntarily adhere to price floors for their products. Enforcement assistance was to be provided by MITI, which agreed to monitor export prices on selected commodity semiconductors. With respect to market access, the STA included commitments by the Japanese government to provide sales assistance to foreign firms and to encourage long-term relationships between Japanese users and these foreign suppliers.

In addition to the explicit terms of the Semiconductor Trade Agreement, a confidential "side letter" was signed by the Japanese government, agreeing to a goal of 20 percent market share for U.S. producers of several semiconductor varieties within five years. Thus, rather than defining "rules" by which trade would be carried out between the United States and Japan, the market-share arrangements of the STA were attempts to manage the "outcomes" of commerce between the two countries. Although the express purpose of this agreement was to increase global competition, by controlling both the prices and market shares of the world's largest semiconductor producers, the STA effectively created an international cartel in these vital commodities.

Was this highly publicized exercise in managed trade successful? The answer, as one might expect, is highly debatable. While market share for American semiconductor producers did increase over the five years following the signing of the STA, the 20 percent goal was not reached until 1996, as shown in Figure 6.1. Arguably, recent gains have been due to American firms establishing a leadership position in highly customized and sophisticated new semiconductor varieties. It is not clear that the blatant pressure applied by the Japanese government on its firms to buy American in the years following the STA had any beneficial effect on the acceptance of U.S. products. Although the Semiconductor Industry Association is still strongly in support of measures such as the STA, the broader impact of this agreement appears to have been largely negative.

Shortly after the signing of the STA, global spot prices for DRAMs rose to as much as six times the long-term contract price, and average selling prices deviated significantly from the precipitous decline typical of the semiconductor life cycle, as shown in Figure 6.2. It is estimated that between $3 billion and $4 billion in excess profits was earned by global manufacturers of DRAMs and EPROMs during this period.[25] Japanese firms received the lion's share of these "bubble profits," but it is important to note that American manufacturers such as Texas Instruments and Micron Technology also benefited greatly from higher prices.

The losers in this case were the users of commodity semiconductor chips. The market for these critical components was characterized by shortages, delays, and rationing during 1987 and 1988. Several U.S. computer manufacturers were forced to delay shipment of new models due to inadequate supply of memory chips and other devices. While it has been suggested that the spike in global semiconductor prices during this period resulted from a surge in market demand, combined with technical difficulties in the production of one-megabyte DRAMs, it seems likely that the STA contributed significantly to soaring prices and shortages.

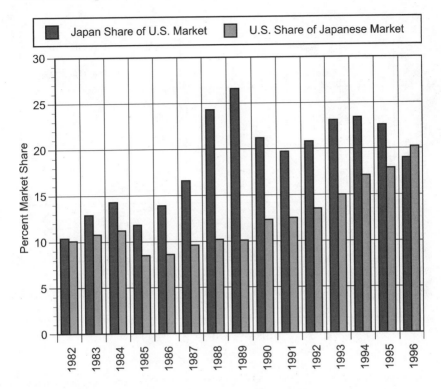

**Figure 6.1:** Reciprocal market shares of U.S. and Japanese semiconductor firms from 1982 to 1996. Source: Semiconductor Industry Association, 1998, *Semiconductor Industry Statistics,* SIA Website.

In the aftermath of the STA, it is difficult to determine whether the recent rise in competitiveness of U.S. semiconductor firms was accomplished through the aid of managed trade actions or in spite of them. The STA did little to protect the industries that were originally targeted; by 1987 only two American firms were still producing DRAMs for export. Meanwhile, the potential for monopoly profits through this type of cartel-like arrangement was not lost on new global competitors. In 1993, the South Korean government and its semiconductor industry association offered to engage in an agreement similar to the STA, in return for the United States dropping antidumping actions.[26]

The moral of this story is that managing to "outcomes" in global markets can have unpredictable and potentially negative effects. Although the STA and its successor arrangements have been strongly supported by the semiconductor industry, the potential loss of competitiveness in related industries must be carefully considered. [27]

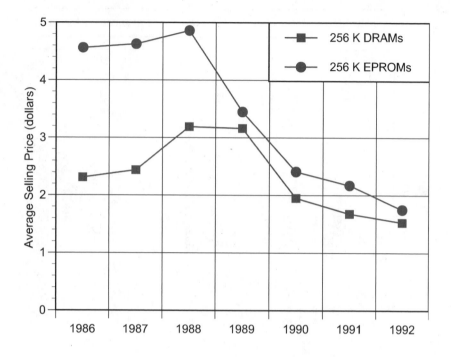

**Figure 6.2:** Average selling price of 256 Kbyte DRAMs and EPROMs from 1986 to 1992. Source: Dataquest, Inc. Data used with permission.

Critics of American trade policy have noted that actions directed at specific markets, such as Section 301 sanctions and antidumping complaints, are focused on achieving "selective reciprocity": the exchanging of bilateral commitments and concessions in an attempt to control or equalize markets.[28] This is a significant departure from multilateral trade agreements, such as the GATT, which attempt to ensure fair market access by all nations over a broad range of products. Again, the equitable imposition of "rules" seems more likely to have a positive long-term impact on American competitiveness in high technology than does the targeting of specific industries or firms.

Ultimately, the best way for U.S. firms to ensure market access in Japan, China, or any other nation is to introduce unique, desirable, and superior products. Perhaps our interest in achieving arbitrary shares of commodity markets is misguided. In the long run, the greatest benefit to be derived from access to global markets is not higher revenues in one or two industries, but opportunities for diverse learning and technological advance in all industries.

# Intellectual Property, Piracy, and Antitrust

There is no issue more central to the welfare of technology-intensive industries than the protection of intellectual property.[29] The cycle of innovation critically depends on the appropriability of sufficient profits from new and novel creations. If this "right to profit" is undermined by imitation or outright piracy, firms will have little incentive to invest in research and development. It is the security offered by patent laws, copyrights, trademark protection, and other intellectual property rights (IPRs) that inspires enterprise to approach the risky frontiers of science. Without an effective system of IPR protection, the "fire of genius" will be starved for financial fuel.

## *The "Golden Age" of Intellectual Property?*

It is easy to assume that the importance of intellectual property is a modern manifestation of information technology. Latter-day giants such as Microsoft and Oracle, along with almost all of the nation's fastest-growing firms, derive their value from intellectual rather than physical capital. Indeed, technology-intensive firms often have little or no tangible assets; their high stock-market valuations are based on an anticipated stream of future revenues from intellectual property.

Although knowledge-based industries have brought IPR protection to the forefront of business strategy and government policy, its importance was recognized by our forefathers. The U.S. Constitution includes a provision to "promote the progress of science and useful arts, by securing for limited times to authors and inventors the exclusive right to their respective writings and discoveries."[30] In fact, Thomas Jefferson was the first patent commissioner, and personally examined every application submitted during the period from 1790 to 1793.

Since that time, the U.S. Patent and Trademark Office (PTO) has granted more than 5.5 million patents. The term of protection was extended in Lincoln's day to seventeen years, a period which was not reevaluated until very recently.[31] In fact, most of the current structure for protection of intellectual property has not changed significantly in more than one hundred years.[32] Recently it has become apparent that this isolated and highly specialized government agency has not responded to changes in the technological environment. With procedures and structures designed to meet the simple needs of the industrial era, the PTO has been rendered woefully inadequate by the advance of technology.

Intellectual property is no longer confined to the familiar categories of patents, trademarks, and copyrights. A more comprehensive listing now includes layout designs of integrated circuits, breeder's rights, trade secrets, industrial design, a broad spectrum of artistic works, and a number of special (and typically unresolved) cases such as software, pharmaceuticals, and genetic innovations, as shown in Table 6.1. Actually, only a small fraction of the proprietary information that is used by industry today is formally protected; most of it falls into the category of secret know-how. This form of tacit knowledge is embodied in people and methods, making it far harder to protect through legal means.

As I discussed earlier, there is a fundamental tradeoff between ensuring sufficient returns to innovating firms and maximizing social benefits through the rapid diffusion of new knowledge. This balance was less important in earlier times, when intellectual property was subordinate to physical capital as the source of competitive advantage. During the industrial era, firms freely shared proprietary knowledge. Today, with the cost of R&D skyrocketing and the government's share of total research funding on the decline, there is increasing pressure on firms to dig deep into their profit streams to fund future innovations. In this modern economic climate, the prospect of cost-free sharing of intellectual property seems unlikely at best.

It is therefore incumbent on both national governments and international bodies to strike a critical balance between appropriability of returns and diffusion of knowledge. In addition to the obvious economic issues, this activity must take into account a number of social concerns. While most constituents would agree that the inventor of a clever labor-saving device deserves the reward of monopoly profits, it seems unlikely that society would tolerate such a monopoly on, say, a cure for cancer. Does this imply that the government must have the right to "buy" socially vital innovations from inventors for some arbitrary fee?

Other important factors that must be reassessed by policymakers include the methods of protection, the types of innovation, the income level of the innovator, the length of patent protection, methods of enforcement and arbitration, and a host of other variables. Unfortunately, these policy decisions will be guided by very little hard data. There have been several studies examining the impact of the length and quality of patent protection on rates of innovation, but these narrow studies suffer from a common weakness: There are significant differences in IPR requirements among industries, nations, and technologies, making extrapolation of these isolated results to national policy recommendations highly suspect.[33]

Establishing equitable protection for software products has been a notable challenge for the PTO in recent years. As a general rule, basic software concepts have not been granted strong intellectual property rights. Hence, the idea of a blinking cursor, the notion of a compiler, or user-interface features such as overlapping windows are not protectable.[34] In a recent infringement suit, Quantel Ltd., a British software firm, claimed that the popular Photoshop software package created by Adobe Systems Inc. had violated five of its patents that cover "painting" with a computer stylus. Despite the PTO's issuing patents on this innovation in

| Intellectual Property Category | Duration of Protection | Description |
|---|---|---|
| Non-disclosure Agreement | Typically 3–5 years | Any confidential information disclosed during an interaction that is not public or received through other legitimate means. |
| Patent | Up to 20 years | Processes, machines, manufactures, composition of matter, original designs, certain agricultural plants. |
| Copyright | Life of creator plus 50 years | Products of the mind that are produced in tangible expressions: writings, paintings, movies, music, sculpture, computer software. |
| Trade Secret | As long as secrecy is maintained | Any commercial formula, device, pattern, process, or information that is secret, substantial, or valuable. |
| Trademark, Trade Name, Service Mark | 10 years for initial filing plus indefinite 10-year renewals | Word, name, symbol, device, numeral, picture, or any combination thereof that represents a unique identity or brand. |

**Table 6.1:** Overview of the various categories of intellectual property and the periods of protection currently provided under U.S. patent and trademark law.

the mid-1980s on the grounds that it was "novel and nonobvious," many software programmers recognized the principle at the time of Quantel's filing. In this case, a jury rejected the claim of infringement, agreeing with the defense's contention that upholding such a broad patent would open the door to suits against hundreds of other software producers.[35]

When IPR protection has been granted to the final products of software development, it has typically fallen under copyright laws rather than patent laws.

There are no clear guidelines as to what constitutes a patentable software innovation. Does it make sense to confer the same level of protection to software inventions as is currently provided to hardware breakthroughs? Given the rapid pace and unprecedented interdependency of software products, the same period and strength of protection may not be optimal. Perhaps a more relaxed standard would accelerate the diffusion of new concepts and thereby maximize social benefit. Clearly, with respect to new knowledge industries such as software, bioengineering, and microelectronics, we are only beginning to understand the issues.[36]

## Piracy and Protection in Global Markets

The importance of IPR protection in international commerce is demonstrated by its impact on the global strategies of multinational corporations. A number of studies over the last decade have examined the dependence of foreign direct investment on the levels of intellectual property protection afforded in various countries. While selecting a location for an overseas subsidiary is impacted by many strategic factors, in the most IPR-sensitive industries protection can play a major role in FDI decisions. In a recent study of ninety-four U.S. multinational firms in six industries, it was found that the impact of IPR protection on FDI was heavily dependent on the type of investment. While the location of R&D facilities was strongly influenced by IPR considerations, decisions to invest in factories or distribution centers were unaffected.[37]

The effect of IPR protection on foreign investment is also heavily dependent on the levels of technology in a given industry. An OECD study recently noted that the lack of software protection was a major disincentive to FDI in that sector.[38] Likewise, the pharmaceutical industry is highly sensitive to IPRs, due to the enormous cost and time required to develop new products and the fact that pharmaceutical firms privately fund 99 percent of the research and development for patented drugs. One survey concluded that more than 60 percent of all new drugs introduced into the market would not have been developed if adequate patent protection had not been obtained.[39]

While IPR protection has received considerable attention in recent multilateral trade negotiations, the costs to American industry of unlicensed usage and outright piracy remain staggering. One well-documented example of the value of intellectual property is offered by Texas Instruments, which has mounted a rigorous campaign since 1985 to appropriate license fees from firms in all nations that utilize its substantial library of patents. The result has been a stream of revenue totaling more than $1.8 billion in the years from 1985 to 1993.[40] In fact, this highly profitable effort now represents one of Texas Instruments' largest sources of revenue.[41]

Infringement on protected intellectual property can be divided into two general categories: unauthorized use of an *idea* and imitation of the *actual product*. In the first case, a competing firm incorporates the protected innovation into its own products, either to avoid the cost of development or to divert some of the monopoly profits granted by patent protection to itself. For example, in a recent suit filed by Digital Equipment Corporation against Intel, DEC claimed that Intel infringed on aspects of its Alpha microprocessor-chip design. Typically, theft of an idea is difficult to prove; often the plagiarism is unintentional, and simply the result of parallel-technology development. In other cases, infringement suits are used purely for strategic reasons, either to issue a "warning shot" to competitors or to bog down a market leader in litigation.

The more blatant form of IPR violation is almost always both intentional and malicious. Piracy of protected products has gone far beyond the fake Rolex watches sold in alleyways. The International Anti-Counterfeiting Coalition has estimated that between 5 and 8 percent of all products and services provided worldwide are counterfeit, amounting to roughly $200 billion in lost revenues annually. In a separate study, the U.S. Customs Service estimated that the United States lost 750,000 jobs to foreign commercial piracy in 1993, with more than 200,000 attributable just to counterfeit auto parts.[42]

Although all forms of IPR are routinely violated by pirates, the theft of trademarks is the easiest to accomplish, and is probably the most costly to American business. There is an entry barrier of sorts associated with the piracy of some forms of intellectual property. The technical sophistication and financial resources required to reverse-engineer and duplicate an integrated circuit, for example, eliminates all but a few potential counterfeiters. With respect to trademarks, on the other hand, this entry barrier is almost nonexistent, resulting in a proliferation of pirated products displaying famous global brands.[43]

Most developing countries have instituted acceptable laws regarding the protection of patents, trademarks, and copyrights; however, enforcement is often grossly inadequate. This problem can be exacerbated by a lack of commitment on the part of local officials; many developing countries believe that the advanced nations are hoarding their knowledge, and see piracy as an acceptable way for their country to transfer desperately needed technology. While this argument may have some moral leverage with respect to pharmaceuticals, for example, it is a poor justification for the counterfeiting of entertainment CDs.

In some of the largest developing nations, piracy has become almost institutionalized.[44] India, for example, is developing a dubious reputation as a center for intellectual property rights infringement. A combination of weak IPR laws, inadequate enforcement, and a technically adept workforce has made India a "center of excellence" for the counterfeiting of high-technology products.[45] China represents an even larger problem, as discussed in Box 6.1. Whereas the culture of the West has a strong tradition of protecting the property and human rights of individuals, no such cultural foundation exists in China. Hence, the world's largest marketplace is becoming a global leader in the pirating of software and entertainment products.

Rather than wait for multilateral treaties on IPR protection to take hold in these developing markets, affected industries are taking action on their own to reduce foreign piracy. The Business Software Alliance (BSA) and the Alliance to Promote Software Innovation (APSI) are examples of a new generation of proactive consortia established to supplement poor foreign government enforcement of intellectual property rights. Recently, the BSA initiated a raid on a leading maker of compact disks in Singapore. The firm, a listed company on the Singapore Stock Exchange, was suspected of producing thousands of illegal copies of Microsoft Windows 95 and other software titles.[47] The cost to American industry justifies the exposure and risk of such aggressive tactics; the BSA estimates that more than 225 million business-software applications were pirated worldwide in 1996 alone, amounting to a revenue loss of roughly $11 billion, as shown in Figure 6.3.[48]

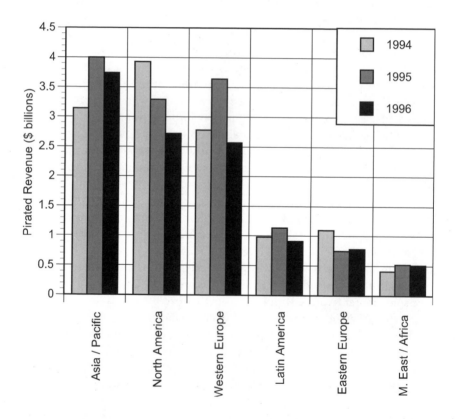

**Figure 6.3:** Results of a global survey of the financial impact of piracy on the software industry in various world regions for 1994 through 1996. Source: *Global Software Piracy Report*, Business Software Alliance Website, 1997.

## Box 6.1: Piracy in China: An Elegant Offense?[46]

Counterfeiting of American music CDs and software titles in China represents a significant obstacle to Western firms entering that enormous emerging market. Weak IPR laws, lax enforcement, and a generally insolent attitude toward the West prompted the U.S. Trade Representative in 1994 to threaten severe trade sanctions under Special 301 of the Omnibus Trade and Competitiveness Act. Trade agreements resulting from these aggressive actions have yielded significant progress toward stronger IPR laws and enforcement methods.

Yet piracy in China is still ubiquitous. Counterfeit CD titles turned up on the streets of Shanghai and Beijing less than two months after the United States and China narrowly averted a trade war over IPR. This frustrating lack of progress may be rooted in fundamental cultural and institutional differences between China and the West, as exemplified by a now-famous Chinese proverb, "To steal a book is an elegant offense." The concept of intellectual property was neither defined nor addressed in the long history of imperial China. In fact, before China opened its markets to the West, the copying of a book was considered to be a compliment to the author. This lack of a cultural basis for IPR protection may significantly reduce the effectiveness of tougher new laws.

There is, however, room for optimism. In 1994, the Walt Disney Company won a copyright infringement suit against a Chinese printer for illegally pirating its cartoon characters. Although the win represented a moral victory, the judgment awarded to Disney was a mere $91. In contrast, a suit filed in 1997 by Koller Group, an American manufacturer of VCR components, resulted in a more satisfying award of $120,000.

Still, the judicial system in China must be treated as a unique and temperamental beast. For example, it is best to announce your intentions to file an infringement suit well in advance, to give the offending firm adequate time to desist. Likewise, officials want to ensure that all parties have a fair chance, so Western firms should not behave as indignant aggressors. Finally, it may be advantageous to hire local private-investigation and legal firms on a percent-contingency basis.

Eventually, China's efforts to meet global standards for IPR protection and enforcement will begin to take hold. In the first five months of 1997, for example, Chinese courts handled 5,296 cases of intellectual-property violation. Ultimately, the Chinese will recognize the true value of IPR protection when its firms begin to develop indigenous intellectual property and are faced with the piracy of their own valuable assets in foreign markets.

The primary coordinating body for international intellectual property rights is the World Intellectual Property Organization (WIPO), headquartered in Geneva, Switzerland. The WIPO is a specialized agency of the United Nations, chartered with administering a loosely connected body of multilateral agreements, as well as providing arbitration and mediation services. More than 160 nations have joined the WIPO since its inception in 1970, but not all members are parties to all treaties. Although participation in the WIPO does represent acknowledgment of the importance of intellectual property rights, it does not guarantee harmonized protection among member nations.

Two multilateral agreements form the foundation for international IPR protection. The Paris Convention for the Protection of Industrial Property (1883) provides a common IPR regime among member states, covering patents, trademarks, and industrial designs. This agreement establishes provisions for *national treatment,* thereby ensuring that each contracting state grants the same protection to nationals of other member nations as it grants to its own citizens.

The second foundational agreement is the Berne Convention for the Protection of Literary and Artistic Works (1886), which establishes minimum standards for protection of copyrighted material. The Berne Convention includes a provision for *automatic protection,* which grants immediate IPRs to copyrighted material in all member countries without the need for compliance with local formalities.

In addition to administering these two basic agreements, the WIPO hosts a number of more specialized accords, as shown in Table 6.2. Several agreements have been implemented in recent times to address rapid advances in technology. The issue of copyright protection for software programs, for example, is dealt with (at least in a preliminary fashion) by the WIPO Copyright Treaty of 1996. Similarly, the Washington Treaty on Intellectual Property in Respect of Integrated Circuits (1989) establishes international guidelines for the protection of semiconductor-device layouts. Both these treaties are open for signature by WIPO member nations, but neither is yet in force as of this writing.

One of the most important treaties administered by the WIPO makes it possible for firms to seek patent protection for an invention in each of several countries simultaneously, by filing an "international" patent application. The Patent Cooperation Treaty of 1970 provides a significant advantage to companies that plan to introduce their products directly into global markets. The potential benefits of this measure are demonstrated by the rapid rise in international patent filings since the agreement went into force: in 1979 only 2,625 applications were submitted, whereas in 1996 this figure had grown to 47,291. The average number of countries listed on applications grew from 6.7 in 1979 to more than 56 in 1996.[49]

Although the WIPO is intended to be a central clearinghouse for global intellectual property rights, it is not the only international body concerned with

| Intellectual Property Treaty | Date | Description |
|---|---|---|
| Paris Convention for the Protection of Industrial Property | 1883 | Applies to industrial property in the widest sense, including inventions, marks, industrial designs, utility models, trade names, and geographic indications. |
| Berne Convention for the Protection of Literary and Artistic Works | 1886 | Protects "every production in the literary, scientific and artistic domain, whatever may be the mode or form of its expression." |
| Washington Treaty on Intellectual Property in Respect of Integrated Circuits | 1989 | Secures protection of original layout designs (topographies) of integrated circuits, whether or not the integrated circuit concerned is incorporated in an article. |
| Patent Cooperation Treaty | 1970 | Makes it possible to seek patent protection for an invention in each of several countries by simultaneously filing an "international" patent application. |
| Madrid Agreement Concerning the International Registration of Marks | 1891 | Provides for international registration of both trademarks and service marks. |
| WIPO Copyright Treaty | 1996 | Provides copyright protection for computer programs, in any mode or expression, compilations of data, and databases. |

**Table 6.2:** Summary of major intellectual property protection agreements administered by the World Intellectual Property Organization as of 1997. Source: WIPO Website.

this vital issue. On January 1, 1995, the Agreement on Trade-Related Aspects of Intellectual Property Rights (TRIPS) came into force after the successful conclusion of the Uruguay Round of GATT negotiations. The TRIPS agreement is administered by the World Trade Organization, and covers much of the same ground as the WIPO treaties. Since there is obvious and substantial overlap between these two institutions, the WTO and the WIPO signed an agreement in 1996, pledging mutual cooperation.

The TRIPS agreement has been described as the "Berne and Paris-plus agreement," since it uses these seminal accords as a starting point.[50] A substantial number of additional obligations have been added to the TRIPS, however, dealing with areas in which the Berne and Paris Conventions were deemed to be inadequate. Thus, provisions for protection of new plant varieties, layout designs of integrated circuits (through adoption of the Washington Treaty), pharmaceuticals, and software have been included in the TRIPS. In addition, strong wording with respect to enforcement is used, requiring member nations to provide "expeditious remedies to prevent infringement and remedies which constitute a deterrent to further infringements."[51]

One substantial benefit of the GATT TRIPS regime is that all members are obligated to adhere to all provisions. Unlike agreements administered under the WIPO, which are not mandatory to all members, the TRIPS represents a first attempt at a harmonized and truly global system of intellectual property protection. To allow time for these mandatory obligations to be implemented by all members, however, Uruguay Round negotiators included a transition period for developing countries. Advanced member nations were given one year to implement the TRIPS disciplines, whereas less-developed countries were allowed up to eleven years to conform. As an additional "equalizing" measure, the GATT TRIPS extends "most-favored nation" obligations to all members.

Despite these promising efforts at global IPR protection, significant problems still remain. The relative security of intellectual property is strongly related to the level of economic development of a given nation. In general, more-advanced countries provide stronger IPR protection, while less-developed nations prefer a shorter period of protection, thereby facilitating rapid inward transfer of technology. This weaker protection is often accompanied by *working requirements* that force foreign firms to manufacture products within the developing nation to receive IPR protection. With respect to critical technologies, such as pharmaceuticals and agricultural chemicals, countries such as Brazil, India, and Argentina have openly refused to offer adequate protection.[52]

Perhaps the most significant obstacle to a truly international IPR regime is the need to harmonize the often conflicting domestic regulations of more than 150 nations.[53] One controversial disagreement centers on the point at which the clock starts on patent protection. Most countries of the world base their patent protection on a *first-to-file* principle, whereas the United States applies a *first-to-invent* rule. An important objection to the first-to-file concept is that it may allow early filing of "fictional" patents: a firm with a good grasp of the technological future within its industry may file a patent disclosure based on extrapolation of current trends. Without holding inventors to high standards of demonstration and proof-of-principle, our technological future could become gridlocked in a maze of overlapping speculative patent filings.

Intellectual property rights laws in the United States must set a global example if we are to secure the strong protection our knowledge-based industries so vitally need. Recent domestic legislation establishing new penalties for copyright violation is a step in the right direction. The No Electronic Theft Act of 1997 provides for criminal action against individuals who copy software or other protected works worth more than $1,000, even if the violators do not profit from their actions. The Business Software Alliance estimates that closing this loophole in U.S. copyright laws will save the software industry more than $100 million annually.[54]

Legislation to reform the Patent and Trademark Office is pending in both Houses of Congress at this writing. The Omnibus Patent Act of 1997 (S 507) would streamline the operation of the PTO, provide for timely publication of foreign patent submissions, and reduce the astronomical cost of patent litigation through the establishment of inexpensive administrative procedures to clarify the scope of granted patents.[55] If enacted, this bill would transform the PTO into a government corporation, with increased budgetary independence and improved industry access. While this is a promising first step, it is only the beginning. The PTO currently represents a significant bottleneck to both the rates of innovation and the appropriability of returns in the United States.

### At the Cusp of Cooperation and Competition

The tradeoffs associated with the strength of intellectual property protection described earlier are mirrored in federal antitrust laws. Furthermore, the goals of these two elements of competition policy are often in direct conflict. Whereas IPR protection seeks to establish monopoly powers for innovators, antitrust laws are intended to ensure healthy competition by breaking up monopolies. The pivotal role that antitrust legislation plays in high-technology competition was made abundantly clear in a contentious suit filed with the Justice Department in 1997 against Microsoft on behalf of Netscape,[56] and in the launching of a broad investigation of Intel Corporation by the U.S. Federal Trade Commission in that same year.[57]

In both these cases, firms that dominate their respective markets were suspected of using their clout to restrict competition, in violation of antitrust laws. It was claimed by their detractors that Microsoft and Intel exerted undue influence on both original-equipment manufacturers and distributors of personal computers to incorporate their products into retail merchandise. In the case of Microsoft, this amounted to requiring producers to "bundle" their version of an Internet browser with other software that is preloaded onto PCs before shipment. For Intel, the continuing investigation is focused on whether it is punishing companies that buy competitors' goods by withholding allotments of its popular microprocessors and limiting its marketing support.

There are some famous historical analogies to this current attack on high-technology giants. In the early postwar years, both AT&T and IBM were investigated under antitrust laws for restricting competition. The settlement of the AT&T suit involved a consent decree restricting the firm's activities to domestic telecommunications and forcing a liberal licensing of electronics breakthroughs such as the transistor.[58] A similar consent decree settled an antitrust suit against IBM by mandating the liberal licensing of its punchcard technology and other computer-related patents at reasonable rates.

These aggressive actions by the U.S. government in the 1950s and 1960s caused American technology industries to limit their cooperative activities, thereby reinforcing inwardly focused strategies such as vertical integration and self-sufficient R&D facilities. On the other side of the Pacific Ocean, however, a more liberal perspective on competition policy was shaping industry in a very different way. Although competition in Japanese industry has always been fierce, the formation of cooperative ventures among firms, and between industry and government, has been actively encouraged.[59] Thus, by the late 1970s there was a stark contrast between the highly integrated Japanese *keiretsu* structure and the insular islands of American industry.[60]

In the early 1980s, U.S. policymakers began to recognize the need for intra-industry cooperation among high-technology firms. Galvanized by the success of overseas competitors, the government quietly began to relax its stand on antitrust violations and to view consortia and strategic alliances as legitimate responses to competitive pressures from abroad. This more liberal attitude was evident in the passage of the National Cooperative Research Act of 1984, which allows pre-competitive collaboration between U.S. firms. Hundreds of alliances and consortia have been registered with the federal government since this bill came into force.[61]

With respect to high-technology industry, the potential conflict between antitrust laws and intellectual property protection can represent a significant obstacle to rapid innovation. Practices such as patent pooling and cross-licensing can offer enormous competitive advantages, provided they do not run afoul of the Department of Justice. In the case of patent pooling, several firms with similar or overlapping patents agree to "share" the rights to their protected innovations, rather than engage in costly litigation. In some cases, the involved firms may choose to establish a separate entity to which they assign or license their patents.

Similarly, cross-licensing of conflicting patents has defused some potentially devastating legal conflicts, to the advantage of both parties. In 1995, for example, two competitors in the networking hardware industry, Bay Networks and 3Com, settled patent litigation by signing a cooperative mutual patent-license agreement. Both firms have since noted that rates of innovation and design flexibility have improved significantly as a result of this agreement. An even more contentious battle between Advanced Micro Devices (AMD) and Intel was settled in a similar way, when these firms agreed to a five-year patent cross-licensing agreement in 1996. As a result, both firms were able to better utilize their production capacity, and Intel received a net royalty income from AMD.[62]

Traditionally, the Department of Justice has been highly suspicious of pooling and cross-licensing agreements, since these strategies can result in price-fixing and other anticompetitive outcomes. In today's rapidly changing technological environment, however, these fears might not be justified. In 1995, the Antitrust Division of the Department of Justice and the Federal Trade Commission jointly issued *Guidelines for the Licensing of Intellectual Property.* These guidelines assert a continuing commitment by the government to pursue anticompetitive practices, but they also represent a significant break from former hard-line views on cross-licensing and pooling. There appears to be a growing consensus among policy-makers that cooperative activities of this sort are both pro-competitive and economically beneficial.[63]

In our innovation-driven world, the lines between competition, trade, and technology policies are rapidly blurring.[64] If American firms are to sustain global leadership, there is a pressing need for harmony among these traditionally distinct fields of governance, on both the domestic and the international fronts. The diffusion of new knowledge must be balanced against the need for incentives to innovate. The benefits of cooperation must be weighed against the risks of collusion and barriers to new entrants. To achieve this delicate balance, industry and government must work together in new and unprecedented ways.

# 7

# The American
# System of Innovation

## A Comparative Look at U.S. Technology Policy

The actions that a government takes to influence the development, commercialization, or adoption of new innovations are referred to as *technology policy*.[1] Ideally, these actions should take the form of a coordinated program to foster all aspects of a nation's technology enterprise, from the funding of basic scientific research to the diffusion of new, productivity-enhancing manufacturing processes. Unfortunately, the fragmented reality of technology policy in the United States is far from this holistic ideal.

The concept of a *national system of innovation* (NSI)[2] extends beyond the limits of technology policy, to encompass all aspects of the business environment that support the development and expansion of innovative activities in a country.[3] In general terms, the government's role in the NSI is to provide ample supplies of human capital, fund basic scientific research, and maintain a robust infrastructure. The private sector plays a complementary role in creating a fertile environment through venture-capital funding, workforce training, supplier development, and intra-industry cooperation. While most of this chapter will deal with the narrower topic of technology policy, it is worthwhile considering this broader perspective, to gain insight into the importance of cooperation between government and industry in the fostering of technology enterprise.

It has been noted that the term 'national system of innovation' embodies two contradictions. First, in today's global economy no system of innovation can be truly national. Many American firms are performing a portion of their research and development overseas, and many foreign firms have established an R&D presence on U.S. soil. Despite this apparent homogenization of innovation, however, there is ample evidence that countries follow unique paths of technological evolution and that firms within nations tend to develop in ways that exploit local environmental conditions.

The second contradiction with respect to NSIs is that the factors which impact innovation within a country cannot accurately be called a "system." In reality, the institutions and structures that create unique competitive environments in a nation are not strategically conceived. NSIs are loosely connected, ad hoc assemblages that are strongly influenced by history, politics, and culture. The challenge for both industry and government is to understand the nature of this ambiguous beast and harness it to create competitive advantage.

In this section, we will take a brief tour of the American national system of innovation, focusing primarily on how U.S. technology policy differs from that of other advanced nations, such as Japan and Germany. The remaining sections of this chapter will delve into specific areas of technology policy, including support for basic scientific research, development of advanced infrastructure, promotion of technology diffusion, and formation of human capital. In each of these areas it will become evident that effective policy must be developed through a partnership between the private and public sectors.

### The Scope of Technology Policy

Perhaps the first question we should consider regarding technology policy is why it is needed at all. It would seem that, at least with respect to technology industries, market forces would be more than adequate to ensure sufficient competition and progress. In reality, however, market forces impact only the commercial tip of the technological iceberg. Firms focus their research and development efforts on creating value for their customers. To achieve this value, they draw on an existing base of scientific and technological knowledge. Some of this foundation is accumulated by individual companies, but the majority of knowledge employed in industrial R&D is derived from a reservoir of public goods.

The problem with this vast body of public knowledge is that it is not of sufficient value to any one firm for that firm to invest in creating it. It is reasonable, for example, that a firm might fund a study of semiconductor etching processes that directly affect its yields and profits. On the other hand, it is unlikely that the firm would be willing to fund a comprehensive investigation of all types of etching under all possible conditions. Creation of this database would not yield a sufficient return to justify investment by a single company. Yet if the knowledge existed in the public sector, all firms could exploit it, resulting in a very high rate of return to society.

Governments justify their investment in certain aspects of technology by citing the need to correct so-called *market failures*. Such investments can be broadly supportive, as in the development of high-volume data-transmission infrastructure, or can be more targeted, as with federal contributions to the SEMATECH consortium (discussed later in this chapter). In all cases, the allocation of public funds is justified by the potentially high social rates of return derived from technology investment.

Indeed, the estimates for economic returns on generic research and development can be very high. The literature valuation of this spillover benefit ranges from 30 to 50 percent, whereas significantly higher estimates have been suggested by recent U.S. government publications.[4] Even if the lower figures prove to be correct, there is still an enormous incentive for government to strategically invest in technology development.[5] The trick is to identify which investments will yield high social benefits without destroying free-market competition.

Some areas that are generally accepted as being within the domain of technology policy are described in Table 7.1. The operative word used to describe these investments is *pre-competitive*.[6] This term is intended to explicitly exclude

| Examples of Market Failures that Justify Government Investment | Description |
| --- | --- |
| 1. Basic Scientific Research | Fundamental research on the nature of things. Principles that underlie applied areas of technology. |
| 2. Pathbreaking Technologies | Areas of research with high potential commercial applicability, but with long payback periods or distributed benefits. |
| 3. Infrastructure | Enabling knowledge and physical structures, including standards and regulations, public goods databases, data transmission networks, etc. |
| 4. High-risk Technologies | Areas of research with high barriers to entry, due to either high initial cost or very high risk. |
| 5. Public-use Technologies | Technologies that directly impact public welfare or safety, such as biotechnology, environmental studies, etc. |
| 6. Strategic Technologies | Technologies that have small or niche markets, but provide a critical capability, including military/defense R&D. |

**Table 7.1:** Overview of several categories of pre-competitive government investment in technology enterprise that have potentially high social rates of return.

the government from "picking winners and losers." The United States has a deeply seated mistrust of *industrial policy*, in which federal funds are used to subsidize specific firms or sectors. Investments in basic science, development of standards and regulations, establishment of infrastructure, and support for the diffusion of knowledge transcend individual firms. While some industry groups favor direct subsidies, it seems reasonable to expect higher rates of return from investments that support a broad and diverse constituency, than from money which is simply thrown at targeted companies.

So what exactly is the scope of technology policy? Clearly, it must reach beyond the arcane world of basic science into a broad gray area that has been dubbed *basic technology research.*[7] Unfortunately, semantics are important here, since debate on technology policy is often reduced to quibbling over the distinction between *basic research* and *applied research*. In fact, these two terms have virtually lost their meaning after years of rhetoric and misuse.

A more useful distinction might be to segregate policy into demand-side and supply-side measures. On the supply side, government directly funds research activities and may also indirectly subsidize technology development through targeted procurement. The demand side is far more passive, using government funding to meet industry needs for highly trained workers, advanced infrastructure, and basic technology. Despite some greatly exaggerated anecdotes from East Asia regarding the reputed success of supply-side policies, I believe that a well-executed demand-side strategy will have a much higher probability of yielding long-term economic benefits.

The current debate in American technology policy is driven by the realities of shrinking federal discretionary funds. Entitlement programs and other fixed obligations are growing more rapidly than the U.S. economy. Pressure to balance the budget, combined with a strong disposition not to raise taxes, makes it unlikely that government investment in technology can be sustained at current levels. Some of the slack must be taken up by private industry. More important, America needs to improve the efficiency with which it employs precious public assets.

Should funding be directed toward creating new science, or toward diffusing existing technologies? Is a static strategy valid in the current technological environment, or should policy be made more flexible and dynamic? Does the current decentralized maze of administrative bodies responsible for U.S. technology investment have the capacity to execute a truly integrated strategy? Each of these issues must be addressed for government to achieve its full potential as a valued partner to American technology industry.

### The Evolution of U.S. Policy

The formation of U.S. technology policy in the postwar years was decidedly bimodal. The perspective of scientists and academics, who saw a minimal role for government in directing basic scientific research, was in stark contrast to America's national commitment to develop the most advanced military and defense capability ever conceived. These conflicting directions for federal technology investment caused a rift between the lofty ideals of academic science and the mission-oriented focus of military technology. For reasons I will discuss below, government support for commercial technology enterprise was marginalized in this political debate.

Shortly after World War II, two competing visions emerged with respect to the future of American investment in basic science. The view held by Vannevar Bush, as articulated in his famous report, *Science: The Endless Frontier,* proposed that scientists should be given a free and open atmosphere in which to study the nature of things. This vision is best exemplified by university scientists, who enjoy substantial intellectual freedom and relatively little accountability for their research activities.

A contrasting view was championed by Senator Harley M. Kilgore, who felt that a more active role for government was needed to ensure high social benefits from science. Kilgore envisioned a system of applied research centers and extension services, with a greater degree of commercial applicability and political accountability. Ultimately, the debate was settled by the formation of the National Science Foundation (NSF) in 1950, which represented a clear victory for the "free-science" perspective supported by Bush.[8]

Whereas the social benefits of basic scientific research were the subject of lively debate in the early postwar years, the military benefits of advanced technologies were abundantly clear. Breakthroughs in rocketry, jet aircraft, radar, and atomic energy, coupled with the success of the colossal Manhattan Project, set the stage for the largest sustained buildup of defense-related technology in history.[9] This period, from the early 1950s through the late 1960s, has often been referred to as the "golden age of American science and technology."[10]

Thus, the paths of U.S. science and technology development were split: a tiny trickle of funds was allocated to basic scientific research, supporting such institutions as the NSF and the National Institutes of Health (NIH), while the vast majority of public funding for R&D went directly to mission-specific defense projects. Even on those occasions when significant investment was made in nondefense technology, it retained the same "big-science" character of the Manhattan Project. The Apollo lunar exploration program, the Space Shuttle, and more recently the Superconducting Supercollider are all typical of America's infatuation with scientific megaprojects.

The rationale behind mission-oriented funding of science and technology was the assumption that there would be a spinoff of benefits to the broader economy. Presumably, the tremendous advances made during the Cold War military buildup and the contentious space race with the Soviet Union would somehow work their way into the commercial sector, enabling the creation of a wealth of new products. The space program in particular was credited with spinning off countless breakthroughs, from advanced materials to powdered orange juice.

This technological trickle-down theory was based on two assumptions. First, it was taken for granted that the technology needed to accomplish specific (often military) missions was applicable to the commercial marketplace. Second, there was an implicit assumption that the spinoff process happened efficiently and cost-lessly, without the need for government action. Both these presumptions have proved to be erroneous.[11]

The foundation for modern U.S. technology policy was established in the National Science and Technology Policy, Organization, and Priorities Act of 1976.[12] This statute defined for the first time the government's role in technology development for purposes other than strategic missions. There was a dawning recognition on the part of policymakers that the huge investments in defense technology were not stimulating the economy, as had always been assumed. Several subsequent measures, including the Bayh-Dole Act of 1980 and the Federal Technology Transfer Act of 1986, were specifically oriented toward encouraging the "dual use" of federally funded mission-specific research by commercial enterprise.[13]

By the dawn of the 1980s, there was clear evidence that other nations were employing industrial policy to promote targeted technology industries. Within the United States, however, the prospect of "picking winners and losers" was still anathema. A decade later, President Bush acknowledged the need for increased co-operation between industry and government, and was supportive of federal fund-ing for generic, enabling-technology development. It wasn't until the Clinton administration, however, that technology policy came into its own.

Within the first one hundred days of the Clinton administration, a position paper was issued that established clear goals for American technology policy. The focus of this proposal was unabashedly commercial, citing the need to "create a business environment where technical innovation can flourish and where invest-ment is attracted to new ideas."[14] This was closely followed by the submission to Congress of the National Competitiveness Act of 1993, which proposed a new and enlightened approach to American technology policy.

Among its many provisions, the National Competitiveness Act called for the creation of a list of "critical technologies." This was by no means a new concept; various government agencies and nongovernment organizations had been creating such lists since 1985. In the past, however, there was a presumption that the lists would provide a focus for government investment. Unfortunately, the criteria for a technology to be ranked as "critical" were often ill-defined; consequently most lists simply documented the obvious. More important, this focus on critical technolo-gies was potentially dangerous, in that the de facto success of technologies in the marketplace was assumed to justify investment in those areas.

Rather than using arbitrary lists to focus federal investment, the Clinton administration called for benchmarking activities to be performed on critical industries, to determine America's relative position in global markets. There was no provision for allocating funds based on the conclusion of these benchmarking studies, but the need for such a competitive assessment to serve as a foundation for policy development is clear.[15]

Today, technology policy in America is administered by a diverse group of agencies, each chartered with a different set of research and development goals. Nominally, the Department of Commerce has the lead in this area, through the auspices of the Technology Administration. Under this umbrella resides the Office of Technology Policy, which is responsible for policy development and sponsorship; the National Institute of Standards and Technology (NIST); and the National Technical Information Service (NTIS). Other agencies that play key roles include the Department of Defense, the Department of Energy, NASA, the National Science Foundation, the National Institutes of Health, and a host of smaller entities.

Although the current organizational structure is far from ideal, U.S. policy in support of technology industries has evolved significantly from the defense-dominated, mission-oriented postwar era.[16] These changes have been made in reaction to world events and economic pressures, however, rather than through a fundamental shift in philosophy. The Clinton administration remains skeptical of government intervention in markets, in the face of growing pressure for America to take aggressive action in response to the economic threats posed by a unified Europe and a maturing East Asia.

Would America be better off if we had embraced a Japanese-style industrial policy? Should we take the final step toward government intervention by attempting to "pick winners"? As I will demonstrate below, the success of advanced nations such as Japan and Germany may well be attributable to their national systems of innovation, but little of their success is directly related to government targeting and subsidy.

### Can Governments Pick Winners?

The discretionary portion of the U.S. budget is expected to be under increasing pressure as aging baby boomers begin to draw heavily on their entitlements. With available funds declining and global competition intensifying, it has become critically important to leverage every public dollar invested in research and development. This harsh reality has caused policymakers to reconsider the laissez-faire assumptions on which American technology policy has historically been based. Has the time finally come for the United States to develop a true industrial policy, in which government funds are used to "pick winners and losers"?

The apparent success of such policies in Japan, South Korea, and even the European Union has prompted an outcry from American technology industries to meet this challenge in kind. First impressions in this case may be deceiving, however. Take, for example, Japan's world-leading energy-conservation industry. It is not clear whether this success was a direct result of Japan's substantial government investment in energy-related research or is simply a reflection of intense domestic interest in this issue. Similarly, the German government has invested heavily in industrial-technology development over the last two decades. Can this investment be credited with *creating* Germany's advanced machinery industry, or is its success simply an outgrowth of the historically close relationship between industry and government in matters of German industrial policy?

The central question here is not whether governments should be broadly supportive of their successful domestic industries, but whether they should engage in overt targeting of individual firms for subsidy. The effectiveness of industry-level government support for R&D is undeniable. The vast majority of world-leading technology industries have received substantial government subsidies. In the European Union, for example, the commercial aircraft industry has been subsidized to the tune of more than $26 billion since the late 1970s, while in the United States there have been huge subsidies for the development of military aircraft.[17] In both cases, domestic governments provided substantial R&D funding and a guaranteed early market for their industries' products. The story is the same in the computer, telecommunications, semiconductor, and biotechnology industries; government support has been persistent and substantial both here and abroad. Does this mean, however, that governments have been successful at creating "winners"?

The prospect of creating a globally competitive industry through targeted government subsidy is analogous to selecting stocks for personal investment. To achieve a high rate of return, the buyer must select stocks that are currently undervalued by the market but which will become highly valued in the future. Although the potential returns for this type of investment may be attractive, the risk to the principal is considerable. After all, there is no proven method for picking stock-market winners, as the "dartboard" challenges conducted by *The Wall Street Journal* will attest.[18]

Identifying which new technology will become the "next big thing" is even more challenging. The same inscrutable market forces that drive the value of stocks make the demand for new technologies highly unpredictable. Moreover, science itself may not cooperate. Despite the enormous potential for high-temperature superconductors, for example, the technical difficulties associated with this new technology have delayed its commercial debut indefinitely.[19]

In the following section, I will compare the American system of innovation with that of Japan. The Japanese government is often cited as being an adept practitioner of industrial policy. In reality, however, its contribution to the competitiveness of Japan's technology industries has been greatly overstated. On the occasions where winners have been created, their success was due to resonance with the entire Japanese system of innovation, rather than due to targeted government subsidy.

## A Comparison of Successful NSIs

The contrast between national systems of innovation in the postwar period is nowhere more dramatic than that between Japan and the United States. While the U.S. scientific and technical community was divided between the creation of world-class basic research and the achievement of mission-oriented military advances, the Japanese thrust was aggressively commercial. Furthermore, Japan's strategy was based on absorption and imitation, as opposed to the American penchant for "breakthrough invention." In a sense, the system of innovation in Japan can be thought of as supporting a fast-second strategy on a national scale.

Beginning in the late nineteenth century, Japan's domestic technology base became heavily dependent on the importation of expertise. The isolation Japan suffered prior to the Meiji Restoration in 1868 had left it far behind the rest of the world. In a concerted effort to catch up, Japan encouraged foreign-born educators and technologists to help develop its industrial and educational systems. The Japanese Ministry of Industries, for example, was established on the advice of a British railway engineer in the early 1870s.

The Japanese government emphasized engineering education at a time when pure science was considered to be of paramount importance by European and American universities. Many of the Japanese engineers who graduated during the early decades of the twentieth century were hired by large, family-owned industrial firms known as *zaibatsu*. These conglomerates often had strong military and government connections. Additional demand for Japanese engineers came from the rapidly growing electronics and automobile industries. Interestingly, virtually all the major firms in these industries were affiliated with Western multinational corporations prior to World War II.[20]

Although government investment in technology development and targeted procurement by the military played a role in Japanese industrial expansion, a more profound influence may have been provided by strong restrictions on imports and foreign direct investment. This left the technologically advanced American and European firms with only one way to exploit their advantage in Japan's domestic market: They were forced to sell their technology through licensing. Indeed, to this day Japan is the world's largest licensee of foreign intellectual property.

This strategy of technology importation became a substantial liability during World War II, when the flow of advanced knowledge came to an abrupt halt. Without this lifeline to support its domestic engineers, technological progress in Japan came to a near standstill.[21] After the war, Japan found itself with a decimated industrial base and a renewed technology gap with the rest of the world. The strategy of imitation and catch-up that had served it well prior to the war was reinstated in earnest.

The success of Japan's technology industries in the postwar period has often been attributed to the influence and support of the nation's Ministry of International Trade and Industry (MITI).[22] The high profile that MITI has held in recent bilateral trade negotiations with the United States has created an exaggerated impression of its importance to Japanese competitiveness. In fact, the role of MITI as the administrator of Japan's industrial policy has been diminishing for some time. In the early postwar years, however, MITI held sway over much of Japan's technological development.

It is a great oversimplification to assume that MITI's role was simply to pick winners and dole out subsidies. Actually, as I have noted previously, the return on MITI's investments in targeted industries was quite poor. Their success was derived from a balanced strategy that included fostering industry cooperation, gathering and diffusing knowledge, and carefully selecting investments to correct "market failures."[23]

The legendary cooperation among Japanese firms has also been misunderstood by most Western observers. It is routinely assumed that leading Japanese companies enter into both tacit and explicit collusion within their domestic markets and readily form price-fixing cartels in international trade. In reality, the rivalry between Japanese firms is intense, with very little evidence of market-distorting collusion. Where it is appropriate and mutually beneficial, however, rivalry is set aside in favor of cooperative action. MITI has played an instrumental role in determining the nature of this cooperation.

Under MITI sponsorship, eighty-seven research associations were formed among Japanese firms during the period from 1961 to 1987. Although cooperation was the motivation for forming these alliances, they actually served as a conduit for MITI's subsidy of promising technologies. This helps explain why only two of these eighty-seven research associations actually established joint research facilities. In all other cases, the participating firms simply took their share of the government subsidy and proceeded with business as usual.

In other cases, however, MITI has been more successful in fostering true cooperative ventures. Recently it has sponsored a number of technology programs, including the Very Large Scale Integration (VLSI), the Fifth Generation Computer Project (FGCP), and the Super-Computer consortia.[24] Although these activities are targeting "critical technologies," they constitute a rather small government investment. In fact, MITI's entire R&D budget is modest, representing only about 13 percent of Japan's total government investment in R&D. Furthermore, few technological breakthroughs have resulted from these ventures. They are more useful in disseminating information and in helping to bring laggard firms closer to the scientific frontier.

Where MITI achieves its greatest leverage is through its highly efficient knowledge-gathering network. The collection and diffusion of information by MITI is achieved through its Agency of Industry, Science and Technology, for technological data, and the Japan External Trade Organization (JETRO), for market reconnaissance. JETRO has been particularly effective as a gatherer of information relating to import systems, price trends, import channels, trading ports, financial arrangements, etc. When combined with other, more informal absorption networks, MITI's agencies represent a powerful resource for Japan's technology industry.

As Japan approaches the scientific frontier in industry after industry, the effectiveness of its "absorb and imitate" strategy has begun to diminish. Today, more than 90 percent of all research and development is performed under contract by private firms, rather than in government-sponsored consortia. This is due in part to the apparent inability of these alliances to perform critically needed basic research. Often the goals of these ventures are corrupted to meet immediate commercial requirements.

Nonetheless, Japan has embarked on an aggressive effort to shift from a strategy of imitation to one of innovation. MITI has responded to the increasing globalization of R&D by sponsoring international consortia, such as the Intelligent Manufacturing Systems Project. Meanwhile, the rest of the Japanese government has begun shifting its focus as well. Recent plans for increased government-funded R&D, coupled with renewed emphasis on developing Japan's basic research capability, should be taken as a serious warning.[25] We cannot assume that breakthrough innovation is our exclusive domain; as the Japanese system of innovation evolves, so will that nation's capacity to advance the frontiers of science.

Japanese technology firms have not become world leaders due to a flood of government subsidy. In fact, the public-sector contribution to total R&D in Japan was only 1.2 percent in 1990, as compared to 33 percent for the United States in the same year.[26] Although MITI and other government research sponsors have played an instrumental role in coordinating industry action and galvanizing technology development, the credit for success must ultimately go to the entrepreneurial spirit of the firms themselves. Government in Japan has made its greatest contribution when it has played a supporting role, by providing a stable economic environment, placing an emphasis on technical education, and sponsoring an efficient system for the absorption and diffusion of knowledge.

### Future Directions for U.S. Technology Policy

In the future, American technology policy must address two fundamental issues: How the government's portfolio of technology investments can be optimized under the realities of budget constraints, and what is the most effective method for administering that portfolio? While the former represents the greater challenge, the latter deserves at least a brief discussion.[27]

Any scheme for administering the tens of billions of dollars in public funding that flows into American technology industry each year must meet some basic requirements. First, the system must be highly flexible and adaptive. This implies that whatever agency structure evolves over the next decade, it must have sufficient resources, both financially and technically, to stay near the leading edge of technical progress. This is not a trivial requirement, given the substantial technology gap that has already developed in many agencies, including the Patent and Trademark Office.

Second, the system must be efficient and well coordinated. This does not mean, however, that it must be *centralized*. In fact, I believe that a distributed structure may be far more expedient. Given the unique needs of various technology industries, and the differing specialties required to provide basic science, infrastructure, education, and technical extension, it seems reasonable that several agencies or levels of government might share responsibility for the execution of technology policy.[28] The haphazard system that has evolved over the last four decades, however, must be substantially overhauled.[29]

Currently, the responsibility for administering U.S. technology policy is divided among three major entities: the Department of Defense, the Department of Energy, and the Department of Commerce's Technology Administration.[30] Each of these agencies has strengths and weaknesses, and it is likely that they will all play a significant role in future technology policy.

The Department of Defense has received favorable marks for the ability of its Advanced Research Projects Agency (ARPA) to select and nurture aggressive new technologies. Perhaps a "civilian ARPA" might be a viable channel for subsidy of promising commercial innovations. Similarly, the flagship of the Technology Administration, the National Institute of Standards and Technology (NIST) is generally regarded as being both capable and effective. Finally, the Department of Energy offers the highest levels of technical competency through its massive network of national laboratories.

While there are substantial benefits to empowering each of these agencies, they all have significant drawbacks as well. None of them, for example, has had sufficient direct experience in industrial research and development. Furthermore, it is hard to imagine transitioning ARPA from a dual-use philosophy to an entirely commercial focus, or partnering the rather stodgy NIST with the wild-and-woolly world of market-driven innovation. Ultimately, each of these agencies will migrate toward a natural and appropriate role. In addition, the final structure may need to be supplemented by an industry-focused, entrepreneurial new unit that is tasked with shepherding the last mile of government technology policy directly into the boardroom and onto the shop floor.[31]

| Technology Policy Goal | Areas of Investment |
|---|---|
| 1. Creation of Knowledge | • Basic scientific research<br>• Basic technology research<br>• Selective industrial "seed money"<br>• Dual-use adaptation and conversion |
| 2. Development of Infrastructure | • Standards and regulations<br>• Knowledge databases<br>• Industrial methods and techniques<br>• Telecommunications and information network hardware and software |
| 3. Diffusion of Knowledge | • Sponsoring of consortia<br>• Absorption of foreign innovations<br>• Technology extension programs<br>• Tacit skills transfer programs |
| 4. Formation of Human Capital | • K-12 educational reforms<br>• Technical-specialist training programs<br>• Industry exposure for advanced degrees<br>• Lifetime learning and skills retraining |

**Table 7.2**: Summary of four critical goals for American technology policy, which collectively provide the balanced government support required for sustainable competitive advantage.

Beyond the two basic requirements given above, I am relatively unconcerned about how technology policy is administered at the highest levels. It should be taken for granted that the U.S. government will not reorganize itself around the needs of science and technology. Hence, whatever structure evolves will, of necessity, resemble in many ways our current decentralized maze. What matters most are the investments themselves and the ways in which they will affect American industry.[32]

There are four reasonable goals for U.S. investment in technology, as outlined in Table 7.2. First, some level of financial support must be directed toward correcting market failures in risky or low-appropriability areas of research through the creation of public-goods knowledge. Since this goal can be easily corrupted into a politically motivated targeting program, it is essential that funding be selectively allocated to universities and industry consortia rather than directly to firms.[33] It is also important that some discipline be imposed on the number of projects that are funded at the proto-competitive level. Critical-mass funding and a high degree of focus are key factors in the success of any technology-development activity.

Second, government must create an infrastructure that can provide the interconnectivity, information flows, and flexibility needed to support knowledge-based industries. This aspect of technology policy covers a broad spectrum, including the sponsorship and creation of standards, provision for high-value database resources, and development of a robust physical layer of telecommunications and data-management capability.

A third goal for U.S. technology investment should be the diffusion of public-goods knowledge throughout American industry. An investment in basic science or technological innovations has little social value if the knowledge is not disseminated to potential users. Small-to-medium-sized firms in the United States are notoriously poor adopters of new technology, and are in danger of being left behind by their more adaptive foreign counterparts.

Finally, and most important, the government must establish a comprehensive program to improve technical education at all levels, from kindergarten through graduate studies. This human-capital development effort should not be restricted to simply placing computers in schools. New categories of education are needed, including technical-specialist training programs that can bridge the gap between a high school diploma and a college degree. As our industrial environment becomes increasingly saturated with technology, the demand for highly skilled workers at all levels will increase. We must make lifetime learning and technical literacy our nation's highest domestic priority.

In the sections that follow, I will delve more deeply into each of these goals for American technology policy. Taken as a whole, they represent a platform upon which our technology-intensive industries can build sustainable competitive advantage. We must use technology policy as a tool to develop our nation into the world's most favorable environment for innovation and advanced research. Although the steps necessary to attain this goal are not yet clear, there is far less danger in "overdoing" technology investment than in allowing our nation to slide into the second tier of industrial economies.

## Building a Pre-competitive Knowledge Base

The creation of new knowledge can be motivated by necessity, curiosity, or a desire for profits. Of these possibilities, the first incentive is by far the most compelling. In developing nations, for example, the introduction of new medical treatments and high-yielding agricultural products is literally a matter of life and death. Even in the more advanced economies, investment in military and defense technology provides essential protection against external threats to lives and property. It is only in a select few nations that the system of innovation is driven not by necessities but by the desire for intellectual fulfillment and economic growth.

It is relatively easy for policymakers to justify the expenditure of public funds on developing a cure for AIDS or assuring a strong national defense. These investments impact a broad constituency and provide a substantial social benefit. Similarly, government support for curiosity-driven research that expands the horizons of knowledge can be defended as creating a desirable public good. On the other hand, the use of government funds to underwrite commercial technology development represents a persistent policy dilemma.

Since the end of World War II, the United States government has funded more than half of all research and development performed in this country, including more than 60 percent of all basic scientific research, roughly 40 percent of applied research, and more than 50 percent of total U.S. investment in technology development.[34] The vast majority of this support has been directed toward achieving mission-oriented military and defense goals, but the outflow of advanced technologies from this massive effort has had a global impact. Military R&D investment by the United States during the 1950s and 1960s was the dominant engine of technology development throughout most of the noncommunist world.[35]

The success of American investment in innovation has been referred to as sustaining a *virtuous circle:* As the nation generates wealth, a portion is reinvested in research and development, which eventually yields still greater wealth. The presumption that public investment in science and technology represents the catalyst for sustainable economic growth has been the underpinning of bipartisan support for such measures over the last half century.[36]

Although military missions have been at the heart of American postwar R&D investment, there have been many less pernicious breakthroughs as well. The National Institutes of Health, for example, have achieved notable success in sponsoring discoveries that have improved both health and the quality of life. Human insulin, the first drug produced through advances in biotechnology, was developed in 1982 under NIH funding. Likewise, vaccines for hepatitis B, production of human growth hormone, and significant progress toward cures for cancer and AIDS have all come about through investment of public funds.[37]

The sources and levels of R&D funding in the United States have undergone a dramatic shift over the last several decades. Defense spending has sharply declined, while investment by private industry has become a dominant force. A new structure for government R&D support that reflects this new environment is needed, one that reinforces our nation's commitment to retain leadership in science and technology.

### America's Standing in the World

When the level of American R&D investment is compared with that of other advanced nations, some interesting contradictions emerge. It is true, for exam-

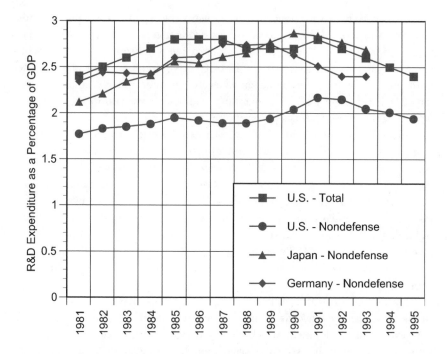

**Figure 7.1:** Total R&D expenditure as a percentage of gross domestic product for Japan, Germany, and the United States, 1981 to 1995. Note that although total U.S. R&D spending is roughly equivalent to the nondefense R&D of the other two nations, our nondefense R&D has consistently lagged behind our overseas competition. Source: National Science Foundation, 1996, *Science and Technology Indicators - 1996,* National Science Board, pg. 155, Table 4-34.

ple, that total R&D expenditures in the United States are currently greater than those of Japan, Germany, France, and the United Kingdom combined. On the other hand, each of these nations expends more than the United States on research in selected commercial fields. Hence, despite impressive total commitments, we may be underinvesting in areas that have the greatest economic potential.

Likewise, although our total expenditures on R&D as a percentage of GDP are similar in magnitude to those of Japan and Germany, our investment in nondefense research and development has consistently lagged behind these nations, as shown in Figure 7.1. Since mission-specific research for military applications has a relatively low impact on economic development, it would seem that from a commercial standpoint the United States is underinvesting in aggregate terms as well.

For a few years in the mid-1960s and again in the mid-1980s, the United States' total investment in R&D approached 3 percent of GDP. In both instances, however, these world-leading levels of investment in research were driven by

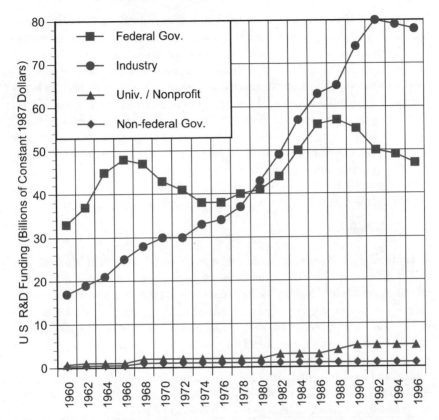

**Figure 7.2:** Contributions to total United States R&D investment by the federal government, state and local governments, industry, and universities and nonprofit institutions, 1960 to 1996, in 1987 constant dollars. Source: National Science Foundation, 1996, *Science and Technology Indicators - 1996*, National Science Board, pg. 105, Table 4-3.

increases in military spending. Private-sector funding for research and development grew at about the same rate as the U.S. economy throughout much of the 1980s, while both Japan and Germany increased their commercial spending far more rapidly. Steep cutbacks in defense spending in the early 1990s prompted some analysts to predict that the United States could be entering a long, slow decline into technological mediocrity. This fear has been abated somewhat by evidence that industry R&D investment has exceeded the growth rate of the U.S. economy for the three-year period from 1995 to 1997.[38]

As defense spending has declined in the 1990s, U.S. industry has taken up the mantle of R&D with encouraging vigor, as shown in Figure 7.2. The turning point for technology investment in America came in 1980, when industry investment in research and development surpassed government investment. The split between public and private funding persisted at about a 50/50 ratio for several years, but since the mid-1980s the gap has widened dramatically. In 1997, the percentage of total R&D investment in the United States that was funded by the federal government dropped to 30.5 percent, its lowest level since World War II.[39]

In principle, this dramatic shift should be a good thing. After all, the most obvious weakness in the American national system of innovation has been its heavy dependence on mission-oriented government programs. Yet what really matters in R&D investment is not the overall magnitude, but rather the efficiency with which the funds are used. In the process of shifting our weight from one focus for R&D to another, we must be sure not to lose some essential element that has kept our virtuous circle intact. Hence, it is even more important in times of tight budgets and steep defense cutbacks that we scrutinize both the individual activities and the overall balance of government investment in science and technology.

### Balancing Domestic Priorities

By any measure, U.S. spending on defense research and development is in a tailspin, as shown in Figure 7.3. While other areas of spending have increased slightly, the overall levels of federal funding for R&D have been steadily declining in real terms.[40] Among the major categories of science and technology investment in the 1997 federal budget, defense received approximately 53 percent of all funding, health received 18 percent, space was allocated roughly 12 percent, general science 5 percent, energy research 4 percent, and the rest was apportioned among transportation, agriculture, and environmental technologies.[41] In terms of trends, investment in health research has grown steadily in recent years, space is down slightly, general science has gradually increased, and energy support is roughly stable.

Another way to evaluate the federal R&D budget is to consider the percentage of funds allocated to basic scientific research, applied research, and technology-development activities. In 1996, for example, 63 percent of government funding for science and technology was expended on activities that are characterized as "development." This might seem an encouraging distribution from the standpoint of commercial enterprise, until it is noted that nearly all of these funds went toward testing and evaluation of weapons systems, and the development of mission-specific tooling and manufacturing methods. The government's investment in commercial technology development has been tiny by comparison.

The remaining 37 percent of the federal R&D budget in 1996 was roughly equally divided between basic and applied research. As I pointed out earlier, it is unclear how these two categories map into an industrial context. Presumably, applied research is oriented toward a targeted application, while basic science is foundational in nature. It seems evident that if the nation's policymakers are to be effective at harnessing public R&D investment to achieve economic growth, some clear connection must be made between categories of research and development and their utility to American industry.[42]

This raises the question of how the U.S. budget for science and technology is actually developed. Given the haphazard administrative structure for government R&D, it should not be surprising that there is no such thing as a "federal science and technology budget."[43] Currently, the allocation of funds is largely driven by agency missions, rather than by national R&D priorities. Tradeoffs between various technology investments are essentially impossible, since the budgeting process for R&D takes place in roughly twenty-six subcommittees of the House and Senate appropriations committees. The only time the total U.S. investment in research and development is pulled together as an integrated plan is after the U.S. budget has been finalized.[44]

The problem with establishing an optimized federal budget for science and technology runs much deeper than congressional procedures. At this writing, only six members of Congress are scientists, two are engineers, and one is a science

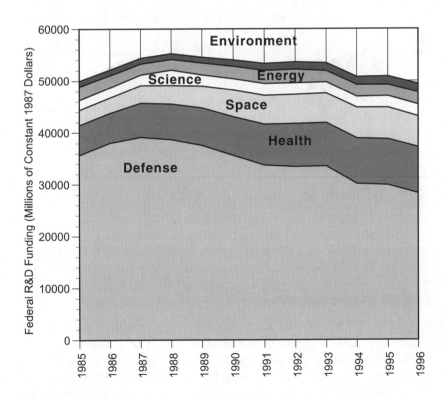

**Figure 7.3:** Federal R&D funding by function, 1985 to 1996, in constant 1987 dollars. Source: National Science Foundation, 1996, *Science and Engineering Indicators - 1996*, National Science Board, pg. 150, Table 4-29.

teacher. Even a casual sampling of congressional debate on technology issues will provide ample justification for a recent comment by Vice President Al Gore that Congress is "approaching science with all the wisdom of a potted plant."[45]

This lack of technological sophistication among policymakers may become a serious obstacle to the future competitiveness of U.S. high-technology industries. Significant reductions in government-funded R&D are planned in both the President's ten-year budget projections and Congress's seven-year outlook. Meanwhile, the European Union has stepped up plans for investment in both higher education and first-class laboratory facilities, and Japan is on schedule to double its 1992 government R&D funding levels by the year 2000.[46] U.S. policymakers will be forced to do much more with less if our government's role in technology enterprise is to remain vital. This will require a level of competence and wisdom with respect to technology policy that is currently lacking in many of our elected leaders.

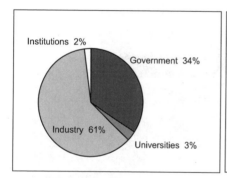
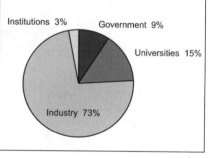

Expenditure by Source – 1996
Total = $184.3 billion

Expenditure by Performer – 1996
Total = $184.3 billion

**Figure 7.4:** Distribution of U.S. R&D investment, by source and performer, for 1996. Source: National Science Foundation, 1996, *National Patterns of R&D Resources: 1996*, National Science Foundation, pg. 3, Chart 1.

### The Roles of Universities, Industry, and Government

Now that we have identified the sources of R&D investment in the United States, it is worthwhile considering where the actual knowledge creation is being performed. Predictably, industry is the dominant performer of research and development, consuming 73 percent of all public and private R&D funding in 1996, as shown in Figure 7.4. In fact, American industry is the primary recipient of government research funding, and retains most of its own R&D investment as well.

From the standpoint of basic research, however, the heavy hitters are the nation's universities and colleges. The majority of funding for academic research flows through the two flagship institutions of America's system of innovation, the National Science Foundation and the National Institutes of Health. Only about 7 percent of funding for university research comes from industry, a figure that is unlikely to grow beyond 8 to 10 percent in the future. Hence, the creation of fundamental scientific knowledge is almost completely dependent on government support.

Since the Vannevar Bush report of 1945, the research programs within the U.S. university system have enjoyed almost complete autonomy. The premise for this high degree of intellectual freedom has been a tacit *social contract* between the scientific community and the taxpaying citizenry. Under this implicit agreement, the scientific community is given the initiative to define research objectives, perform peer review of promising proposals, and allocate basic research funding. In return, the scientific community assures society that the economic benefits of its work will far exceed the financial cost of public support.[47]

The lack of government oversight of academic research has been called into question in recent years. There is growing skepticism over the validity of the social contract in today's environment of global competition and shrinking national R&D budgets. Rather than working toward closer partnerships with American industry, some research scientists have voiced strong opposition to "targeted" research, and have demanded that their efforts remain commercially "unusable."[48] This position has resulted in an unprecedented level of disenchantment with, and resentment of, university scientists in both government and industry.

Until the mid-1970s, for example, many universities viewed patents as obstructions to the free flow of information. By the 1980s, most major research institutions had developed programs for intellectual property protection and had become more focused on commercial needs. Yet a vocal minority of scientists still display an almost contemptuous attitude toward "useful" targeted research. Given current economic and fiscal realities, this has caused some policymakers to view the university research community as just another politically biased special-interest group.[49]

Despite its reluctance to surrender intellectual independence, the American university system still provides higher value to industry than is the case in most other advanced nations. German industry, for example, relies heavily on its system of publicly funded research laboratories, the Max Planck Institutes, which are dedicated to basic research, and the Fraunhofer Laboratories, which perform primarily applied research.[50] Likewise, Japanese universities have historically received little public funding for research and have relatively weak links to industry.[51]

In the future, university research programs will be under increasing pressure to demonstrate commercial relevance. However, we should be careful not to demand too much from these institutions. There is a danger that American universities may prove to be ineffective at truly applied research and will cease to do what they do best: creating a world-leading foundation of basic scientific knowledge. The key to avoiding this mishap is to improve the linkages between universities and industry, through cooperative research activities and other long-term collaborative efforts.

A number of universities have successfully implemented Centers for Manufacturing Extension and Centers for Industrial Business Education and Research. A recent Ford Foundation study identified more than one thousand such centers nationwide, with a total annual budget of $4.5 billion.[52] According to this study, 13 percent of all Ph.D.s who graduated in 1992 had some direct contact with one or more of these centers.[53]

The U.S. government is helping to foster university/industry cooperation through several programs, including NSF's Engineering Research Centers and the Industry/University Cooperative Research Centers. In addition, there are numerous state-government-funded extension and industrial research facilities throughout the country. Although progress has been slow, there are encouraging signs that American industry is placing higher value on university research collaboration. Both General Motors and IBM have recently changed their approach toward university R&D collaboration and are actively seeking to partner with faculty who are working on relevant research projects.

The other important performer of government-funded basic and applied research is the system of Federally Funded R&D Centers (FFRDCs). This group includes the laboratories of NIST and NIH, the Naval Research Laboratory, and several massive energy and weapons laboratories, including Los Alamos National Laboratory, Oak Ridge National Laboratory, the Sandia Corporation, and Lawrence Livermore Laboratory. With recent dramatic shifts in America's defense priorities, this system of more than seven hundred research facilities has been scrambling to find new missions in the post-Cold War era.

The capability of the FFRDCs is truly impressive. With an aggregate budget of more than $20 billion and some of the most advanced facilities and talent in the world, they represent both a tremendous national asset and a troublesome policy dilemma. While these centers have been effective at executing some "big-science" projects, such as the Department of Energy's Clean Car Initiative and the Human Genome Project, they appear to be ill-prepared for a transition to commercial technology development.

Recent administrations have placed considerable emphasis on the use of Cooperative Research and Development Agreements (CRADAs) between federal laboratories and industry as a means to harness the innovative power of these institutions. Unfortunately, there is evidence that CRADAs may be less effective than other forms of consortia at promoting technology transfer and development.[54] More important, simply "counting CRADAs" does not provide evidence of successful laboratory/industry partnerships.

There are several alternative ways in which industry can exploit these vast centers of scientific excellence. It may be possible, for example, for industrial consortia to locate cooperative research facilities at an appropriate national laboratory. This would be cost-effective and provide excellent opportunities for the exchange of tacit knowledge between the laboratory scientists and industry technologists. Another possibility would be for national laboratories to identify their commercially valuable core competencies and market themselves as "R&D for hire."

As federal priorities shift from geopolitical concerns to a focus on economic growth, this huge agglomeration of talent and resources will drift well outside the mainstream of national needs. Some serious restructuring will be needed to achieve a reasonable economic return on this substantial public investment.

## Government Sponsorship of Commercial Development

In addition to supporting basic research in universities and mission-oriented development at national laboratories, the U.S. government has recently begun to implement programs that directly subsidize commercial innovation in the private sector. These include the Advanced Technology Program (ATP) and the Manufacturing Technology Centers, administered by NIST, the Small Business Innovation Research (SBIR) program, and ARPA's Technology Reinvestment Project (TRP).

The ATP provides cost-share funding to commercial firms and industry consortia, with a focus on commercial technology development. To avoid the prospect of politically motivated targeting of individual firms, the ATP accepts unsolicited proposals for specific activities and subjects them to a review for technical excellence, commercial viability, and return on investment. Although results from this program are still preliminary and therefore difficult to evaluate, one recent study concluded that the ATP program is having a significant positive impact on the pace of technology development.[55] Beginning with a budget of $47 million in 1992, the funding for this promising program grew to a requested spending level of $431 million by 1995.

One of the most useful programs for encouraging small, technology-intensive start-up firms has been the awarding of Small Business Innovation Research grants. Initiated in 1982 under the Small Business Innovation Development Act, this program requires eleven federal agencies to invest at least 1.25 percent of their contract R&D in small businesses. In 1992, $450 million in federal funds was expended on SBIR grants, with firms of ten or fewer employees receiving more than one-third of all awards. Follow-on legislation signed in 1992 will double the size of the program by the year 2000.[56]

With respect to military technology development, the United States has been forced to accept a new geopolitical role following the collapse of communism. As part of this transition, the mission of the Defense Advanced Research Projects Agency (DARPA) was forced to change as well. With the demand for strategic weapons systems diminishing in the early 1990s, this pioneer in defense technology was renamed ARPA, and given responsibility for the development of *dual-use technologies*.

The term 'dual-use technology' is something of an anachronism. When the concept was originally proposed in the mid-1980s there was still the possibility of the American military retaining leadership in key strategic technologies such as telecommunications and information processing. The premise of dual use was that technologies could be developed that would provide both advanced military capability and competitive commercial products. Unfortunately, unanticipated difficulty in transferring existing defense technology to the private sector, combined with explosive technical progress on the commercial side, has proven dual use to be a rather hollow concept.

Nonetheless, the success that ARPA has achieved in bringing high-risk advanced technologies to fruition hints at a new and more promising role for this agency. In 1993, ARPA announced an aggressive new commercialization program, the Technology Reinvestment Project, with an initial budget of $472 million in reprogrammed Defense Department funds.[57] Although the concept of dual use was still embedded in the TRP mission statement, the structure of the program was broadly supportive of commercial industry needs. Nearly half of the funding was targeted toward establishing a Manufacturing Extension Program and other diffusion-oriented activities. The remaining funds were allocated to investment in dual-use technology development and providing manufacturing education and training. With ARPA's unique project-management structure and its knack for supporting pathbreaking technologies, this program may prove to be a model for future government support of private industry.

Despite federal budget-tightening and a disjointed administrative system, I am cautiously optimistic about America's future R&D enterprise. Industry research funding has been rising sharply over the last several years, hopefully as both an acknowledgment of, and a response to, a diminishing government role in technology development. More important, funding levels alone are not a particularly strong indicator of a successful system of innovation (although this kind of one-dimensional quantitative metric tends to be irresistible to policymakers and executives).

What truly matters in the development of technology is not the magnitude of investment alone, but rather the synergy between activities that are funded and the time frame over which the investment is sustained. For the first time in the 1990s a general consensus has begun to form over the proper role for government in support of high-technology industry. The following sections describe critical elements of an integrated American technology policy that will serve as a catalyst for private-sector R&D investment, thereby enabling far higher returns and world-leading rates of innovation.

# Developing a Technology Infrastructure

Until fairly recently, the term *infrastructure* evoked images of power plants and highway systems, the underlying structures that enable industrial and social development. Indeed, this traditional definition still applies in many countries of the world. In the advanced economies, however, the availability of such things as ample power and clean water is no longer a serious obstacle to commerce. In a knowledge-based economy, a new and more challenging layer of infrastructure is needed to enable the efficient production and flow of information.

Defining a plan for development of physical infrastructure is a relatively straightforward task. Planners consider a "model factory" that produces generic physical products, and identify all the inputs and outputs of that facility. Infrastructure is then designed to perform three functions: Allow for the efficient transport of inputs and outputs, provide regulation and organization where needed, and create generic underlying industrial technologies that support the consumption of inputs and the production of outputs.

The planning of a technology infrastructure should follow a similar strategy, by considering a typical technology-intensive firm and determining how public funds can be employed to perform the same three functions. Technology infrastructure must provide for the "transport" of knowledge inputs and outputs, perform regulatory activities where needed, and supply an underlying base of generic technologies upon which a knowledge-based economy can build and expand.

## *The Scope of Technology Infrastructure*

The concept of an enabling infrastructure has evolved from rather unglamorous beginnings to become the buzzword of choice in many policy discussions. Along with this elevation in status has come a muddling of scope and meaning, to the point where technology infrastructure in many instances has become virtually synonymous with the national system of innovation.[58] It is therefore worthwhile considering just what is meant by a technology infrastructure and how the current administration proposes to fulfill this need.

In its broadest sense, the goal of U.S. technology infrastructure should be to leverage private-sector investment, reduce commercial risks to individual firms, and improve the productivity of the national technology enterprise.[59] According to the Department of Commerce, technology infrastructure includes protection of intellectual property rights, a national information infrastructure, management of the radio-frequency spectrum, development of an environmental information system, and development of standards, measurement techniques, and test methods.[60] While this list is a good starting point, it represents an enumeration of what is already being done, rather than what will ultimately be needed.

If we look at this from an industry perspective, the list of needs becomes more focused and a bit more self-serving. The semiconductor industry, for example, has expressed long-standing concern over the weakening of the domestic base of small suppliers for processing equipment and high-quality raw materials. In fact, the SEMATECH consortium was initially chartered to address these issues. Thus, according to the Semiconductor Industry Association, a broad and robust domestic base of suppliers is an important component of America's technology infrastructure.[61]

Not surprisingly, both the governmental and industrial perspectives are valid, yet neither clearly defines the scope of technology infrastructure. I believe that the best way to resolve this issue is to return to our "model" technology firm and consider what would be required to enable rapid growth in a highly competitive global marketplace, as summarized in Table 7.3.[62]

The most obvious candidate for government intervention is the development of the world's finest telecommunications capability. Although this activity begins with a traditional hardware focus, including the laying down of fat data pipes and the launching of high-bandwidth satellites, there is much more to telecommuni-

| Component of Technology Infrastructure | Examples |
| --- | --- |
| Information Infrastructure | • The "Information Superhighway"<br>• Data transmission trunk and branch lines<br>• Communications satellites<br>• Information display devices<br>• Data interchange protocols |
| Infratechnology | • Design and automation tools<br>• Processing methods<br>• Materials and other databases<br>• Specialized test equipment and facilities<br>• Specialized supplier base |
| Standards and Regulations | • Spectrum management<br>• Quality standards<br>• Process standards<br>• Interoperability standards<br>• Safety regulations |

**Table 7.3:** Summary of three important areas of technology infrastructure that enable increased rates of innovation and facilitate accumulation of technical knowledge.

cations than just the hardware layer. Data transmission requires several additional layers of control and interchange software, standards to ensure connectivity, regulation of transmission frequencies, and so on. The difficulties in transporting physical objects pale in comparison to the challenges of creating a truly effective information infrastructure.

Another area that has great potential for positive government involvement is the creation of a solid base of facilitating knowledge. This is not the same as the creation of commercial innovations. Instead the focus here is on ensuring that the corporations at the end of the economic food chain (i.e., system integrators and original-equipment manufacturers) have an extensive base of public knowledge and qualified subcontractors to accomplish their commercial goals. This could mean the development of a database of semiconductor material properties by a government laboratory such as NIST, or providing local government support to create a fertile environment for specialized suppliers. In either case the essential goal is the same: to create a broad platform of technical capability on which innovative firms can launch new and expanded lines of business.

Finally, there is a growing need for cooperation in the development of global standards, particularly with respect to data interfaces and product compatibility. Information currently comes in a variety of flavors, as anyone who owns a cellular phone can attest. In an ideal world, private industry would work cooperatively to establish standards for such things as high-definition television (HDTV) or digital videodisks. Where the need for standards overlaps with federal regulation (as is the case with the radio-frequency spectrum) an active government role is appropriate. However, even in purely commercial situations, government facilitation of standard-forming industry bodies may be beneficial.

In each of the above cases, government involvement can help rectify situations in which market forces are insufficient to ensure adequate private investment, or in which a lack of intra-industry cooperation may prove to be costly to consumers. Since the development of a robust technology infrastructure often involves the creation of non-rival goods (such as standards and databases), government investment benefits from increasing returns to scale, making this a highly leveraged component of technology policy.[63]

### Information Highways and Byways

The scope of information technology has become truly breathtaking. Even in its infancy, it spans virtually every aspect of commercial activity, including publishing, entertainment, broadcasting, telecommunications hardware and software, computer hardware and software, and much more. This ubiquitous presence has made the so-called "information superhighway" an accepted focal point for public investment.

At a minimum, information infrastructure includes the physical hardware required to produce, transmit, receive, and display digital information, the software and standards required to interconnect this hardware, and the knowledge content that creates value in the system.[64] Many of these needs are being met with great enthusiasm by private industry, but there are still excellent opportunities for the public sector to play a positive role. Issues such as universal access, protection of privacy and intellectual property, and transaction security are prime candidates for cooperation between industry and government.

It is well known that the Internet had its origins as a defense-communications backbone funded by ARPA, and was later extended to the university system through investments by the NSF. Today, the Internet bears little resemblance to this early incarnation in either scale or structure. It has become a *metanetwork* of international autonomous networks that share a common level of functional interoperability, reaching into homes and businesses in every corner of the planet. There is no human creation that is more thoroughly global in nature than the Internet.

Should national governments play a role in the regulation and development of what has become a global information infrastructure (GII)?[65] Many believe that market forces alone will guide us to an optimal solution, leaving governments little to do but watch from the sidelines. As with most technology-driven developments, however, there are a number of potential market failures, along with some significant social issues, that will require either national or multilateral intervention.

Basically, the GII will perform two distinct functions: It will provide a vital link between neighbors and nations as a conduit of free information to all who desire it, and it will become a vast and powerful economic force through electronic commerce. These disparate roles demand distinct policy measures. In the case of electronic commerce, a minimalist role for government has already been endorsed by both the United States and several multinational bodies, as discussed in Box 7.1.[66] With respect to the role of the GII as a channel for knowledge and a tool of learning and growth, however, the U.S. government has several vexing challenges.

The first goal of national policy for technology infrastructure must be to ensure that all who desire connection into the global information network have the opportunity to do so at a fair and reasonable price. This is one of the most clear-cut cases of market failure, in that the cost of delivering adequate telecommunication services to remote locations may be far too high for individuals to bear. The same type of adjustments will be required for information access as have already been implemented for electric, phone, and television service. As both learning and commerce become increasingly dependent on access to cyberspace, the need for universal access will become compelling. Without government intervention, we are in danger of creating a second class of citizens who are barred from membership in the global information community.

A second critical area for government involvement is in the protection of individual privacy and public safety. To address issues of privacy and security, the

Clinton administration has proposed several "privacy principles" that address such issues as mandatory disclosure by gatherers of private data and methods for individuals to limit the use and reuse of personal information. If these and other privacy concerns are not adequately addressed by industry through self-regulation, governments will be forced into greater levels of intervention to safeguard the public.[67]

The dramatic advance of wireless telecommunication technology has brought about a final important challenge for national governments. Historically, governments have taken a commanding role in the development of telecommunications infrastructure, and in particular the allocation of bands within the radio-frequency spectrum. Even today a majority of nations have not fully privatized their telephone and broadcasting monopolies. The United States has played a leading role in achieving multilateral acceptance of open competition, through the successful conclusion of the WTO's Basic Telecommunications negotiations and the signing of the Information Technology Agreement in 1997. Despite this high-profile endorsement of global competition and deregulation, however, the United States has much room to improve on the domestic front.

Although opinions on the success of the Telecommunications Reform Act of 1996 are decidedly mixed, it is safe to say that competition has not reached a fever pitch, particularly with respect to local phone service. Meanwhile, the Federal Communications Commission (FCC) has become a reactive agency, torn between congressional and court actions, while slipping further behind the volatile edge of telecommunications technology.[68] Attempts to encourage the development of HDTV by the FCC have been abortive, and the critical need for a national cellular phone standard has been largely ignored. Promulgating erratic policy and ineffectual regulations, as has become common in American telecommunications policy, is the antithesis of an effective technology infrastructure. What is most important to the expansion of technology enterprise is long-term regulatory stability and a light governmental touch.

### Building a Knowledge Platform

Between the commercial realm of product development and the esoterica of basic science lies the great gray expanse of applied technical knowledge. For new innovations to reach the marketplace, this rift must be crossed, either through public or private funding. Residing in this domain is knowledge that is either too broad to be commercially valuable or too specialized to have a sufficient market. Again, this represents a case of potential market failure: individual companies may find the development of such knowledge for their exclusive use to be prohibitively expensive.

Consider for a moment the full body of technical knowledge required to produce a personal computer. This case is particularly illustrative because even though the knowledge content of PCs is exceedingly high, virtually every small

## Box 7.1: A Global Free-Trade Zone?[69]

The Clinton administration has been forced to face the realities of the technology age. The U.S. government lacks the economic resources, political influence, and technological savvy needed to control the Internet, so it has endorsed the next best thing. In mid-1997, the Office of the President issued a document entitled *A Framework for Global Electronic Commerce*, which proposes that the Internet be made a global "free-trade zone." This laissez-faire policy is a reflection of government being forced to play catch-up with the rapidly expanding global metanetwork. Vice President Al Gore described the proposed policy as "a digital Hippocratic oath—first do no harm."

There are still important responsibilities for government, however, including the establishment of a "uniform commercial code" to guide the formation of contracts in cyberspace, enforcement of IPRs, and assurance of privacy and security. This last consideration has been the subject of lively debate. Payment methods represent the weak link in the electronic commerce chain. Secure and reliable methods for validating personal identity and transferring funds are central to building consumer confidence. Proposed identification techniques include credit cards, electronic signatures, and smart cards, all of which depend on the use of some type of secure cryptographic key. A number of issues surround the development and proliferation of so-called *strong-encryption* technology, including the alleged need for governments to have access to private transactions for national security reasons.

This final concern has caused heated debate, with industries pressing to export their most powerful encryption products, and the U.S. government attempting to restrict global access to these technologies. Early attempts at an encryption standard based on the "clipper chip" are now obsolete. The administration is currently working with industry and international bodies to develop market-driven standards, public-key management infrastructure, and key-recoverable encryption products.[70]

Industry analysts predict that electronic commerce on the Internet could reach more than $200 billion by the year 2000, and perhaps $1 trillion by 2010. One of the architects of the U.S. policy, presidential advisor Ira Magaziner, put the importance of electronic commerce in perspective: "We think it will be the engine of growth for the world economy well into the first quarter of the next century."

town has a local firm capable of producing them. To create the motherboard, specialized components must be procured from various suppliers, each of whom has established a fairly narrow core competency. The design of the board will require computer-aided design (CAD) tools developed by other specialized firms, which capture schematics and verify board layouts. Finally, the assembly of the motherboard, fabrication of the enclosure, and final system testing might all be performed based on standardized interfaces, using processes developed by industry consortia or government-sponsored testing laboratories.

Evidently it takes very little resident knowledge for a firm to manufacture PC clones. The reason for this is not that the technology embodied in PCs is trivial, but rather that there is a rich infrastructure of specialized suppliers, industrial tools, and process standards available to enable such an enterprise. This base of *infratechnology* represents a tremendous lever for increasing total-factor productivity, and is an essential element of a successful national system of innovation.

In the past, it was possible for major multinational corporations such as General Electric, AT&T, and RCA to perform virtually all the research necessary to produce new products internally. The powerful central laboratories of these industrial giants transcended the gray expanse of applied technology, allowing them to perform everything from world-class basic research to the development of new industrial processes.

Today, the central laboratories of large MNCs have given way to multiple islands of product-specific research that depend increasingly on the specialized R&D capabilities of their supplier firms. This horizontal industry structure eliminates market failures through achieving higher levels of technical specialization. A semiconductor-manufacturing equipment supplier, for example, may produce only one or two products for a short list of customers. They can survive, however, by dominating their narrow niche and capturing monopoly profits commensurate with their high levels of specialized capability.

One of the most important functions of a national technology infrastructure is to provide a platform of infratechnology upon which firms can rapidly build their enterprise. This platform consists of two categories of knowledge: that which is embedded in specialized supplier firms, and that which is noncommercial, but enabling. The first category requires government action to create a fertile economic climate for small start-ups and technical specialists, while the second demands direct public support for the creation and dissemination of enabling knowledge.

The benefits of government-funded infratechnology were evident in the years following World War II. The National Advisory Committee for Aeronautics (NACA, the predecessor to NASA) provided valuable enabling technologies such as wind tunnels, test methods, and safety standards for America's promising aeronautics industry. Since that time, the United States has led the world in performing this type of generic research, including the characterization of material properties, development of design tools and simulation models, creation of quality control and production process methods, and publication of engineering handbooks.[71]

Among the hundreds of government laboratories that perform basic and applied research, NIST stands out as a center of excellence for the creation of infratechnology. Its activities include the development of standardized testing and measurement methods, creation of vast databases of material properties, and providing leadership for broader programs of government-funded research. NIST has consistently received high marks from industry for contributing to productivity and reducing the transaction costs between buyers and sellers of technology.[72]

Another approach to government funding of infratechnology is exemplified by large-scale demonstration projects, such as the Clinton administration's Clean Car Initiative. This program, which includes participation by the Big Three automakers and several second-tier suppliers, is mission-specific and unambiguously commercial. Yet there is an underlying goal to develop a base of generic technologies that will help fight air pollution, improve transportation safety, and increase energy efficiency. In a sense, these unspoken externalities justify the expenditure of public funds on a mission to improve auto-industry competitiveness.[73]

The creation of valuable infratechnology alone is not sufficient to reap productivity gains for American industry. This platform of knowledge must be made available to every enterprise at minimal effort and cost. This goal has been made infinitely more achievable with the advent of the World Wide Web. In my personal search for knowledge while performing research for this book, I discovered that virtually every government and industry site that I visited held vast storehouses of data, presented in clear, usable, and often downloadable form. The U.S. government has been aggressively loading the resources of national laboratories and federal departments onto the Web. There could not be a better expenditure of public funds than to expand this program to subsidize site development for every source of infratechnology, whether public or private.

The need for access to this base of generic knowledge is most pressing for America's small-to-medium-sized firms.[74] Although these firms make up more than 90 percent of all industrial enterprise in this country, their rate of adoption of new technology has been perilously slow. Often these companies lack the depth of resources necessary to be good absorbers of external knowledge, and they may be isolated from industry and public sources of new technology. As our dependence on specialized suppliers increases, significant government action will be needed to create a more favorable environment for America's smaller firms.

Creating an "incubator" for small technology-intensive firms presents a multifaceted challenge. Changes in intellectual property protection and antitrust laws may be needed to ensure early profitability for new start-ups. Other measures, including tax incentives and investment guarantees, may be appropriate, provided that they are doled out in an equitable fashion to all industry, whether low or high technology. More important, the venture-capital markets in the United States must continue to reward innovation. The shortsightedness of private sources for venture capital is often an obstacle to pursuing long-payback opportunities.[75] A more "patient" investment-capital system is needed to inspire researchers to look at longer-term projects with potentially high social payoffs.

Anyone who has attempted to establish his or her own firm knows just how much knowledge is required. It is not sufficient to be an expert in your field and possess a basic understanding of business principles. Successful proprietorship requires a knowledge of accounting, finance, marketing, export processing, human resources, and manufacturing. Failure to grasp any one of these essential skills can cast even the most promising innovations into the oblivion of business failure.

The need for a holistic approach to mentoring small businesses has been recognized in several other nations. Germany, for example, has initiated the Baden-Wurttemberg program, which includes both business and technical assistance for start-up firms. Brazil has launched the SABRAE program, which addresses the "total" needs of small enterprise. While the U.S. government has established some highly successful manufacturing and technical extension programs, it has been left to the individual states to provide complementary business-consulting services. If we are to continue to lead the world in technical innovation, programs of this sort must be broadly institutionalized, so that barriers to business entry are reduced and the risks of failure for small firms are minimized.

### Managing the Interfaces

The days of stand-alone products are over. Today, virtually every commercial innovation must interface successfully with existing products and infrastructure, and be compatible with myriad government regulations. This high degree of interdependency manifests itself at times in obvious ways, such as the requirement for interoperability of computer hardware and software products, or it may have a more subtle impact, as in the case of electromagnetic-compatibility (EMC) regulations. Thus, managing interfaces and attributes among disparate products has become an important enabler of technological expansion.

The weapons of choice in the battle for interoperability are technical standards. Although the need for standards is most apparent in information and electronics-based industries, the ubiquitous nature of these technologies has extended the need for interface management into many nontechnical fields as well. Standards can accelerate adoption of new technologies, aid in the dissemination of information, and form the basis for many types of industry cooperation. On the other hand, standards are increasingly employed as a component of aggressive competitive strategies, at times resulting in obstacles to new innovation.

The U.S. government has played a long-standing role in facilitating the development and diffusion of industrial standards. In fact, this has been a central mission for institutions such as NIST (which was revealingly named the National Bureau of Standards until recently). In the European Union, the government-funded ESPRIT program addresses the reduction of entry barriers to regional markets, and supports joint research on cooperative technical standards.

Despite a relatively high-profile mandate for government facilitation, the vast majority of standards development takes place in the private sector. This is fortunate, given the dubious history of direct government involvement in technical standard-setting, particularly at the international level. Attempts by the United States, Japan, and the EU to position their "domestic champion" standards for high-definition television provide a dismal example. Although the outcome of this battle is still in doubt, it is likely that the world will be saddled with another several decades of incompatible television standards.[76] Actually, the entire question may be moot, since the delays associated with this contentious standards debate may have already rendered several of the proposed technologies obsolete.

On the other side of the spectrum (pardon the pun) is the ongoing battle over cellular phone standards in the United States. The Federal Communications Commission has dragged its feet on mandating a nationwide standard for digital-wireless phone systems. Rather than stepping into what it considers to be a market-driven contest, the FCC has allowed the U.S. market to be split among three different (and not surprisingly, incompatible) standards. Meanwhile, in Europe, the global system for mobile communications (GSM) standard has been adopted across the entire expanse of the EU, allowing a cell-phone user to cross a dozen national borders while making calls with the same handset.[77]

Despite an obviously disruptive and inefficient current situation, it is not clear that the FCC should step in to establish a national cell-phone standard. There is a significant danger in empowering governments to pick technologies, rather than allowing market forces to decide the issue. Although Europe is indeed homogeneous with respect to cellular technologies, the GSM standard that has been adopted is now essentially obsolete. A changeover to newer technologies will be costly, and may lag behind the more competitive U.S. market. Ultimately, situations of this type are reduced to a tradeoff between the benefits of allowing competition to sort out the best and brightest technology, and the network externalities that can be gained by having broad interoperability of products.

In general, the most effective method for developing commercial standards is essentially from the bottom up. The technical issues associated with interoperability and compatibility are usually far too complex for nonspecialists to grasp. Hence, industry working groups consisting of developers and users of a new technology are most effective at laying the groundwork for standards. Organizations such as the American National Standards Institute (ANSI) serve as a coordinating body and clearinghouse for such industry-developed standards. In addition, ANSI represents the United States in international forums to negotiate global technical standards. This process of bubbling up standards from the user community has been highly successful in rapidly changing technology industries such as Internet hardware and software.[78]

The U.S. government has played a beneficial role in this arena by funding standards research, both in federal labs such as NIST and within private-industry consortia. An excellent example of a cooperative venture that may offer high productivity gains is the ANSI National Standards Systems Network on the World Wide Web. This effort was undertaken by ANSI under funding by ARPA's Technology Reinvestment Project, and was provided with technical support by NIST. The system, which is available to any firm with Web access, provides overviews and sources for more than one hundred thousand military and commercial standards.[79]

As technology markets become increasingly global, there is a growing need for multilateral coordination of both product and interface standards. Organizations such as the International Standards Organization (ISO) and the International Electrotechnical Commission (IEC) have made an immeasurable contribution to global standardization. Despite their best efforts, however, there are still enormous barriers to the development of "world products."

In most cases, these obstacles are simply bureaucratic. Minor differences among national standards, combined with a lack of interest and participation by the firms themselves, can result in drawn-out development exercises and slow rates of adoption. The IEC, for example, has been developing a comprehensive set of global standards for the electrical safety of medical devices for more than twenty years. The IEC 601 series of standards is now the criterion of choice in most of the world's important markets. Yet, due to minor differences between these standards and the accepted American standard, UL 544, developed by Underwriters Laboratory, the IEC guidelines have been slow to take hold. Even though recent negotiations have eliminated virtually all the differences between the documents, the Food and Drug Administration (FDA) has not yet adopted IEC 601-1 as the American national standard.[80]

Adherence to international standards is critically important to industries that are highly specialized. Often there is insufficient demand even in the huge U.S. domestic market to offset the cost of product development. Without prompt access to global markets, many firms will have insufficient incentives to innovate. This is even more critical for information-based technologies that depend on rapid deployment to recoup their R&D investments. Software firms, for example, have reported that they expect to receive up to 75 percent of their revenues from new products within the first two years after introduction.

The global harmonization of technical standards has been prominently featured in recent WTO multilateral trade negotiations, and has been pivotal in market-opening bilateral discussions with China and Japan. As tariff barriers dwindle in both magnitude and importance, non-tariff barriers have become the protectionist measures of choice in many countries. China, for example, had implemented more than seventeen thousand product standards by 1995, with only four thousand of them being aligned with international guidelines.[81] Although the development of technical standards should be driven by industry and market forces, national governments must be aggressive in promoting harmonization and adoption of these standards in the global marketplace.

It should be clear from this discussion that managing the interfaces in a technology-driven economy represents a critical factor in the economic growth of both firms and nations. Tremendous gains in efficiency and productivity can be attained through cooperation between industry and government. At times, however, this cooperation can be undermined by the use of standards as strategic weapons for competitive advantage. In some industries, such as computer software, the conflict over standards has expanded into a battle for market domination, with network externalities and consumer preferences tending to drive these markets toward virtual monopolies.

As information technology becomes a dominant factor in global competition, it is likely that the strategic use of standards will become central to achieving competitive advantage. Already there are a number of subtle applications for product and interface standards that can give firms an important edge, as shown in Table 7.4.

Although every CEO dreams of exploiting first-mover advantage to become "the standard" in an industry, recent experience in both the computer and telecom industries suggests that more is needed than just being first to market. Sun Microsystems' recent attempt to have its Java language become established as an international standard provides an excellent example. In this case, it was Sun's reluctance to release control of its source code to an impartial body that caused their appeal to be rejected.

Becoming a standard requires broad acceptance both by the customer community and by the myriad interested parties whose products depend on such a standard. In the future, no firm will be allowed the privilege of "standard status" without first demonstrating its willingness to cooperate with all vested parties.

## Fostering Diffusion and Absorption of Knowledge

American industry has shown a remarkable lack of interest in adopting productivity-enhancing technologies. Inadequate R&D spending by U.S. firms on process innovation has become legendary. Moreover, domestic rates of adoption of commercially available process improvements has been lagging by world standards. This problem is particularly acute for smaller firms, which lack the

| Role of Technical Standard | Examples |
|---|---|
| 1. As a bridge between distinct product types or technologies. | • Engineering Design Automation (EDA) tools.<br>• Factory automation hardware and software.<br>• Plug-and-play test equipment. |
| 2. As a bridge between two components of infrastructure. | • Linking of major infrastructure functions, such as the merging of the National Information Infrastructure with the Intelligent Highway System.<br>• Standardization of formats and templates for Electronic Document Interchange (EDI). |
| 3. As a way to create new markets. | • Collaboration between fifteen of the world's largest electronics firms on a standard for new magneto-optical (MO) data storage disks. |
| 4. As an assurance of performance or quality. | • ISO 9000 / QS 9000 quality standards.<br>• International standards for surface mount fab., silicon wafer geometry, EMI / EMC, etc. |
| 5. As a way to lock out competition. | • Sony Corp., Philips Electronics NV, and HP have each decided to make their own version of digital videodisks (DVDs). |
| 6. As a way to invade new markets. | • Microsoft, Intel, and Compaq have proposed a new TV standard that merges computer monitor and TV functions. The goal is to "merge" markets for entertainment & computer products. |
| 7. As a way to capture competitors' market share. | • Sun Microsystems Inc. has petitioned ISO to adopt its Java programming language as an intl. standard. Thus far, the request has been rejected due to complaints by Microsoft and others that Sun wishes to retain control and ownership, while reaping the benefits of being designated a global standard. |
| 8. As a way to build barriers to foreign imports. | • Most developing nations, including China, Brazil, etc. use technical standards to protect domestic industries.<br>• Japan uses "type exceptions" and technical standards to impede entry of foreign imports. |
| 9. As a way to gain network externalities. | • NEC recently announced that it would abandon its proprietary system in favor of the ubiquitous "Wintel" standard. |

**Table 7.4:** A sampling of the many roles that standards play in optimizing infrastructure and achieving strategic advantage. Sources: *Business Week,* various articles, *International Herald Tribune,* various articles, and Branscomb, L. M and J. H. Keller (1996).

resources and external connections needed to efficiently access new productive technology.[82]

The insular attitude of industry has been reflected in U.S. government policies toward technology diffusion and adoption. Until recently, the vast majority of federal funding has flowed into research, with little attention paid to fostering higher levels of adoption and utilization. Our entire national system of innovation has been geared toward creating the next great invention, rather than addressing the more mundane necessities of maintaining a competitive manufacturing capability.

Clearly, we cannot expect to lead the technological pack into the twenty-first century while being handicapped by obsolete processes and low-productivity methods. Despite the recent attention paid to improving American manufacturing techniques, there is still evidence that advanced technologies are being adopted at a faster rate, and with greater effectiveness, by foreign firms. Since nations that foster the highest rates of technology diffusion will tend to have the most rapid growth in productivity, the need for a change in attitude on the part of the U.S. technology enterprise is apparent.

### Lagging U.S. Technology Adoption

Not only do American firms underinvest in developing advanced manufacturing processes, but they appear to be delinquent in adopting proven industrial technologies as well. Surveys of small-to-medium-sized enterprises (SMEs) have shown that productivity-enhancing methods such as computer-aided design and robotic automation have penetrated fewer than one-third of these businesses, as shown in Figure 7.5. Even the use of relatively mature technologies such as numerically controlled machines has been embraced by less than 40 percent of SMEs.[83]

This presents a worrisome problem for America's technology enterprise. As production strategies become more horizontal, the value chain for technology products will be dominated by the productivity of specialized suppliers. It is therefore critical that advanced methods and processes penetrate beyond the major original-equipment manufacturers (OEMs) into the vast supplier network that supports them. Both government and industry must work together to ensure that the barriers to technology adoption are lowered for America's SMEs.

There are a number of reasons for the low rates of diffusion and adoption observed among U.S. firms. The first obstacle is a lack of awareness. Many small businesses are highly focused on a narrow specialty and have little exposure to unrelated industrial technologies. With limited resources available to research options and perform trade studies, a small firm might be reluctant to make a substantial investment in advanced processes. Likewise, the added support costs and learning curve associated with the adoption of any new technology may appear to be prohibitive.

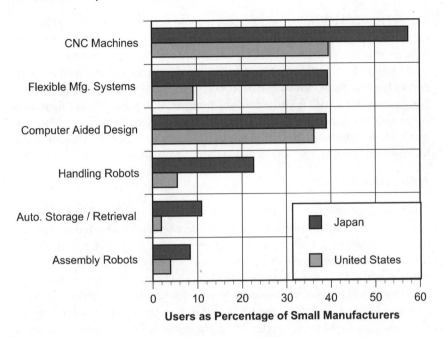

**Figure 7.5:** Comparison of the percentage of small manufacturers in Japan and the United States that employed advanced manufacturing technologies as of 1988. Source: National Academy of Engineering (1993, Table 2.2, pg. 39).

More important, there is often a mismatch between the needs of small production operations and the solutions being offered in the marketplace. The purchase of an enterprise resource planning (ERP) software package by a small business, for example, can be a frustrating exercise. The available products are feature-rich, expensive, and require huge up-front installation efforts and continuing maintenance. These systems are typically not designed for a firm with fewer than fifty employees, and can actually *subtract* productivity in such an inappropriate application.

The same is true for other advanced industrial technologies. The U.S. government, for example, established the Automated Manufacturing Research Facility to develop and disseminate state-of-the-art process techniques. Unfortunately, it has been estimated that only about 4 percent of all manufacturers believe that the automation and robotic technologies being developed there are applicable to them. One of the primary reasons why America's small manufacturers are not rapidly adopting productivity-enhancing processes is that these technologies have not been adapted to their specific needs.

Achieving higher rates of diffusion of productive technologies will require action on three fronts. First, a new awareness of the needs of small manufacturers must be instilled into the producers of advanced process technologies. This will require a shift in emphasis on the part of both government and industry away from seeking "breakthroughs" and toward identifying pragmatic solutions to real-world productivity problems. Second, small firms must be enfranchised as a vital part of the American industrial network. This aspect of the problem has become far more tractable with the advent of the Internet, which allows small firms easy access to a global pool of information. Finally, American firms (both large and small) must understand the critical role of productivity in global competition and become well versed in the ways in which it can be enhanced.

### Lessons from Abroad

American reluctance to adopt new manufacturing technologies is deeply rooted in our industrial history. Our major industries were built on leadership in product-related technologies, while foreign competitors have based their strategies on imitation and recurring-cost reduction. Again, we have been slow to recognize a fundamental shift in the commercial landscape. The center of gravity for global competitiveness has moved toward emphasis on state-of-the-art process technology, with leadership in this arena firmly residing on the other side of the Pacific. Even when advanced process technologies are developed within the United States, we are slow to embrace them. According to University of Michigan faculty, for example, both German and Japanese automakers frequently apply the results of U.S. university process research sooner than the Big Three U.S. automakers or their suppliers.[84]

One of the most important contributions that government can make to domestic technology enterprise is the development of effective diffusion strategies. Although the U.S. government has paid increased attention to industrial extension programs and other mechanisms in recent years, we are relative latecomers to technology diffusion. In 1989, the United States allocated only 0.2 percent of its total R&D budget to "industrial-development" activities, which included extension programs, consortia participation, pilot plants, and demonstration projects. By comparison, Japan spent more than 32 percent and Germany invested more than 20 percent of their respective R&D budgets on economic development programs, with 5 to 13 percent going toward industrial development specifically.[85]

The Japanese government has been particularly aggressive in establishing diffusion programs, including activities tailored to the needs of smaller firms. A network of 170 small-business industrial-extension centers has been deployed under Japan's *Kohsetsishi* organization, with an annual budget of $500 million. With such ample funding, these centers can support roughly six agents per one thousand companies, as compared to the single agent per one thousand firms provided by the best American programs. This financial commitment reflects a recognition of the infra-structural role of smaller manufacturers in advanced industrial economies.[86]

In addition to extension programs, Japan has implemented subsidies and favorable tax treatment for the adoption of advanced manufacturing-process equipment. Several public corporations have been formed by the Japanese government to procure domestically produced robotics and computer equipment and lease these capabilities to firms at modest rental rates.[87] Collectively, these programs increase the awareness of new industrial technologies and reduce the financial barriers and risks associated with their adoption.

Although these activities provide explicit mechanisms for assisting SMEs, the most important channel for diffusion of technical knowledge in Japan is through collaborative partnerships. In the United States we tend to view consortia as the means to achieve a mission-specific end. The Japanese consider the cross-fertilization and communication links created by consortia to be an end in themselves. The primary goal of cooperative organizations is often to close the gap between the average and the most advanced of the participating firms.

In the West, we have frequently judged Japanese consortia such as the Fifth Generation Computer Project (FGCP) as having failed to produce commercial results. Although this appears to be a fair judgment, we may be using the wrong criterion to evaluate their success. The FGCP, for example, was intended to develop Japan's basic research capability and to create communication links among industry, universities, and government laboratories. Hence, these spillover benefits should be included in our rather nearsighted assessment.[88]

Similarly, collaborative agreements such as ESPRIT and UNIDATA in the European Union have not had a great impact on industrial competitiveness, but they have facilitated communication among technical firms from fifteen different countries. The most important lesson to be learned from the experience of other advanced nations is that consortia are often more effective at building cooperative, information-sharing relationships than they are at creating new commercial technologies.[89]

### U.S. Industrial Extension Programs

Since the mid-1980s, the U.S. government has become an active partner in the diffusion of productivity-enhancing technology. State and federal governments have long funded such programs in agriculture, but recent concerns over global competitiveness have caused a shift in technology policy toward increased support for the adoption of advanced industrial processes.

Technology-diffusion programs in the United States are divided into three categories. First, several parallel efforts are underway to establish industrial and technology extension centers in regional locations. Industrial extension has proven to be successful in Japan, Germany, and Scandinavia, and offers potentially high economic returns for relatively low levels of public investment. Second, the U.S. government has enacted policies to promote collaboration among universi-

## Various Forms of Industrial Extension

- **Business Assistance**
  Consultation on business practices, funding sources, human resources and management methods.

- **Incubators**
  Provision of facilities and equipment to start-up firms.

- **Research Parks**
  Industrial clusters planned specifically to develop a horizontal base of technology firms, and often centered on a university.

- **Seed Capital**
  Grants and research subsidies for commercialization.

- **Technology Assistance**
  Programs designed to increase rates of adoption of productive technologies.

- **Technology / Research Centers**
  Centers for collaborative research among universities and industry.

**Table 7.5:** Several forms of industrial extension, all of which facilitate the transfer and adoption of new productive technologies. Source: Adapted by the author from Chapman et al. (1990).

ties, federal laboratories, and industry. Finally, the formation of industrial consortia to perform demonstration projects or to develop pre-competitive technology has received direct government support and participation in recent years.

The purpose of manufacturing and technology extension services is to transfer advanced process knowledge to smaller firms and provide assistance in implementing these improvements. This assistance can take on several forms, including management consultation, technology transfer, and direct financial support for start-ups, as shown in Table 7.5. The benefits to small business are obvious: higher productivity at reduced cost and risk.

The Manufacturing Technology Centers (MTCs) program, established under the Omnibus Trade and Competitiveness Act of 1988, specifically targets small business for assistance. These large regional centers, which are administered by NIST, were originally chartered to stimulate the introduction and use of advanced productive technologies. After several years of experience, however, NIST has modified its definition of "advanced" to include primarily commercial off-the-shelf (COTS) solutions. What began as a delivery system for supply-side technology breakthroughs has transitioned to a more practical, demand-side mechanism for meeting the needs of small business. NIST's current plans call for establishing thirty MTCs by 1999, along with nearly one hundred smaller Manufacturing Outreach Centers.

The National Science Foundation has developed its own brand of diffusion program, which involves the formation of Science and Technology Centers (STCs) at selected U.S. academic institutions. After ten years of operation, twenty-four centers remain active, spanning the spectrum of important new technologies from biomedical to computer and information sciences. Funding for the creation of ten new centers has been budgeted beginning in 1998.[90] In addition, the NSF has established Engineering Research Centers in seventeen states. The mission of these facilities is to perform cross-disciplinary and systems-oriented research, to provide education and outreach, and to facilitate collaboration and technology transfer.

The diffusion of industrial knowledge is one aspect of technology policy in which state governments have made a major contribution. Currently, forty states have implemented some form of manufacturing extension service, often in conjunction with business and financial-assistance programs. These state-level activities are typically focused on small enterprise development and offer the significant advantage of regional sensitivity and flexibility. In fact, many of the state-funded programs are more innovative than their federal counterparts, causing national policymakers to consider these new concepts for federal adoption.

This multilevel approach to diffusion policy avoids the "one size fits all" dogma typical of centralized federal programs. A recent study of thirty-eight extension programs in twenty-five states noted that, on average, states were spending $5,000 to retain or create a job, and that the state received at least one dollar in increased tax revenue for every dollar spent.[91] In California, facilities such as the California Manufacturing Technology Center provide assistance to SMEs through the deployment of proven technologies and standards. Similarly, Texas has established a network of forty-seven Small Business Development Centers, which includes several Technology Assistance Centers to facilitate the commercialization of new and innovative technologies.

In some cases, states have attempted to take up the slack caused by waning federal R&D support. In 1996, California launched an industry/university/state partnership designed to promote workforce development and pathbreaking research in economically important industries.[92] The Industry/University Cooperative Research Centers program, administered by the University of California, provides seed grants to investigators from both universities and the private sector, and supports a training environment that promotes research experience in economically relevant fields. One noteworthy goal of this program is to build a "better" Ph.D.

In the best of all worlds, government at the state and federal levels would work cooperatively to achieve an optimized diffusion network. Such an arrangement would demand an equitable cost-sharing arrangement to ensure that all states have equal opportunity to deploy these assets. Agreements such as the U.S. Innovation Partnership may provide a template for state and federal cooperation on infrastructural technology policy. Ultimately, the state governments, and even regional and local entities, must play a larger role in tailoring diffusion programs to their constituents' needs. Meanwhile, the federal government's efforts should be directed outward, to serve both as a filter for new breakthroughs and as a harvester of advanced productive technologies from all nations.

### Diffusion Through Cooperation

The only drawback to technology-extension programs is that they are based on a one-way, top-down flow of information. These programs offer few opportunities for isolated small firms to become connected into a cooperative industrial network. This aspect of technology diffusion is best accomplished through participation in collaborative research partnerships among federal labs, universities, and individual firms.[93]

Since the passage of the National Cooperative Research Act of 1984, firms have been allowed to register the formation of R&D consortia with the Department of Justice, thereby securing some protection against antitrust actions. Subsequently, more than six hundred R&D joint ventures have been registered, spanning every form of collaboration, from basic research to joint production. Due in part to a concerted lobbying effort by industry, the original legislation was recently extended and expanded through the National Cooperative Research and Production Act of 1993.[94]

The primary vehicle for collaborative ventures among government laboratories, universities, and industry is the Cooperative Research and Development Agreements (CRADAs). The use of these structures has been growing rapidly in recent years, particularly as a means for the transfer of dual-use technology from federal labs.[95]

The primary benefit of CRADAs is the opportunity they provide for firms to acquire tacit skills and form cooperative linkages with other participants. Similar advantages are derived from participation in the Industry/University Cooperative Research Centers (I/UCRCs) program, funded by the NSF. These centers are typically started by a small grant from the NSF to an individual university faculty member. Once industry involvement and support has been demonstrated, the faculty member can submit a proposal for continued funding of the center. Other universities often join the consortium to lend additional synergy and capability to the group. I/UCRCs have been established to develop advanced manufacturing technologies, to study nanotechnology fabrication methods, and to investigate revolutionary new materials.[96]

Despite the proliferation of CRADAs, the effectiveness of these entities is in some doubt. Many firms are skeptical of the cultural differences among government laboratories, universities, and industry. Noncommercial institutions are notorious for displaying a "cost-plus" mentality, implying that control of budgets can be very difficult. There are also concerns over governance; federal laboratories are accountable to federal departments, which can impose bureaucratic inefficiencies. The Department of Energy, for example, has a reputation for micromanaging its laboratories from Washington.[97]

Finally, firms have expressed worry over increased government intrusion into their operations as a result of participation in a CRADA. The success of some CRADAs sponsored by the National Institutes of Health, for example, has prompted Congress to urge the NIH to influence the pricing policies of CRADA partners. If such attempts to extract *quid pro quo* from CRADA partners continue, these collaborative structures will become far less attractive to firms in the future.[98]

At least in theory, firms should expect to gain economic benefits from participation in collaborative R&D. These gains may come in the form of reduced financial risks, shared non-recurring costs, pooled expertise and resources, or more-rapid development cycles.[99] It is worth reiterating, however, that collaborative projects are best at fostering diffusion and utilization, rather than advancing the state of the art. Firms should keep this in mind when considering membership in a consortium, regardless of whether it includes government, university, or exclusively industry participation.

The benefits of collaboration tend to be long-term in nature and are quantifiable primarily in terms of systemic improvements and tacit learning, rather than being based solely on discrete commercial outputs.[100] Firms should avoid justifying participation based on expectations of patents, papers, or prototypes. The transfer of explicit knowledge will tend to be a minor part of the total value of collaboration.[101]

The U.S. government has attempted to emulate successful strategies in Europe and Japan with respect to diffusion and technology transfer. Unfortunately, the policy results have fallen short of the superior models on which they were based. This is largely a result of inertia: it is very difficult to establish a new federal program that fails to fit within existing structures and organizations. Since the magnitude of a policy change is often inversely proportional to its probability of adoption, new solutions always tend to resemble those already in place.[102]

If government is to play the role of catalyst for industrial R&D collaboration, it must commit patient financial resources (meaning that multi-year funding is essential), and must avoid reactionary swings in technology policy to build industry confidence. There is no question that federal policy should favor the formation of industrial R&D consortia. Whether government should play an active role in industrial partnerships is dependent on enacting stable policy measures that respect the boundaries of commercial enterprise.

There are several ways in which collaborations can form, including any combination of government agencies and laboratories, universities, and firms. Some consortia have been founded on the basis of a specific mission. Such demonstration projects have helped to accelerate commercialization and adoption of new technologies when technical uncertainties were low and cost-sharing arrangements could be established between all participants. Examples include energy demonstrations, waste disposal, recycling, and other "pilot scale" facilities.[103]

In some cases, government agencies can be instrumental in forming collaborative agreements, such as the Partnership for a New Generation of Vehicles. In this case, the original emphasis was on securing participation of the Big Three automakers. Recently, as an acknowledgment of the growing importance of SMEs, an effort has been made to include first and second-tier suppliers in this project as well.

The National Electronics Manufacturing Initiative (NEMI) is another example of government serving as a catalyst for cooperative research and development. The NEMI was created to ensure sustained growth and competitiveness in the U.S. electronics industry, and includes in its membership OEMs, suppliers, and key government agencies. Activities have included the creation of technology roadmaps, technology-gap analysis, and specific research on such important technologies as flip-chip packaging and energy-storage systems.[104]

Although government sponsorship of consortia is often a necessary first step, it is essential that industry recognize opportunities for collaboration on its own. Early examples of privately initiated R&D consortia include the Microelectronics and Computer Technology Corporation and the Software Productivity Consortium. Diffusion of standards has been facilitated by such industry-specific organizations as Bellcore, and the general level of manufacturing expertise has been raised considerably by such diffusion-oriented groups as the Association for Manufacturing Excellence (AME).

The available evidence suggests that cooperative R&D is a critical factor in long-term technology leadership, but it is not clear how best to instill such a culture. Is it appropriate for public funds to be allocated to overtly commercial activities, even if the economic justification is compelling? What criteria should be used for funding decisions? Who should be allowed to participate, and how can "free riding" be avoided?

There are no easy answers to these questions. There is, however, at least one example of a government-funded industry consortium that may offer some insights. SEMATECH is by far the most studied and scrutinized collaboration in history. Although controversy abounds, some important lessons can be derived from this groundbreaking experiment in U.S. industrial policy.

### The SEMATECH Experiment

The semiconductor industry has truly been a pioneer in developing collaborative ventures. The establishment of the Semiconductor Industry Association in 1977 represented a landmark in technology-intensive enterprise; for the first time a high-tech industry overcame its fragmented structure to speak with a common voice. From this initial cooperative effort flowed other collaborations, including the Semiconductor Research Corporation, which was chartered to perform long-range research and develop highly qualified technical personnel.[105]

In 1987, fourteen U.S. semiconductor firms established the Semiconductor Manufacturing Technology (SEMATECH) consortium, funded in equal parts by industry and government. The initial agreement with Congress allocated $100 million per year for five years, after which the group was to proceed entirely on industry funding. Although it took somewhat longer than originally planned, the consortium informed the U.S. government in 1996 that it would no longer require federal funding.[106]

The SEMATECH consortium was founded around three strategies: to improve manufacturing processes, to enhance industry infrastructure, especially the supply base of equipment and materials, and to improve the management of semiconductor factories.[107] To extend the benefits of SEMATECH to a broader group of equipment, materials, software, and service suppliers, a parallel consortium called SEMI/SEMATECH was formed to serve as a link between SEMATECH member firms and the semiconductor industry. Almost half of SEMATECH's funding has gone toward aiding the diverse community of small infrastructural suppliers, rather than directly benefiting the fourteen member companies.[108]

One of SEMATECH's primary goals was to purchase semiconductor-chip manufacturing equipment from domestic suppliers, establish a leading-edge wafer-processing facility, and subsequently transfer the knowledge gained from this experience to members and the industry at large. This objective was subverted somewhat in the early 1990s, with SEMATECH becoming the "buyer of first resort" for many U.S. equipment manufacturers. In 1991, for example, expenditures on supplier equipment totaled more than $130 million, well over half of SEMATECH's annual budget.[109]

During its first few years of operation, SEMATECH was in organizational disarray. Issues of leadership, direction, priorities, and methods virtually paralyzed the consortium. In a real sense, the founders of SEMATECH were breaking new ground. It is a testament to the strength of early leaders such as Robert Noyce that the agreement did not collapse in those early years. What ultimately prevailed was a renewed sense of commitment; despite many obstacles, members recognized the critical importance of their mission, and aggressively sought ways to cooperate effectively.

It is safe to say that after the first five years of operation, a majority of outside observers (and even several members) judged SEMATECH to be a "qualified failure." Some noted economists asserted that by excluding the membership of foreign firms, SEMATECH was adopting a dangerous mercantilist philosophy. Others suggested that SEMATECH had become no more than a channel for subsidy to equipment manufacturers, who despite massive support showed few signs of recovery in the early 1990s. These critics, however, were missing two vital elements: perspective and patience.

As of this writing, SEMATECH appears to have been at least partially vindicated. Consortia take time to develop; members must learn to work together, and leaders must explore options and grow from their mistakes. To expect a billion-dollar organization of powerhouse corporations to come together successfully in a few short years is unrealistic. Now that the price of initial learning has been paid, however, SEMATECH should prove to be an invaluable mechanism for attacking future competitive challenges.

According to the current director of SEMATECH, William J. Spencer, "Our goal was that semiconductor manufacturers would use at least 50 percent domestic equipment, and today it's well over 60 percent."[110] In 1996, member companies reported a 400 percent return on their investment. As a result, the current membership has voted to increase dues by 30 percent. One SEMATECH member noted that it paid more than four times the government's share of funding in taxes that same year.[111]

These glowing results are due in large part to a booming economy and an industry-wide recovery, but member firms appear committed to a sustained relationship. In fact, the SEMATECH structure has prompted imitation by consortia involved in textiles, automobiles, and batteries. The Japanese have adopted many of the same objectives and methods for use in their Sortec consortium, as well as several other collaborative agreements.[112]

The most controversial issue regarding industrial consortia involves the dangers of *free riding*, either by smaller firms or by foreign participants.[113] SEMATECH has never allowed membership by foreign firms, ostensibly to create a competitive advantage for American industry. However, as William Spencer observes, "The issues we face today, whether related to economics, the environment, or health, are planetary issues." Indeed, there are several opportunities for cooperation between SEMATECH and foreign counterparts in areas such as standards for twelve-inch silicon-wafer equipment, and environmental issues involving the disposition of solvents used in the manufacturing process. Perhaps these initial forays will eventually lead to a "global SEMATECH," with participation open to firms from all interested nations.[114]

The Japanese have made significant moves in this direction. The Intelligent Manufacturing Systems Project was intended from its inception to be an international consortium. Interestingly, the Japanese are no strangers to the problems of free riding. They have had experiences similar to those of U.S. consortia, including members who contribute second-rate researchers and equipment, and resistance to full disclosure of corporate technological know-how.[115]

The evidence from SEMATECH suggests that industry consortia can play a valuable role in encouraging cooperation and increasing technological competitiveness, provided that time is allowed for the membership to learn and grow together. Governments can play an instrumental role in reducing the barriers to the formation of consortia, through temporary allocation of matching funds. More important, however, government can offer assistance in controlling opportunistic behavior in collaborative R&D, thereby mitigating the single greatest deterrent to cooperation among competitors.[116]

## Forming Productive Human Capital

As the nature of products evolves toward higher levels of knowledge content, industry strategies must shift from an emphasis on plants and equipment to the development of *human capital*. Economists coined the term 'human capital' to demonstrate that the skills, education, and creativity of individuals contribute to productivity in much the same way as physical capital does. Yet describing such assets as tacit knowledge and lifelong experience in sterile economic terms belies the complexity and subtlety involved in forming productive human capital.

This challenge is heightened in technology-intensive enterprise. Public understanding of issues related to science and technology is surprisingly poor. In Carl Sagan's book *The Demon-Haunted World: Science as a Candle in the Dark*, he wrote, "We've arranged a global civilization in which most crucial elements profoundly depend on science and technology. We have also arranged things so that almost no one understands science and technology."

This dichotomy between an expanding need for skilled technical workers at all levels and a public that appears to be ambivalent toward fulfilling that need is most evident in the United States. When American twenty-three-year-olds were rated for their performance in mathematics and science against Koreans, Canadians, and Europeans of the same age, they finished last on every test.[117] Without an ample supply of trained minds, the U.S. technology enterprise is in danger of suffering the intellectual equivalent of diminishing returns to scale.

### The Importance of Human Capital

The first step toward understanding the tremendous importance of human-capital development is to consider what it means to live in a technology-driven economy. On the demand side, there is ample evidence that technology-based products are being pulled into the marketplace by broad-based consumer preference. Ironically, Americans seem far more comfortable with the benefits of innovative products and services than with the growing technical content of their jobs. This raises concerns that the supply side of our technology-driven economy may soon be starved for qualified human capital.

The most common fallacy with respect to this "new economy" is that it is fueled by an elite cadre of technological experts. At the cutting edge of innovation this assumption may be true, but for America to achieve sustained, broad-based growth, virtually every worker must develop new skills at higher levels of technical sophistication. Clever inventions represent the seeds of economic expansion. For growth to be achieved, however, a nation's workforce must be capable of adopting and utilizing these productive innovations. A personal computer, for example, provides little economic benefit without the learning of new technical skills by secretaries, engineers, and executives.

Today, machinists must not only understand production processes, but must also be adept at computer numerically controlled (CNC) machine programming and perhaps even understand computer-aided design and computer-aided manufacture (CAD/CAM). Likewise, shipping clerks, accountants, shop-floor workers, truckers, inspectors, and a host of other previously "nontechnical" labor categories must now contend with rapidly changing technology.

Given the enormous attention that has been paid to technology industries over the last twenty years, it is surprising that Americans have not taken up science and engineering in larger numbers. A recent study of twenty-four-year-olds from several countries indicates that while the United States leads the world in the percentage of first-level college degrees awarded, it lags behind at least seven other countries in the percentage of this age group with degrees in technology-related fields, as shown in Figure 7.6.

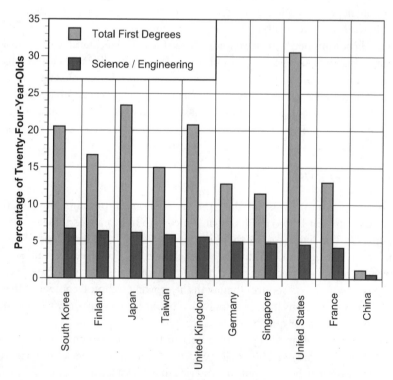

**Figure 7.6:** Percentage of twenty-four-year-olds with a first-level college degree for various countries in 1992. Source: National Science Foundation (1996, Appendix Table 2-1).

The situation is more optimistic with respect to doctoral degrees, as might be expected given America's historical lead in advanced scientific education. The United States produces more Ph.D.s than any other nation, and more science-related doctorates than its three nearest rivals combined, as shown in Figure 7.7. Yet embedded within these apparently positive data are grounds for long-term concern. The number of degrees awarded by American universities includes a far higher percentage of foreign students than any other country. In fact, almost one-third of all science and engineering doctoral degrees completed in 1992 were awarded to foreign citizens, as shown in Figure 7.8.

Furthermore, universities in the newly industrialized nations are producing a higher percentage of technical degrees than American institutions. Although their numbers are still relatively small, this emphasis on technology education as the pathway to economic success has become deeply instilled in their cultures. As high-growth nations such as Taiwan, South Korea, and even China expand their economic influence, they may come to dominate the world's supply of technically skilled workers as well.[118]

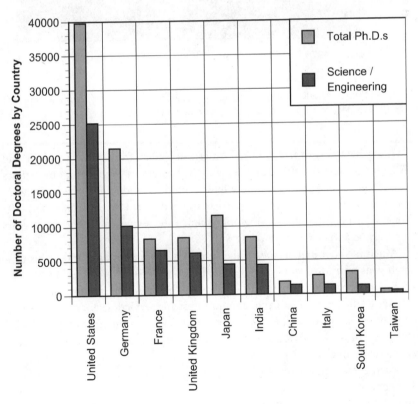

**Figure 7.7:** Number of doctoral degrees earned by country in 1992. Source: National Science Foundation (1996, Appendix Table 2-31).

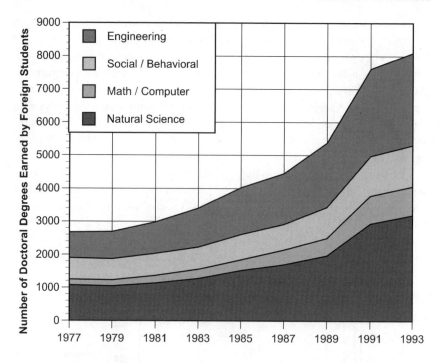

**Figure 7.8:** Number of doctoral degrees earned by foreign citizens in the United States in various fields, 1977 to 1993. Source: National Science Foundation (1996, Appendix Table 2-29).

A recent survey of five hundred American leaders from both the public and private sectors revealed that education was their foremost concern for our nation's future.[119] Since the early 1990s there have been dire predictions of a nationwide shortage of highly skilled workers, particularly in the fields of electronics and software.[120] The persistently low unemployment rates experienced in recent years seem to substantiate these concerns.

Much of the initial debate on the subject of labor shortages and "skills mismatch" was prompted by a report entitled *Workforce 2000: Work and Workers for the Twenty-First Century.*[121] In this document, the "glut-versus-shortage" discussion was framed in the context of a rapidly growing services sector and a continuing drain of scientists and engineers into the defense industry. Indeed, roughly 18 percent of America's technological workforce was consumed in jobs directly related to national security in 1992.[122]

Furthermore, more than half of U.S. industrial science and engineering employment is in service-related jobs. With the services sector expected to expand rapidly, particularly in technology-intensive industries, predictions of a skilled-labor shortage are becoming increasingly credible.[123]

Interestingly, the development of human capital is the most "national" facet of technology policy. With raw materials and investment capital becoming increasingly mobile, skilled labor now represents a truly advantageous endowment for any nation. Moreover, the economic and social benefits of public investment in education and training are unparalleled.

In the days of American hegemony in both scientific education and technology industry, foreign nationals who received degrees from U.S. institutions could often be enticed to remain in our workforce. Today, there is a rising nationalism among expatriate students from Taiwan, Korea, and other industrializing nations. Not only are these graduates being drawn to opportunities in their homelands, but an increasing number of U.S. citizens are being attracted overseas by exciting opportunities as well. This "reverse brain drain" could become the most contentious aspect of American technology policy.[124] Is it reasonable for industry and government to invest in the technical training of foreign citizens? Likewise, should we be considering policy measures to help retain highly skilled American citizens in domestic employment?

As our technology-driven economy matures, the desire to capture economic spillovers from U.S. university education and industry training may drive us toward a new form of protectionism: the creation of barriers to the migration of human capital.

### Teaching the Right Skills

With few exceptions, the American system of education has remained essentially unchanged for more than a century. Although primary and secondary education are ubiquitous, the K-12 system in the United States is designed to impart only generic skills in reading, writing, and math. A high school diploma is perfunctory, offering little in the way of marketable skills or specialized training.

Conversely, college degrees are treated with disproportionate respect by employers, and have become essentially mandatory for any serious career in technology. As a result, a caste system has developed in our workforce that elevates the stature of college-level education and relegates the lowly high school graduate to menial duties and limited responsibilities. This two-tier structure retains a striking resemblance to the scientific-management philosophy of Frederick W. Taylor, in which the white-collar elite directed industrial activities, while the blue-collar masses performed repetitive tasks. Indeed, our current educational system predates this outmoded model of industrial organization.

Clearly, it is time for America to reevaluate its processes for human-capital formation, in light of the dynamic nature of our technology-driven society. We can no longer be satisfied with minor "tweaks" to the current system; major adjustments are needed to the types of educational programs that are available and the fields of study that are emphasized.

For our economy to continue its current expansion, the needs of industry and the skills acquired through formal education must converge. In Germany, for example, the growing demand for highly skilled technologists is serviced through a wide variety of institutions, ranging from polytechnics (known as *Fachhochschulen*) to vocational academies and technical trade schools. The duration of these programs varies according to the level of skills they impart. An average of 4.5 years of post-secondary education is typical for the polytechnical institutions, while the trade and technical schools offer a pragmatic three-year program as an alternative to upper secondary education. Graduates from formal educational programs in Germany are often placed directly into industry, where they receive further apprenticeship training.[125]

The primary benefit of this broad range of educational options is that students with differing abilities can be guided toward high-productivity employment. Those individuals with a poor affinity for numbers can still be given high levels of technical training, although at an applied level. Meanwhile, students with a facility for theory can advance through more traditional degree programs.

Similarly, Japan has supplemented its traditional four-year and advanced university degree programs with a network of technical colleges and junior colleges that provide the equivalent of a two-year associate degree. Once graduates find industry positions, their learning continues through rotation and mentoring programs. The Japanese have long recognized the importance of tacit learning, and emphasize the transfer of "wisdom" from senior managers and technical specialists to freshman employees.[126]

In the United States, there has historically been a stigma attached to skilled workers with "vocational" training. It is important that we overcome our fixation on traditional college degrees and embrace a spectrum of options for specialized technical education. Moreover, individuals who receive this training should be granted both the compensation and respect warranted by their valuable contribution to industry.[127]

Even within the established college curriculum there is a growing need for broader advanced education in support of technology enterprise. Corporate law practices are unable to meet the demand for specialized legal expertise in patent and copyright law, high-technology mergers, and product liability. Firms are snapping up graduates who express an interest in these fields, and are increasingly being driven to provide high-quality post-degree training in these specialties for their new associates.[128]

College degree programs have become highly specialized, particularly in the sciences and engineering. A physics student, for example, will likely have little exposure to the more "practical" engineering disciplines. Likewise, among the engineering fields it is rare for a student's education to span more than one specialty.

As technology-intensive industry evolves, there will be a growing need for technologists who are skilled in multiple disciplines. Many recent innovations are at the crossroads of several advanced technologies, and the ability to partition systems and perform tradeoffs among these fields will be highly desirable. Although industry can provide opportunities for multidisciplinary training, it is appropriate for universities to encourage this expanded form of education as well. Some important areas that are not given adequate attention include manufacturing and industrial engineering, concurrent product development methods, process engineering, design for manufacture, reliability engineering, design for unit cost, and design tradeoff analysis.

Moreover, every engineer or applied scientist who graduates from an American university should be required to participate in some form of hands-on training. Human-capital development must emphasize the acquisition of tacit skills in parallel with exposure to a structured curriculum. Work-study programs, industrial-liaison positions, summer-hire programs, and industrial-rotation opportunities are examples of existing mechanisms that could be expanded in the future. By incorporating practical experience into a formal education program, all parties will benefit. Industries will be forced to develop better relationships with local institutions, educators will be incentivized to make their teaching more relevant, and students will have the opportunity to explore career options in a supportive environment.

The final issue that must be considered as we retool our educational system is the manner in which we measure academic performance. There is a growing controversy over the use of educational standards for evaluating individual performance. While the need for feedback during the learning process is vital to successful education, there is a growing sentiment among technical leaders that the way we currently evaluate students is virtually meaningless. Rather than testing for cognitive reasoning and problem-solving skills, students are currently subjected to a barrage of multiple-choice questions that require memorization and word association. Since these examinations are used as important criteria for college entrance, are we selecting our next generation of knowledge leaders based on test-taking ability rather than thinking ability?[129]

### Providing a Lifetime of Productive Learning

Who is responsible for the formation of human capital in the United States? Historically, the federal government has accepted this role willingly, providing for K-12 education for all citizens and sponsoring the proliferation of universities and colleges to all fifty states. Yet the government typically draws the line at specifying what should be taught, leaving the development of curricula to the educational institutions themselves. Unfortunately, American universities have only recently begun to recognize the shortfall in their programs with respect to the needs of industry.

It should not be surprising that the responsibility for formation of human capital must be shared cooperatively by industry, universities, and government. The goal of this collaboration is straightforward: Develop the American workforce into the most productive, creative, and versatile in the world. Identifying the steps necessary to achieve this objective, however, presents some interesting challenges.

One of the reasons why the issue of human-capital formation has become so critical in recent years is that the rate of change of technology is accelerating. By comparison, education and training move at a frustratingly slow pace, being inherently limited by an individual's ability to absorb new knowledge. As I indicated in Chapter 4, a mind that has been "preconditioned" to learn can absorb new material much more rapidly. Hence, the early and continuous development of learning skills must be a central theme of U.S. education and training programs.

There is no aspect of technology policy that is more appropriate for government involvement and public investment than the expansion of our base of skilled technology workers. In particular, the U.S. government has proved to be effective at encouraging broad participation in advanced education and training initiatives. After World War II, the federal government undertook several programs to democratize higher education. The G.I. Bill, for example, provided substantial financial subsidies to all veterans who wished to attend college-level institutions. This program represented history's largest single investment in human-capital development, and has been given credit for fueling much of America's stunning postwar economic boom.[130]

In more recent times, the Clinton administration has been consistent in its support of new education and training programs. The Goals 2000: Educate America Act of 1994 establishes targets for increased competence in science and mathematics, along with programs for adult education and retraining. The School to Work Opportunities Act of 1994 supports the founding of state and local programs that provide work experience and apprenticeships for younger students. Finally, the Clinton administration has actively supported the creation of an educational information infrastructure under the High Performance Computing Act of 1991. This bill authorizes the establishment of a National Research and Education Network (NREN), which would be cooperatively funded by the public and private sectors and would serve as a knowledge-delivery system and research asset for schools, universities, government, and industry.

Although the enthusiasm displayed by the current administration is admirable, some of the targets for the American educational system may be unrealistic. One of the stated goals of the Clinton technology agenda is to achieve world leadership in basic science, mathematics, and engineering. Given the current condition of our K-12 educational system, and the declining performance of our young college graduates in science and math, it would seem more realistic to aim for *broad-based competence with areas of technological excellence.* We should avoid hyperbolic political rhetoric in this vital matter, and admit that the United States cannot lead every nation in every category of human-capital development. If we fail to focus our efforts on those fields of knowledge that best serve the demands of industry, we are in danger of developing a population of workers who are mediocre at everything.

While government can play a major role in reducing the barriers to higher education, the nature and quality of that education is clearly the responsibility of our universities and colleges. America has the most diverse and capable network of universities and colleges in the world, serving every populous region of the nation. With this immeasurable asset at our disposal, we are well positioned to achieve the educational goals described above.

This is not to say that our system of higher education is without problems. There is still a certain "priest-class" mentality on the part of some professors that limits their receptiveness to more applied forms of learning. Although the link between education and research is stronger in U.S. universities than it is in Europe, the potential to produce "irrelevant Ph.D.s" is still very real. The ambivalence displayed by some educators could be diminished by instituting post-tenure review programs for all senior faculty, in which their ability to produce productive and employable doctoral candidates would be reviewed.

Furthermore, an increased emphasis should be placed on master's degree programs in science and engineering. Unless graduates plan to spend their lives performing laboratory research, the master's degree provides all the formal learning needed to succeed in the most demanding technology career. Work-study master's programs are ideal for transitioning an advanced student from the arcane world of academia to the practical realities of gainful employment. Similarly, participation by graduate students in university-based engineering centers can provide the balanced perspective necessary to excel in commercial industry.

While joint industry/university research programs have helped to reorient the curricula of advanced education, the direct role of industry in the development of human capital cannot be overstated. Formal education can train the mind, but it cannot instill the specific and detailed knowledge necessary for a worker to become an integral contributor to a firm's success. Hence, industry training represents the last, and perhaps the most neglected, aspect of America's human-capital development system.

Firms in the United States spend between $30 billion and $40 billion annually on training programs to upgrade worker capability and productivity, as shown in Figure 7.9. These investments range from incidental skills development, such as electrostatic discharge (ESD) practices and safety training, to advanced (and sometimes even accredited) formal coursework, such as that offered by the Motorola Executive Institute and the Motorola Training and Education Center.[131]

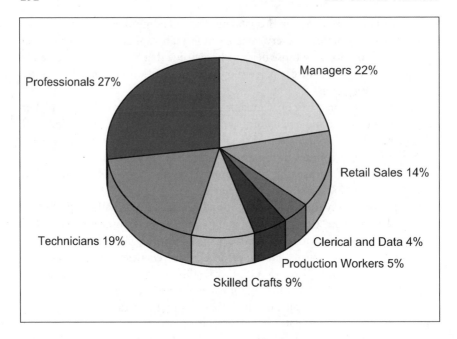

**Figure 7.9:** Distribution of U.S. private-sector expenditure on formal training in 1990. Source: National Center on Education and the Economy (1990, pg. 49).

High-technology companies lead the field in mid-career training, spending an average of $911 annually per employee.[132] Although this is the highest expenditure of any U.S. industrial sector, it is not nearly adequate to keep pace with the rapidly changing technological and social environment. As a percentage of employee wages, the current levels of investment represent only a 1 percent commitment to human-capital improvement, far lower in relative terms than firms spend on maintaining factory equipment or upgrading computer systems.[133]

American firms must assume responsibility for continuing the learning process initiated by public schools and the university system. In-house education must be elevated from an incidental activity to a strategic mandate; every long-term action of the firm should be considered in the context of learning opportunities and competency development. The accumulated knowledge of a firm's workforce is both vital and disturbingly transitory. Human capital depreciates rapidly; without constant renewal, the very lifeblood of technology-intensive industry will fade into obsolescence.

# 8
# Regional Advantage

## The Sources of Regional Advantage

Today, enterprise stands at the boundary of two economic epochs. The industrial age of the nineteenth and twentieth centuries is rapidly giving way to a new era, in which bricks and mortar are being supplanted in strategic importance by intellectual capital and technological capability. This industrial transformation calls into question some of the most fundamental tenets of business strategy.

In the past, large-scale factories were located near natural resources, transportation links, or sources of cheap labor, while R&D activities were centralized and kept close to corporate headquarters. With the recent explosion of telecommunications and information technologies, however, it has become fashionable to think in terms of a "borderless world" in which economic activity occurs on a global basis, with little regard for geographic proximity. This perspective is strategically naive.

As other factors of production become more equalized among the advanced economies, regional advantages may actually be increasing in strategic importance. Investment capital, transportation systems, industrial equipment, and low-wage labor are available on every continent, whereas specialized knowledge and tacit skills are becoming highly concentrated and localized. The ability of firms to access and exploit these clusters of technological sophistication will become a determining factor in the competitions of the next century.

### Why Do Industries Cluster?

The tendency for firms in like industries to gather in close geographic proximity was noted as early as the late nineteenth century. In *Principles of Economics,* Alfred

Marshall observed that positive externalities could be gained from locating factories in "industrial districts" where "the secrets of industry are in the air."[1] Today, the agglomerations of textile mills and steel refineries of Victorian England have been supplanted by the intense regional concentrations of technology industries in California's Silicon Valley, North Carolina's Research Triangle Park, and Texas' Telecom Corridor. The strategic motivations, however, are the same. Like moths drawn to a bright light, firms at the frontier of innovation instinctively draw close to create a critical mass of specialized, industry-specific factors.

Regional advantages span a broad spectrum from the relatively mundane to the exotic, and can be divided into factors that are governmental and geographic in nature. A tax-incentive structure is an obvious example of a governmental factor that can impact a firm's decisions on facility location, while proximity to a respected university is an example of a purely geographic consideration. The list of potential benefits is long and varied, as shown in Table 8.1. The common element is that each regional factor imparts a small but persistent competitive advantage to the firms that partake of it.

At the heart of regional advantage is the potential to increase rates of innovation and productivity growth. Although it is possible to perform transactions with firms located anywhere in the world, it is incrementally easier to execute these transactions when firms are nearby. This benefit may be insignificant when engaging in "turnkey" transactions such as procurement of commodity materials, but it becomes vitally important during the rapid development of a complex, technologically sophisticated product. The competitive environment in global technology markets is a "game of inches" in which the frictional losses associated with collaboration across time zones, cultures, and geographic distances can mean the difference between being a first mover and a not-so-fast second.

The notion that industries with related needs for infrastructure and specialized factors will gather together in close proximity has prompted a new area of economic research known as *cluster theory.*[2] The formation of *technology clusters* was first studied in detail by a French research group, which suggested that so-called *filiere* were a result of competitive forces in a highly innovative environment.[3] Over the last decade the literature has expanded considerably, motivated in part by the massive study of national and regional advantages described in Michael Porter's *The Competitive Advantage of Nations.*

The most fundamental observation that can be made about the concept of technology clusters is that they are self-reinforcing. Once established, firms from other regions will begin to recognize that they cannot compete without equal access to the complex "stew" of specialized resources available in a thriving cluster. As the cluster expands, it will begin to segregate into finer degrees of specialization. This will tend to improve the aggregate productivity of all firms participating in the regional economy. Hence, a positive feedback loop is established, wherein

## An Overview of Regional Advantages

### Increased Rates of Innovation

- Close proximity to suppliers and customers
- Optimal environment for design collaboration
- Regional rivalry provides creative tension
- Technical problems can be solved through "informal network"
- Rich pool of contract professionals, specialty suppliers, etc.

### Improved Appropriability of Returns

- Opportunities to adapt products to new markets or applications
- Potential for supplier collaboration on risky R&D activities
- Participation in local alliances to protect intellectual property
- Horizontal industry structure can reduce capital outlays and other non-recurring costs of product development
- Possible tax incentives and other local inducements
- Concentration of venture capital and "seed money"

### Enhanced Absorption and Accumulation of Knowledge

- Close proximity enables transfer of tacit knowledge
- Skilled human capital from other firms or regional institutions
- Opportunity to access "regional informal network"
- Opportunities to monitor related industries
- Potential to form information-sharing alliances

**Table 8.1:** Overview of several competitive advantages that may be gained through proximity to a regional technology cluster.

firms will demand increased rates of innovation from suppliers and strategic partners, which will drive all participants to higher levels of creative efficiency.

The competitive advantages of technology clusters can be divided into four categories: human resources, technology infrastructure, knowledge resources, and capital resources.[4] The availability of specialized human capital may be the most important single benefit of physical proximity to related firms. As I have emphasized previously, the accumulation of the unique tacit knowledge of experienced workers can provide a sustainable advantage for firms. Moreover, these specialized resources are far less mobile than other assets.[5] Hence, it is highly advantageous to locate a facility near a spawning ground for technically excellent human resources.

There is a countervailing concern that firms located in close proximity to their rivals may lose valued employees. Indeed, the turnover of human resources in technology regions such as Silicon Valley is significantly higher than for firms located in relative isolation. However, these high rates of turnover provide a unique environment for the development of human capital, through the varied experiences gained by these itinerant professionals. Although the retention of valued employees must be given extra attention in such hotbeds of competition, all companies will gain from the accelerated pace of tacit-skill development.[6]

The second category of regional advantage addresses the availability of technology infrastructure that can support rapid innovation. Access to physical assets such as specialized laboratory equipment and testing facilities can be an expediting factor in the product development process. More important, however, is the presence of a pool of talented suppliers available for collaborative engineering projects. Most advanced products embody the innovative contributions of several specialty suppliers, which are coordinated through the system-engineering function of the end-product manufacturer. During the early phases of product design and process development, close proximity can reduce the time and risk associated with design reviews, specification changes, prototyping, qualification testing, and process validation.

Many of the technology clusters that have evolved over the last three decades have formed around the knowledge resources of major research universities. Silicon Valley was built on the intellectual outflows from Stanford and the University of California at Berkeley. Likewise, the clustering of high-technology firms along Boston's Route 128 was seeded by early collaborations with (and many spinoffs from) the nearby Massachusetts Institute of Technology (MIT).

Once a technology cluster is established, access to spillover knowledge from nonacademic institutions becomes significant. Industry associations, regional trade bodies, and local chapters of professional societies all become hotbeds of intellectual interchange. Generic problems can be rapidly solved through this network of like minds, and needed assets and capabilities can be readily located. Although this may seem to contradict "old-school" thinking with respect to protecting proprietary information, the benefits of sharing high levels of technological expertise far exceed the risks.

Finally, the availability of investment capital within a technology cluster can be a critical factor in rates of innovation and aggregate productivity. Venture capital provides the fuel for technological progress at its most sensitive point: the inception of a new idea. Regional advantage works in two synergistic ways to provide this vital seed money. First, by nurturing a sea of specialty suppliers, contract professionals, consultants, and equipment manufacturers, a technology cluster dramatically reduces the cost of entry for new entrepreneurs. Second, the high rates of innovation present in a healthy cluster will attract the favorable attention of investors from around the globe.

When taken collectively, the competitive advantages associated with participation in an appropriate technology cluster are undeniable. What is not clear is the evolutionary life cycle of these regions. How are they born? Can they be created through the actions of governments or industry? Finally, and most important, can the positive feedback last indefinitely, or do technology clusters by their very nature contain the seeds of their own destruction?

### The Evolution of Technology Clusters

Of all the possible locations that could have supported the generic requirements of high-technology industry, why did California's Santa Clara Valley give birth to the world's premier center for technological excellence? The available evidence suggests that the formation of technology clusters requires a "seed crystal." This might take the form of a major university, a large and successful industrial firm, or a government laboratory.

Regardless of its form, the seeding institution contributes to cluster formation in two ways. First, it creates a pull effect, drawing industries in related fields into close proximity to bask in its knowledge resources and economic opportunities. Second, it accelerates the horizontal specialization of the region by spinning off start-up firms in large quantities. Provided that the seeding institution has sufficient "energy" to get the process started, a cluster will begin to snowball on its own.

In the case of Silicon Valley, the seeding institution was initially Stanford University, followed in later years by industry giants such as Fairchild and Hewlett-Packard. For the Route 128 region of Massachusetts, similar roles were played by MIT, Digital Equipment Corporation (DEC), and Raytheon. Each of these entities meet the two criteria described above: They attracted industrial activity through their preeminence in specialized technologies, and they aggressively cast off smaller start-up firms headed by entrepreneurial professors and industry executives.

The development of a technology cluster can be characterized by two stages: convergence of related firms and institutions, and decentralization of activities into higher levels of specialization.[7] This process can take years to mature and, much like biological evolution, can be easily disrupted. Sustained market demand for the region's specialty products is an essential element in cluster growth. It is also essential that the technology upon which the cluster is based remains at the core of multiple technology sectors. Semiconductor devices, for example, represent a pivotal technology for a plethora of unrelated industries.

In recent years, the dramatic increase in strategic alliances among firms has been a catalyst for the formation of industry clusters. This effect is particularly important in explosive sectors such as information technology and telecommunications. A recent study of alliances in information technology demonstrated that firms in this industry derived a significant competitive advantage from forming complex networks of cooperation among rivals. The benefits of these collaborations included the ability to capitalize on economies of scope and increased access to diverse technical specialties.[8] Although these "clusters" of alliances were not always geographically concentrated, the growing importance of cooperation and collaboration greatly facilitates the formation of regional clusters.

As a technology cluster matures, the availability of specialized factors will tend to broaden and deepen. Local universities, community colleges, and trade schools will begin to tailor curricula to the needs of the cluster. Similarly, service providers will emerge to handle the unique requirements of the cluster's core technologies. The design, fabrication, maintenance, and support of cleanrooms (which provide a dust-free environment for semiconductor production), for example, have become a minor industry cluster in its own right in the Santa Clara Valley.

It is this deepening of specialization that could hold the seeds of destruction for technology clusters. As long as markets can continue to support such a complex network of interdependent enterprises, the cluster will achieve world-leading productivities. If the demand for end products falters, however, the result can be catastrophic.

Technology clusters require enormous economic "energy" to sustain their highly specialized structures. Larger original-equipment manufacturers (OEMs) can typically weather the cyclic nature of economies and markets, but the small, highly specialized supplier firms are at far greater risk. Even a brief loss of momentum can induce a cash-flow crisis among this vital infrastructure of suppliers. A major downturn could cause an irreversible collapse.

When observed in the long view, technology clusters may ultimately be destroyed by the very innovation that feeds them. The Route 128 cluster of minicomputer companies (including DEC, Data General, Prime, and others) was undermined by the microcomputer revolution. Likewise, mighty Silicon Valley was critically wounded by fierce competition in commodity semiconductors from the Japanese and South Koreans.

Yet both of these regions have rebounded.[9] The elements that allowed their formation are still present, even if the direction of technology has taken an unexpected turn. Some suppliers, and even a few major corporations, might not survive such dislocations, but the valuable core of human capital and technology infrastructure will not be denied. New entrepreneurs will sense opportunities, and declining clusters will reinvent themselves. A regional cluster of tacit skills and accumulated knowledge is far too rare a resource in our technology-driven world to be allowed to languish for long.

### Can Governments Create Clusters?

If there were such a thing as a fad within global technology policy, it would be the attempts by governments to create technology clusters. This is not so much the case within the United States, but the promotion of science parks and industrial regions has become a major thrust among newly industrializing countries, as well as within many of the advanced economies.

The reasons why national governments would adopt such policies are clear. Technology clusters offer an excellent opportunity for nations to appropriate a disproportionate share of economic spillovers and other positive externalities. The intensive competition and cooperation present in a cluster creates a vortex for specialized knowledge. Firms in close physical proximity will tend to share information and resources at a much higher level than similar firms that are geographically dispersed. Like a magnet for technological competence, a vibrant technology cluster will draw talent and investment into its domain, thereby guaranteeing high levels of economic benefit for the entire region.

Given the obvious economic desirability of technology clusters, it is understandable that many countries are attempting to seed them within their borders.[10] These attempts are based on the presumption that clusters will form in regions that provide the specialized infrastructure required for technology industry to flourish. The operational model for such efforts can be traced back to such early examples as the Stanford Industrial Park, which established a critical connection between a "seeding" university and the business community during the formative stages of Silicon Valley.[11] The legendary success of this model has inspired widespread imitation in regions as diverse as Singapore, Scotland, and Israel.

In some cases these attempts at emulation have been successful. Singapore, for example, has exploited its investment in the Singapore Science Park to attract badly needed human capital from other nations. Expatriate Chinese researchers, in particular, have been targeted; these individuals have often acquired advanced degrees in the United States or Europe, and are not yet ready to accept China's less-than-lucrative opportunities.[12]

When taken as a whole, however, it is not clear whether government sponsorship has had a generative impact on the formation of technology clusters. It is overly simplistic to assume that a successful cluster can be launched by simply providing a tract of industrial space juxtaposed with a local university or government lab. Likewise, an agglomeration of high-technology firms on its own may not provide sufficient fuel to light the fire of economic synergy. The ingredients that must be present for a cluster to ignite are not fully understood, and many of the essential elements are not within the sphere of government control.

Yet many developing nations have committed to budget-busting projects aimed at creating the next Silicon Valley. In some cases, such as the Hsinchu Science-Based Industry Park in Taiwan, these investments have paid huge dividends, as described in Box 8.1. Conversely, the high-profile attempts by Malaysia's government to "construct" a technology cluster near Kuala Lumpur have yet to generate the faintest economic spark, as discussed in Box 8.2.

## Box 8.1: A Silicon Valley of the East[13]

The Hsinchu Science-Based Industry Park (HSIP), located an hour's drive southwest of Taipei, Taiwan, stands out among the plethora of science parks and technology clusters sprouting up in Asia's industrializing nations. A brief review of the companies it hosts reveals that the majority of resident firms are homegrown. Such rising stars as Winbond and Taiwan Semiconductor Manufacturing Corp. have carved out a place in the global semiconductor market, and play an increasing role as foundries for Silicon Valley's advanced chip-design houses. Although there are many similarities between HSIP and its legendary ancestor, their histories could not be more different. Whereas Silicon Valley congealed out of natural technological and market forces, HSIP was created as a deliberate act of Taiwanese technology policy.

HSIP was founded in 1980 for the express purpose of attracting the vast numbers of overseas Taiwanese scientists and engineers away from jobs in American and European high-tech firms and back to their homeland. The park encompasses more than five thousand acres of government-owned land in close proximity to two technically oriented research universities, National Chiaotung University and National Tsinghua University, and in the immediate vicinity of the Industrial Technology Research Institute (ITRI).

By the end of 1992, the government of Taiwan had invested roughly $500 million in land acquisition, while the number of participating firms grew from seven in 1980 to 148 by 1994. During its first ten years of existence, the park's number of facilities and employees grew at double-digit rates. In 1992, HSIP's total output of $3.35 billion represented one-third of the total output of Taiwan's computer, semiconductor, and telecommunications industries.

HSIP is the first technology cluster outside of the advanced nations that has been acclaimed as a model for other developing countries. This success was achieved through a unique combination of government oversight, accumulation of human capital, attention to demand-side factors, and a strong national spirit of entrepreneurism. As is the case with Silicon Valley, it is unlikely that such a subtle recipe can be easily copied. In the words of two well-known observers, "It is for each nation, each region, each city, to work out an appropriate strategy, with as much vision and as much imagination as it can conjure up."[14]

## Box 8.2: The Multimedia Supercorridor[15]

The idea of the Multimedia Supercorridor (MSC) is attractive: a sprawling regional technology cluster that combines world-class infrastructure with the lush equatorial settings of central Malaysia. The proposed industrial zone stretches from the current capital, Kuala Lumpur, to a vast new airport site thirty miles to the south. At the heart of the project is a 2.5-to-10.0-gigabit optical-fiber network, managed by a paperless bureaucracy. Ultimately, the development will include a new capital city named Putrajaya, and a "technology city" called Cyberjaya.

The MSC is the most recent brainchild of Malaysia's charismatic prime minister, Dr. Mahathir bin Mohamad, who has become a legend in Southeast Asia for driving his country into the modern age. The MSC is just the cornerstone of a grander scheme called Vision 2020, a plan to transform Malaysia into a global technology leader within two decades.

Ascension to "MSC status" is granted to firms that address seven broadly defined categories of "flagship applications." Thus far the irresistible incentives being offered, including a ten-year tax holiday and the waiving of selected import duties, have attracted more than forty companies, including Oracle, IBM, Sun Microsystems, and Microsoft.

Amid the media hype (everything about the MSC is either "smart," "super," or "excellent"), there has been little discussion of the challenges of creating a hub for the booming Asian telecom market. The vast majority of construction contracts for the MSC have been awarded to Malaysian companies on a "sole-provider" basis. This hardly inspires confidence, considering the dubious reputation of firms such as Malaysia TeleKom, the sole provider of the region's fiber-optic links. Perhaps the biggest risk facing the MSC is the seemingly irresistible tendency of developing countries to engage in cronyism and political favoritism.

Despite the recent nosedive of the ringgit, the $20 billion project has been given the nation's top funding priority. This substantial investment is justified by the promise of technology transfer, an essential ingredient if Malaysia is to become an innovation center in the twenty-first century. Unfortunately, in an all-out effort to attract the world's most powerful technology companies, Malaysia has imposed few mandates to ensure that firms will share their technology.

This problem is further aggravated by a lack of information-technology-literate workers prepared to absorb the technologies that are being so vigorously pursued. In the near term, this issue has been addressed by eliminating restrictions on the immigration of foreign scientists and engineers. In the longer term, Malaysia must create a pool of human capital capable of benefiting from proximity to the best and brightest.

Although governments are well positioned to encourage the rapid expansion of technology clusters, it is unlikely that they have either the resources or the knowledge needed to create them. This situation is similar to government attempts to "pick winners" through subsidy of domestic industries. In both cases, supply-side policies should be supplanted by well-conceived demand-side measures. Governments and firms should work together to define and develop regional infrastructure and human capital. Similarly, firms within a region must establish communication links at several levels, both to foster higher degrees of specialization within the cluster and to ensure its sustained growth.

In the sections that follow, I will describe a variety of technology clusters that are representative of the complex milieu that surrounds these unique economic entities. What will emerge is a common theme: Successful clusters are based on high levels of cooperation and synergy among firms, and between industry and public institutions. Each actor in this communal effort plays a vital role. Firms provide the focus for innovation and economic gains, institutions offer forums for cooperation and knowledge exchange, and governments fulfill the needs of industry through sensitive (but noninvasive) demand-side policy.[16]

## A Global Survey of Technology Clusters

The best way to understand the competitive advantages of technology clusters is to consider them from a purely human perspective. Unlike the industrial agglomerations of an earlier age that formed around valuable natural resources or major transportation links, today's clusters are founded on human factors and human interactions.

Nuclear physics provides an excellent (albeit somewhat unlikely) analogy for the explosive nature of technology clusters. The nuclei of radioactive material emit subatomic particles at a steady rate, due to the natural process of atomic decay. As long as there is not much of this material around, the process will benignly continue until the end of time. If, however, a *critical mass* of such material is brought together in close proximity, a startling event occurs. The subatomic emissions of decaying atoms collide with nearby atoms, causing a chain reaction of rapid and highly explosive decay.

Similarly, if firms with common interests and related technologies are widely dispersed, their spillovers of human capital and specialized knowledge are scattered to the four winds. Once these firms draw into close proximity, however, a critical mass is achieved, wherein the majority of spillovers are captured by neighboring firms. Thus, explosive economic growth is achieved, provided that an adequate source of "fuel," in the form of human capital and technical knowledge, can be maintained.

## Proximity as a Catalyst

Human-capital availability and knowledge spillovers are the primary motivations for firms to locate in close proximity.[17] This conclusion is borne out by strong evidence that the "emissions" from technology-intensive firms are geographically localized.[18] Physical location within a cluster, however, does not guarantee that firms will benefit from spillovers of knowledge and human capital. Likewise, a group of firms in close proximity does not automatically transmute into a latter-day Silicon Valley. Human factors are the source of economic growth within clusters, so it is logical that human factors also play a determining role in a cluster's economic vitality.

The gains from regional proximity can be achieved only through appropriate behavior on the part of both the "senders" of knowledge (in both tacit and explicit forms) and the "receivers" of that knowledge. An open and externally focused attitude on the part of firms, and more important, their employees, is critical. The most successful clusters have actively fostered a culture of cooperation, both within the regional firms themselves and among competitors, suppliers, and public institutions.

For a cluster to reach critical mass, a pervasive climate of trust, fairness, loyalty, and reciprocity must be present. Such a culture evolved to high levels in Silicon Valley, for example, but languished in Massachusetts' vertically structured Route 128 region. As a result, firms such as DEC (located in the Route 128 cluster) have noted that they have been able to establish closer relationships with Silicon Valley institutions and firms, located thousands of miles away, than they were able to achieve with their local counterparts.[19]

The most obvious human factor that impacts firms in technology clusters is that workers are relatively immobile. The availability of a pool of qualified, experienced, and tacitly skilled professionals is an almost irresistible attraction for many firms. Although turnover within a cluster may be high, employees have greater opportunities to find the "right fit" for their skills, thereby increasing both the productivity of firms and the satisfaction of workers. Clusters offer a unique opportunity for highly skilled individuals to continue their professional development without sacrificing the comforts of a stable personal life.

Beyond the human-capital-enhancing power of the major seeding institutions within a cluster, there is a wealth of alternative institutions for knowledge interchange and skills development. Mentoring and networking among professionals in a technology cluster are common, and often extend well beyond the boundaries of individual firms. There are also many social opportunities available, including charitable organizations, sporting events, clubs, and most important, active chapters of professional associations. Technology clusters are a magnet for talented people and a catalyst for human-capital development.

Firms within a cluster gain significant strategic advantage from close proximity to suppliers and rivals. Transaction times and costs are reduced across the board, providing a crucial edge in time-to-market. The ability to exchange information rapidly, and often face to face, can facilitate decision-making and avoid catastrophic misinterpretations of specifications or interfaces.

Firms that develop complex products benefit most from close proximity.[20] Sun Microsystems, Inc., a Silicon Valley resident, depended heavily on specialized suppliers to create its wildly successful SPARC workstations. After launching its enterprise through the innovative use of regionally available, commercial off-the-shelf components, it proceeded to fund the development of a custom chip set, including a unique RISC-based microprocessor. This work was performed in such close cooperation with Silicon Valley-based Cypress Semiconductor and Weitek Corporation that it was hard at times to identify the boundaries between the companies.[21]

Such close cooperation between suppliers and original-equipment manufacturers (OEMs) during advanced development is characteristic of the most successful technology clusters. Potential users of a new innovation can be brought into the development cycle early, both to contribute to the embodied form of the technology and to become presold on its advantages. As larger firms are attracted by the possibilities of co-development, a region will evolve from a source of low-level components to a center for complex, end-user products.[22] Silicon Valley has made this transition, originating as a semiconductor cluster and evolving into a center for advanced computer and networking hardware.

Small firms within a healthy technology cluster are provided with additional benefits. Many smaller companies lack the resources to purchase state-of-the-art design tools or specialized production equipment. Close proximity to firms in a similar situation offers the possibility of pooling important capabilities such as CAD/CAM systems. By sharing in both the cost and utilization of these important tools, small firms can compete more effectively with industry leaders.[23] Likewise, the ability for entrepreneurs to "one-stop shop" for experts, equipment, suppliers, and facilities significantly reduces the barriers to new and aggressive entrants.

Unfortunately, there are limitations to the growth of technology clusters. As a region spreads geographically, the cost of housing and other essentials will tend to rise. Although high rent is a small price for firms to pay, it can be a deterrent to inflows of new human capital. Just as with any other economic system, technology clusters are subject to diminishing returns to scale. Without an ample supply of investment capital, human capital, and supporting infrastructure, the synergy of a cluster may be broken. It is the vital role of government to help satisfy these demands, and thereby retain the critical mass needed for continued technology-driven growth.

## The Archetypical Technology Cluster

The popular business press is fond of describing the phenomenal success of Silicon Valley in terms of economic trivia. Admittedly, I find it hard to resist the temptation myself. For example, fully one-fifth of the world's top one hundred high-tech companies have located their headquarters in this ten-by-twenty-five-mile stretch of Northern California. The top five firms collectively boasted revenues of $40 billion in 1996, with a market capitalization of more than $250 billion.[24] In that same year, on average, one Silicon Valley company went public every five days, creating an average of sixty-two millionaires every twenty-four hours. Moreover, the spectacular economic performance of this region appears to be accelerating; in the first quarter of 1997 alone, 150 new start-up firms were recorded.[25]

Dazzled by these amazing statistics, it is easy to neglect the hard battles that have been fought against global competitors and fail to recognize the remarkable recovery that these recent figures represent. In fact, Silicon Valley can be thought of as two separate phenomena, the first having been based on narrow specialization in semiconductor devices, and the second representing a broader industry mix, involving almost every facet of the information-technology sector.[26] Between these two stages of development, the region suffered through a tumultuous period in which the competitive advantage of this powerful cluster was temporarily lost.

During Silicon Valley's formative years, major firms such as Fairchild Semiconductor became a spawning ground for an impressive list of spinoffs, most notably the 1968 founding of Intel Corporation by Robert Noyce, Andrew Grove, and Gordon Moore. This "start-up" was funded by a New York investment banker, Arthur Rock, who became one of the Valley's first venture capitalists. Additional sources of seed money for promising start-ups included successful regional firms and even Stanford University, which frequently carved out portions of its endowment to sponsor new enterprises.[27]

Another vital contributor to Silicon Valley's early expansion was the unsurpassed quality of human capital available in the region. The vast academic resources of Stanford and Berkeley contributed a steady flow of world-class technologists, and provided ongoing opportunities for collaborative research. By the early 1960s, however, demand for scientists and engineers had outstripped supply. Local educational institutions began to offer multiple levels of technical training, ranging from vocational programs for technicians and data-entry clerks to a stream of advanced degrees from the major universities. This, combined with the constant interchange of experienced personnel between firms, created an exceptionally productive and skilled regional workforce.[28]

Then, as now, the dynamism of Silicon Valley could be attributed to a willingness by entrepreneurs and executives to take chances and make mistakes. A tolerance of failure, combined with ample rewards for success, made Silicon Valley an ideal proving ground for aggressive and ambitious talent. Daredevil risk-taking was encouraged, while failure was tolerated to the point of being considered an advantage from an experience standpoint.

The region's venture-capital community played an important role in balancing risks and rewards. Unlike the traditional bankers of New England, the California breed of investors were often experienced executives and entrepreneurs themselves. Hence, venture-capital firms would often encourage start-ups within their portfolio to work with each other cooperatively. Experienced investors might even step in themselves to correct obvious strategic or organizational weaknesses.[29]

During the Valley's first two decades, an extensive informal network of professionals developed. Personal relationships between peers were often based on years of collaboration, and could command more loyalty than the firms that employed them. As the tides of fortune ebbed and flowed, it was not uncommon for a valued mentor in one professional setting to become the apprentice in another. Thus, the hierarchical concepts of position and title became almost irrelevant. What mattered most in Silicon Valley was experience, knowledge, and, of course, success.

There is a good deal of folklore regarding the exchange of secrets and the cutting of deals at such colorful Silicon Valley watering holes as the Wagon Wheel Bar and the Roundhouse.[30] Although these and other hotspots were replete with techno-gossip, they served more as a metaphor for the culture of cooperation that evolved in Silicon Valley. The true power of this informal network resided in its ability to connect needs and capabilities with great efficiency. The fortunes of many foundries, contract manufacturers, marketing consultancies, and service providers were built on the word-of-mouth recommendations that flowed freely throughout this network.[31]

The vitality of regional networks in Silicon Valley during the 1960s and 1970s stood in sharp contrast to the vertical isolation of technology firms along Massachusetts' Route 128 corridor. The rather stodgy culture of New England permeated the management philosophies and organizational structures of industry giants such as DEC and Raytheon. These firms grew to global leadership by retaining command and control of virtually every aspect of their operations. DEC, for example, performed all functions related to the development of its industry-leading minicomputers in-house, from the design and layout of custom integrated circuits to the final assembly and test of finished products.

Whereas Silicon Valley was open, horizontal, and collaborative, Route 128 firms were typically secretive, territorial, hierarchical, and antagonistic to outside institutions.[32] There are few folksy stories about mutual support and cooperation associated with the Route 128 technology cluster. Instead, there is a legacy of conflict among individual titans, such as the epic battles between Ken Olsen, the founder of DEC, and Edison DeCastro, who left DEC to found rival Data General in the late 1960s. During this period, Route 128 firms were more like autarkic islands than members of a vibrant technology region.

These cultural differences manifested themselves in the competitiveness of regional firms. Time after time, firms within the Route 128 cluster lost market opportunities to Silicon Valley firms, or were blindsided by major new technological advances. The competition between Apollo Computer of Route 128 and Sun Microsystems of Silicon Valley, both promising start-ups in the early 1980s, offers an excellent example.

These firms competed in the newly developing workstation market at a time when the decline of the minicomputer had already begun. Apollo was a pioneer in the workstation industry, and was initially very successful. Even after Sun Microsystems entered the market several years later, Apollo retained a commanding lead, based on what was generally recognized as a superior product. Despite Apollo's first-mover advantage, however, Sun rapidly gained ground by exploiting the rapid innovation rates and flexibility that the Silicon Valley environment allowed. By the time Apollo was purchased by Hewlett-Packard in 1989, it had fallen to fourth position in the workstation market, while Sun had become an industry leader.[33]

Even the rich collaborative network of Silicon Valley, however, was susceptible to unanticipated attacks by overseas competitors. As I discussed in Chapter 7, it had become apparent by the mid-1970s that Japanese semiconductor manufacturers were achieving superior yields and significantly lower costs than their American rivals. The Japanese strategy was simple: Focus on ultra-high-volume devices such as DRAMs, and steadily refine manufacturing processes until the lowest marginal costs in the industry were achieved. Japanese firms adopted a vertical structure for their organizations, primarily because this was the only way to ensure the exceedingly high quality of raw and intermediate materials needed to fabricate superior devices.

The response to this threat by leading U.S. manufacturers, including Intel and National Semiconductor, was to abandon the horizontal, networked structure that had served them so well, and mimic the mass-production mentality and vertical orientation of the Japanese. Unfortunately for the American firms, however, the high-volume devices upon which the Japanese had built their industry were perfectly suited to their culture of continuous improvement, and benefited both from higher operational efficiencies and (at the time) lower average wages. Hence, the giant American firms were forced to abandon the commodity semiconductor market first to the Japanese and later to the South Koreans.

Could the severe economic slump that followed this thrashing have been avoided? Only in retrospect is it obvious that we stood little chance of beating the

Japanese at such a cost-sensitive game. Once this uncomfortable reality was recognized, however, Silicon Valley firms adopted a more suitable strategy. By developing a deep understanding of the unique needs of their customers, firms such as Cypress Semiconductor, Cirrus Logic, and Maxim Integrated Products began to manufacture small lots of specialized devices, thereby fragmenting the commodity market held so firmly by the Japanese. Not only did this strategy undermine the market for "generic" devices, but these customized solutions also offered greater value to customers and allowed significantly higher profit margins for manufacturers.

Today, Silicon Valley firms dominate global markets in highly specialized and customized semiconductors. More important, however, the region has expanded in scope to include higher levels of system integration. The region now hosts world leaders in Internet software development, networking hardware, disk drives, semiconductor manufacturing equipment, and electronics contract manufacturing. The old-guard firms such as Intel, Hewlett-Packard, and IBM have been joined by a new breed of high-growth powerhouses, including Cisco Systems, Novell, Netscape, Oracle, Intuit, and Adobe.

The recent resurgence of the Silicon Valley cluster is a direct result of firms recognizing the value of tacit design and customization skills, and understanding the importance of close customer collaboration. Although industry structure in the Valley is more fragmented today than in its earlier heyday, regional leaders are well aware of the need for intra-industry collaboration and industry/government cooperation. By developing lines of communication at many levels, the firms within Silicon Valley may now be better situated to anticipate and prepare for the inevitable threats to their economic future.

### Technology Clusters and National Specialization

Before our very eyes, the technological world is accreting into clusters of specialization. Whether they form spontaneously, due to natural-factor advantages, or are promoted into existence by government action, these centers of excellence represent the most powerful current trend in technology enterprise. Although seeding institutions often provide an early technology trajectory, the specialization that evolves within a given cluster is greatly influenced by the cultural, educational, and political environments of the host nation.

One of the most important determinants in the development of technology clusters is the proximity of the host nation to the scientific frontier. The technological sophistication of the workforce and seeding institutions within a nation impacts the way in which the structure of the cluster will evolve. In general, the level of capability of a cluster will be limited by the capability of the local workforce. Areas such as Germany's Baden-Wurttemberg, a center for advanced machine tools and electronics, benefit from a world-class talent pool, enabling firms in the region to innovate at technology's leading edge.

In other cases, however, technology firms gather geographically not to innovate, but to take advantage of a local abundance of semi-skilled and cost-effective talent. These industrial clusters are actually high-tech manufacturing and assembly centers, and might fail to capture the higher levels of design and development work needed to generate spillovers between firms. Examples include Scotland's lowland region (referred to as "Silicon Glen") and similar areas within Wales.[34]

Despite severe infrastructure limitations, the region surrounding Bangalore, in southern India, has become a global center of excellence for software development. India's software exports have expanded from $125 million in 1991 to more than $850 million in 1996, with projections of growth to an astonishing $10 billion by 2002.[35] Major multinationals have begun to outsource software to the Bangalore region, due to the relatively low labor costs for highly skilled software designers and programmers, and investment-friendly policies toward foreign firms.

The cultural differences among nations can have a profound effect on the focus and structure of technology clusters within their borders. Japan's culture, for example, promotes a rivalrous environment, in which open competition is encouraged and firms are systematically ranked and rated in the business press. This love of competition, combined with Japan's impressive technology base, has allowed firms with such specialties as industrial robotics to achieve global leadership. Hundreds of Japanese firms compete in the robotics industry, many of which have clustered near the base of Mt. Fuji, including pioneers such as FANUC, Matsushita, Komatsu, and Toyota Machine Works.[36]

In Israel, a technology cluster is forming around Internet-related products and services. The region surrounding Jerusalem, which is cleverly referred to as "Silicon Wadi," has produced more entrepreneurial start-ups in this industry than anywhere else in the world outside of Silicon Valley.[37] As an indication of the intensity of interest in global networking within Israel, this technologically isolated Middle Eastern nation played host to the Internet World conference in 1997.[38]

Although there has been a trend in recent years toward industry-seeded clusters, technology regions are still forming around major universities. Microsoft recently chose a site in close proximity to Cambridge University in England to build an $80 million European Research Laboratory, its only electronic research base outside of the United States. Incidentally, England currently hosts several established science parks that have shown strong growth over the last several years.[39]

One of the earliest examples of an industry-seeded cluster within the United States formed in Southern California between 1940 and 1960. By the late 1980s, this region supported a huge and diverse concentration of more than two thousand aerospace and defense firms, employing more than half a million workers. Industry giants such as Rockwell International, Lockheed, Northrop, TRW, and Hughes Aircraft Company had major facilities in this region. The primary causes for the formation of this cluster were opportunities to draw on skilled human

capital, a postwar housing and infrastructure boom, good weather for flight testing, and proximity to institutions such as the California Institute of Technology
(Caltech), the Jet Propulsion Laboratory, and Edwards Air Force Base.[40]

The rate of industry-based cluster formation in the United States has been
accelerating in recent years. A multinational cluster of telecommunications leaders has formed near Richardson, Texas, allowing firms from several nations to
benefit from location within the highly sophisticated and competitive U.S. telecom market. The early seeding institutions for this "Telecom Corridor," Texas
Instruments and Rockwell Collins Radio, have been joined by more than five
hundred other technology companies, including Southwestern Bell, Alcatel, and
AT&T. Hundreds of small start-ups have been launched in this area since the
passage of the Telecommunications Reform Act of 1996.[41]

Within the United States today, virtually every major growth industry has a
technology cluster associated with it. Advanced materials are the focus of a cluster
of firms in the region surrounding the Corning Glass Works in New York state,
collectively referred to as "Ceramic Valley." Japanese industry leaders Kyocera
Corporation and Toshiba have joined 110 other advanced ceramics companies
in this cluster, to take advantage of local spillovers and the human capital being
created by several excellent regional universities, including Cornell.

Other areas within the United States that show promise of birthing vibrant
technology clusters include North Carolina's Research Triangle Park, which is a
center for biotechnology research by both U.S. and foreign firms,[42] and the areas
surrounding Orlando, Florida, Huntsville, Alabama, and the Raleigh-Durham
region.[43] Every state in the nation has some form of program to promote technology clusters, an obvious indication of the potential benefits that these economic
wellsprings offer to the regions that host them.

## Strategies to Exploit Regional Advantage

The strategic benefits to firms located within an appropriate technology cluster
occur on many levels. A given cluster might incorporate specialized institutions,
an innovative local culture, a highly specialized supplier base, or access to pools of
human capital. However, not every cluster will have all of these factors in the same
relative abundance. This represents a dilemma for firms: Implementing strategies
to exploit these multifaceted resources can stretch even the most flexible organization to the breaking point.

If a firm specializes in microwave radio equipment, for example, it might
choose to locate a facility near a university that is performing pioneering
research, or select a site that is within a telecommunications-industry cluster,
or even position itself in close proximity to an appropriate supplier network.
Depending on the competitive strengths and weaknesses of the firm, any of the

above options might be desirable, but it is unlikely that all can be achieved at a single location.

Since the cost of geographic expansion is substantial, some difficult choices must be made. Does the groundbreaking science and specialized human capital offered by a university research cluster provide a greater competitive advantage than sharing industry gossip with your customers in an OEM cluster? Are there greater strategic benefits to be derived from proximity to suppliers or to customers, during new product development? The goal of this section is to shed some light on the critical elements of a location strategy, and offer some examples of the geographic expansion strategies of successful technology-intensive firms.

### Key Elements of an Expansion Strategy

Technology clusters are capable of rapidly adapting to changing market and technological conditions, but the orientation of the cluster will determine its sensitivity to these external factors. A region that is focused on the development of leading-edge commercial products, for example, might not provide useful spillovers in basic research. Likewise, a university-based cluster that is dedicated to expanding the scientific envelope might be insensitive to market demand, and could easily become sidetracked into activities that are less than commercially viable. Hence, one of the first decisions a technology firm must make in planning for geographic expansion is to determine what aspect of its business needs the most external reinforcement.

At the most basic level, there are obvious logistical advantages to locating a production facility near suppliers and distributors. Several firms within Silicon Valley, for example, have noted that the vast majority of their suppliers are located within a one-hundred-mile radius.[44] The logistical benefits associated with co-location of suppliers and production activities, however, are far outweighed by the strategic benefits of knowledge spillovers within a technology cluster.[45]

For firms in technology-intensive industries, the most important location decisions involve regional headquarters and R&D centers, rather than production facilities. This does not mean that the location of manufacturing and distribution functions is not important, but from the standpoint of long-term market leadership, the real leverage comes from gaining access to the world's premier technology base. Thus, location decisions must be made in a global context.

In the past, firms established international facilities primarily to benefit from low-wage, semi-skilled labor, or to gain access and credibility in local markets. Today these motivations must be expanded to include proximity to specialized technologies and highly developed intermediate suppliers. In other words, technology transfer is no longer a one-way street.

Selecting an international site for a regional headquarters or R&D center is a far more complex decision than choosing an equivalent domestic location.

A number of nation-specific factors must be considered, including local political and monetary stability, cultural differences, trade policies, etc. Siemens Corporation, for example, currently considers the rising protectionist sentiment within both the United States and Europe to be one of its prime considerations when selecting locations for R&D activities.[46]

In the 1960s and 1970s, several U.S. technology firms, including Fairchild Semiconductor, established overseas operations in Hong Kong. These decisions were based on Hong Kong's relative political stability, open financial system, easy repatriation of profits, excellent telecommunications and air/sea transport facilities, and the availability of a small but talented core of skilled technicians and scientists.[47]

These initial forays into geographic decentralization were focused primarily on reducing production costs, with little emphasis on accessing technological specialization or gaining knowledge spillovers. Today the picture is quite different. Major multinational corporations are basing their international siting decisions on technology-driven factors, and some have even relocated their corporate headquarters to other nations to take advantage of critical regional advantages.

The fact that established firms would uproot their "home base" to gain a competitive edge serves to underline the strategic importance of regional advantage. The Canadian telecommunications giant, Northern Telecom, recently shifted its headquarters for digital central-office switching equipment from Canada to a 2.2-million-square-foot facility in North Carolina. Its motivations included access to sophisticated software engineers and world-class university research programs, and the opportunity to conduct its innovative activities within the intensely competitive and sophisticated U.S. telecommunications market.

Similarly, Hyundai relocated its home base for personal computers from South Korea to Silicon Valley, after discovering that it could not keep up with the dynamic PC industry from within its home country. With all of its competitors exploiting the same global market for low-cost components, it was impossible for Hyundai to gain a significant cost advantage. The only way for it to achieve a competitive edge was through rapid new product introduction. A Silicon Valley location provided the specialized infrastructure needed to accelerate product development and allowed direct access to the world's richest distribution network.[48]

Although some firms have chosen to move their headquarters to new countries or regions, it is far more common for companies to establish R&D centers or business-unit headquarters in remote locations. The "neuron model" for R&D strategy discussed in Chapter 5 provides an excellent organizational framework for this type of expansion. In this model, headquarters facilities for each of a company's product lines are located in the region that provides the best market advantages, while satellite R&D centers are established within technology clusters that possess the appropriate specialized factors.

This model has been successfully implemented by Hewlett-Packard, the world's largest developer of advanced electronic test equipment and a global leader in the PC and workstation computer markets. Almost 50 percent of H-P's assets and employees are located outside the United States, including sales and support offices in 110 countries. Although much of this international activity involves production and distribution functions, an increasing amount of its advanced R&D is being performed overseas as well. Plants in Singapore and Malaysia have been empowered to conduct process research and development, and software coding is taking place in human-capital-rich regions of India, China, and the countries of the former Soviet Union.

A number of other major U.S. multinational corporations have located R&D centers in proximity to foreign technology clusters. Ford Motor Company, for example, has located a research facility near the University of Aachen in Germany, to take advantage of the world-leading automotive technologies being developed there.[49] Motorola now has fourteen research facilities in seven countries, and Bristol-Myers Squibb has a dozen R&D centers in six different countries.[50] The extensive resources of these giant firms allow them the luxury of establishing specialized facilities to access basic research. Small-to-medium-sized firms, on the other hand, must base site selection for R&D facilities on more pragmatic considerations.

One way to define an optimal strategy for R&D siting is to consider two broad categories of research centers. An R&D facility can be thought of as either 1) providing a feed of absorbed knowledge back to a central location, or 2) a technology-support facility for in-country product development and manufacturing.[51]

A *knowledge absorption* facility should be located in proximity to technology clusters that perform cutting-edge research in the desired fields. Industry leaders such as Siemens, NEC, Matsushita, and Toshiba, for example, operate R&D sites near Princeton University and Lucent Technologies' R&D laboratories (formerly Bell Telephone Laboratories). The goal of such absorption centers is to track the leading edge of technology development, take part in regional collaborations, and gain access to unique tacit skills.

The *technology support* facility is the more common approach for expanding the R&D capabilities of small-to-medium-sized firms. In this case, an already existing manufacturing facility might be selected for upgrading to a process development center. Alternatively, a small industrial engineering and design group might be established at a remote location to tailor products for local markets. Although these examples represent somewhat modest beginnings, the seeds of innovation are still sown. Given the availability of sufficient human capital and specialized technological resources, these remote islands of R&D activity can be expanded to form an important node in a firm's innovative network.

## A Spawning Ground for Start-ups

The brightest stars in a regional technology cluster are typically the start-up firms. During periods of strong economic growth, start-ups burst forth from virtually every institution to explore the leading edge of the technological frontier. Start-ups are almost entirely vision-driven, and are propelled by the willingness of their founders to risk abject failure in the pursuit of their innovative dreams. Unencumbered by the inertia, bureaucracy, and risk aversion of larger firms, entrepreneurial ventures are the ultimate iconoclasts, dedicated to shattering market preconceptions about the technological future.

Despite their notorious fragility, start-ups represent an essential source of new competition in industries that have become stagnated. Rather than simply extending the existing path, they enthusiastically explore new technologies and search for "jump-shift" breakthroughs. Such creative destruction is possible only when innovators have little to lose and everything to gain; established firms typically have far too much at stake to willingly destroy their own markets.

One of the most economically important categories of start-up firms is the specialty supplier. By providing highly concentrated expertise within a narrow range of competencies, the specialty supplier can enable major OEMs to achieve significant productivity improvement. Andy Grove, chairman and CEO of Intel Corporation, once said, "Anything that can be done in the vertical way can be done more cheaply by a collection of specialist companies organized horizontally."[52] By promoting horizontal delegation of tasks, specialty-supplier start-ups encourage the formation of standardized interfaces and product configurations. Whereas only a single technical option may be available within a vertically structured firm, the presence of design and interface standards allows firms to access a spectrum of viable technologies from several specialized suppliers.

Collaboration with start-ups can reduce the risks of innovation for larger firms, by sharing the cost of specialized R&D and the building of proof-of-concept prototypes. Moreover, their presence allows larger firms to maintain two or more parallel approaches with different suppliers, thereby mitigating the risk of a supplier failing to deliver. If a region becomes too dependent on a single supplier, market forces will encourage the entrance of a "second source" competitor. Often such a new entrant is guaranteed an initial market share, due to the order-splitting strategies preferred by many large firms.

The environment within a technology cluster provides a broad safety net for entrepreneurs contemplating the founding of start-up firms. Rather than being saddled with developing an entirely new product, the rich supplier network present within a cluster makes it possible for an entrepreneur to outsource many of the specialized capabilities needed to execute a new technological vision. The availability of contract professionals spanning many fields, plus the potential use of regional contract manufacturers, further reduces barriers to market entry.[53] If the cluster is sufficiently deep, an entire "virtual enterprise" can be formed with minimal initial investment in either facilities or human resources. In this way, the cluster environment significantly reduces the height of that "first step" for prospective entrepreneurs, from secure employment to a risky new venture.

Perhaps the most generative factor for new enterprises within a technology cluster is the presence of savvy, risk-taking investors. Venture capitalists are attracted to clusters, primarily because they are often hotbeds of economic growth.[54] For the right entrepreneur in the right industry, there are ample opportunities to trade a chunk of equity for enough seed money to bring a new innovation to life.

The reduction of entry barriers for start-ups, however, provides little benefit without the most vital ingredient, a supply of visionary entrepreneurs. Technology clusters provide many role models for aspiring entrepreneurs, and can offer opportunities to learn from the successes and failures of colleagues. Clusters promote diversity in organizational thinking, technologies, methods, and cultures. In a sense, they are like a vocational school for future industry leaders.

The typical cluster-seeded start-up is formed by a group of friends or colleagues with an innovative idea that does not fit within the constraints of their current employment. Clusters provide a fertile pasture for the formation of such groups, by playing host to an interconnected network of large and successful firms, government laboratories, think tanks, and major universities.

Leading think tanks, such as SRI International (formerly Stanford Research Institute) and Sarnoff Corporation (formerly David Sarnoff Research Center), are a potent mixture of entrepreneurial talent and leading-edge technology. These two centers alone spun off twenty new ventures between 1993 and 1996, including start-ups specializing in advanced materials, medical instruments, communications components, and video-display technologies.[55]

Although many of the critical ingredients are present, some technology clusters fail to provide a fertile spawning ground for start-ups. Whereas Silicon Valley is legendary for proliferating spinoff enterprises, the Route 128 region has been far less conducive. It appears that the most important indicator of a cluster's "fertility" is the degree to which it has evolved into a horizontal network of specialty suppliers and producers.

Silicon Valley was built on the organizational visions of thinkers such as Robert Noyce and other Fairchild and H-P alumni, who viewed the world of commercial technology as a "community of equals."[56] Such a democratic culture can make the prospect of business ownership seem tangibly real. It is a small sideways step for an aspiring entrepreneur to move from employee to employer in such an egalitarian environment.

On the other hand, few executives left DEC and other major Route 128 technology firms to form their own start-ups. Ken Olsen of DEC, for example, took the vertical organization to such an extreme that it functioned as an independent sociological unit, an island unto itself. Olsen openly discouraged his executives from founding their own ventures, and aggressively attempted to destroy those who tried. In this hostile environment, the launching of a new start-up could mean professional suicide.[57]

Today, the entrepreneurial climate along Route 128 is changing for the better. Silicon Valley, however, still receives at least three times the amount of venture capital, and gives birth to far more start-up firms. Evidently, a technology cluster can be either a fertile valley or a harsh wasteland for new ventures, depending on the abiding attitudes and cultures prevalent in the region.

### The Informal Network

Productive innovations are not necessarily the result of methodical research and development. Often they are born of chance and serendipity, the kind of free-association thinking that is the defining characteristic of creative minds. Such thinking is best performed against an intelligent sounding board. Hence, the need to discuss, debate, argue, and dissect technological problems draws innovators together into informal networks that defy the arbitrary boundaries of firms and institutions.[58]

The interactions within a technology cluster's informal network are of a uniquely personal nature. They are driven by curiosity, frustration, ambition, and a natural human desire to collaborate on challenging problems. In such networks the velocity of information flow is very high. Rather than dealing with the inefficiencies of formal alliances and company structures, valuable knowledge is freely exchanged among trusted peers.

The informal interactions between collegiate professionals have a high level of tacit content; solutions are described in detail, experiences are shared, "rules of thumb" are offered, and personal recipes are disclosed. Although this type of interaction can permit the loss of proprietary information if not carefully executed, the benefits often far outweigh the risks. Employees in such an environment should be educated in the "rules" of informal interaction, and then encouraged to engage in it at every opportunity. All workers within a firm should develop and maintain a network of "experts" in their areas of interest who can assist them in finding resources, solving problems, or identifying unanticipated risks.[59]

Unfortunately, not every technology cluster supports a healthy informal network. These tenuous structures can form only under favorable conditions, including a regional culture of openness, a horizontal industry structure, and the presence of active venues for the exchange of knowledge and ideas.

High employee turnover in a region can facilitate the evolution of informal networks. Professionals will often stay in touch with respected former colleagues throughout their careers, and form additional long-term relationships with valued suppliers, educators, and consultants. In this way, informal networks are created by the independent actions of individuals; each professional establishes personal linkages that eventually merge into a regional continuum.

Formal alliances between firms within a cluster can play an important role in the formation of informal networks. Anytime talented people from disparate backgrounds are brought together, new linkages will be formed that can last well beyond near-term activities. The informal network becomes stronger and more diverse as a result.[60]

Gaining access to an informal network requires a conscious effort, particularly for new firms within a technology region. Trusted personal relationships take a lifetime to build, a process that cannot easily be short-circuited. Superficial contacts will tend to be self-serving, with each participant trying to manipulate the interaction to follow his or her own agenda. Only those relationships based on shared history, respect, and mutual interests will result in high-value informal collaborations.

The first step for firms attempting to merge into an informal network is to encourage a company culture that is compatible with the intellectual openness of the region. Many foreign laboratories within the United States have adopted cultures and organizational structures that are patterned after U.S. laboratories, rather than following their home-country norms.[61] This is often done in a conscious effort to enhance the absorptive capacity of these facilities. To become accepted into an informal network, new participants must "fit" within the social structure of the region.

One of the best ways for a firm to become connected is to encourage employees to volunteer for local industry associations, standards bodies, committees, etc. Similarly, regional special-interest groups and clubs provide an excellent venue for relationship-building and resource-sharing. Another way for firms to assimilate into a regional network is to join, or better yet become a sponsor of, an industry association or organization.

Most technology industries have special institutional needs that cannot be affordably addressed by individual firms. Examples include specialized technical education, assistance to start-up firms, global market research, technology roadmapping, political action, regional promotion, etc. These opportunities for cost and risk-sharing among several firms offer a primary economic benefit, as well as a secondary benefit of establishing interpersonal relationships that strengthen the informal network.

A good example of this type of cost- and risk-sharing alliance is the pooling of information and legal strategies by eight semiconductor firms in Silicon Valley. In 1990, Cypress Semiconductor, LSI Logic, Sierra Semiconductor, and several other custom integrated circuit suppliers agreed to cross-license each other's technologies and form a united front against litigation brought about by their larger competitors.[62] Similar relationships have been formed between smaller firms to share unique and expensive capital equipment, engineering design resources, and pre-competitive market research.

In the early days of Silicon Valley, the informal network was nurtured by such regional institutions as the American Electronics Association (AEA) and the Semiconductor Equipment and Materials Institute (SEMI). These groups sponsored trade shows and seminars, hosted dinners, provided political and competitive information, and helped establish industry standards, such as specifications for the silicon wafers used in semiconductor processing. More recently, a broad cross-section of Silicon Valley policymakers, executives, consultants, and educators have formed "Venture: Silicon Valley," with the goal of promoting the region's image and the effectiveness of inter-industry communication and collaboration.

Ultimately, the responsibility for the formation and maintenance of informal networks is shared among individuals, firms, and regional policymakers. While government can serve as a facilitator and catalyst, it is up to the leaders of industry to suppress their mistrust and competitive paranoia, and encourage an increased level of sharing and collaboration with other firms within their region.

# PART 3

## Strategies for Sustained Market Leadership

# 9
# Negotiating the Technology Rapids

## Dynamic Strategy Formation

If we set aside banal excuses for the decline of American competitiveness vis-à-vis our international rivals in the 1970s and 1980s, what remains is a stark truth. Despite our leadership in mass production and breakthrough innovation, American firms were repeatedly outdone in the strategic management of technology enterprise. From an "internal" standpoint, our shortfall was evident in the lack of emphasis on process technologies, a reluctance to collaborate, and the failure to rapidly commercialize important innovations.

The story was no better with respect to "external" strategy formation. American firms repeatedly underestimated their foreign competitors, failed to recognize fundamental shifts in the competitive playing field, and held on to outdated planning tools long after the pace of technology change rendered them useless. Despite recent gains in global competitiveness by America's technology industries, the lack of proven theories, tools, and methods for technology management remains a persistent weakness.

The goal of this chapter is to survey the best available thinking on competitive strategy in technology-intensive industries. Competitive strategy is the branch of strategic management theory that is concerned with how external factors can shape the future prospects of a firm. Ideally, we desire an integrated theory for strategy formation and implementation that accommodates rapidly evolving technologies, chaotic changes in market conditions, and the complexities of global competition. Unfortunately, no such integrated framework exists. Instead, we must sift through a collage of traditional static theories, along with some partially developed yet potentially powerful new concepts.

### Do Static Management Theories Apply?

Not only is the competitive environment for technology industry changing, but the rate of change is increasing as well. As recently as the 1980s, competition in high-growth sectors such as telecommunications and information technology was still dominated by traditional leaders: IBM and DEC in computers, Motorola in wireless telephony.[1] Beginning in the early 1990s, these giants became embroiled in a tumult that humbled even the most visionary executives. By 1992, prices and profits had collapsed for IBM and DEC, while the "new guard" of Intel and Microsoft saw their prospects soar. More recently, Motorola's seemingly invincible lead in cellular phone technology has been undermined by foreign usurpers such as Nokia and Ericsson. Evidently market power, vast financial resources, and world-class talent are no longer sufficient to ensure sustainable advantage.

Against such an onslaught of change and uncertainty, our current understanding of strategic management theory stands pale and weak. While concepts for a "theory of the firm" are still being debated in the literature, technology industries have evolved into networked structures that blur the boundaries between firms, rendering such narrow insights largely irrelevant.[2] Likewise, the disagreement over "structure follows strategy" versus "strategy follows structure" is marginalized by the growing necessity for rapid adaptation.[3] Under such constraints, both of these statements miss the point: Because the formation of strategy is inexorably linked to the capacity of a firm to execute, strategy and structure must develop concurrently, synergistically, and with constant feedback.

The questions that consume strategic management scholars today are important as a basis for theoretical understanding, but are of little practical value to technology firms. How do firms behave? Why are firms different? What purpose does "headquarters" serve?[4] Even if definitive answers to such questions were forthcoming, they would be outdated before they could be of strategic value. Without knowing exactly how firms behave, one can reasonably conclude that they will behave differently under changing stimuli. Similarly, the differences between firms will not remain in stasis awaiting leisurely analysis by business scholars, nor will a definitive role for headquarters be either discernible or predictable to a useful degree in the future.

Today many observers of industry performance doubt that conventional strategic management principles can measurably boost a firm's productivity.[5] Widely accepted tools such as *strategic planning* have fallen into disfavor, largely due to their inability to remain valid long enough for the ink to dry.

Even powerful industry-analysis frameworks such as Michael Porter's "Five Forces" model offer only snapshot insights into evolving systems.[6] The bargaining power of suppliers, for example, will change as value-producing innovation moves up and down the supply chain. Likewise, the bargaining power of buyers will evolve in lock step with applicable technology "s-curves," creating a buyer's market for mature technologies and monopoly power for firms that invent the next big thing. Threats of substitutes and new entrants will rise and fall as well; breakthrough technologies enable giants to be toppled and kingdoms to be won. At the nexus of these forces, the rivalry among competitors will be pulled by changing market demand, pushed by new ideas, distorted by swelling R&D costs, and constantly reshaped by alliances, mergers, and acquisitions.

The unique demands of technology-intensive industries spawned a new field of business study during the 1980s: the management of technology (MOT). Although MBA courses have been developed and thousands of papers have been published, this field still lacks coherence and tends to be dominated by fads. Much like the early investigators of subatomic particles, MOT researchers are dazzled by the complexity of what they observe, and are prone to quickly jump to "silver bullet" explanations. Texts on technology management often read more like random opinion surveys than integrated works, leading to the conclusion that there is still no consensus model for the formation of effective technology managers.[7]

Rather than further bemoan the underdeveloped nature of strategic technology management, I will take a reductionist approach in the sections that follow, highlighting those insights that appear to have practical utility in gaining and retaining productivity leadership.[8] My criteria for the selection of topics include: 1) the incorporation of time as a driving factor, 2) accommodation of technological change, risk, and uncertainty, 3) a "holistic" approach to the formation and execution of strategy, and 4) the ability to measurably enhance innovation, appropriability, and productivity. What remains after this filtering has been performed is a concentrated essence that offers hope of longevity for those firms that partake of it.

### Adding the Dimension of Time

Applying the criteria described above to the full range of strategic options yields the list of factors summarized in Table 9.1. The most important single contributor to sustainable market leadership is the decision-making ability of a firm. In a slowly changing competitive environment, strategic decisions (i.e., those that directly impact the future of the company) are made infrequently, and are followed by long, stable periods of implementation and refinement. As the pace of change quickens, however, this methodical approach becomes untenable.

Instead, continuous decision-making must be the central focus at all levels of the organization. Significant environmental changes mandate immediate action, tactics of rivals demand a prompt countervailing response, shifts in customer preference become the catalyst for new product innovations, and the impending shockwaves of breakthrough technologies demand structural "bracing" in anticipation. In each case, organizations must make timely choices based on the best

| Factors in Sustainable Strategy Development | Description |
| --- | --- |
| Decision-Making Process | Dynamic strategy is developed through a continuous process of decision-making at all levels of an organization. |
| Systems Thinking | In rapidly changing industries, all functions within the firm should be thought of as an enterprise system, which must adapt its strategic behavior to the external competitive environment. |
| Organizational Factors | Strategy formation and implementation are a continuous, cyclic process. Firms that can rapidly adapt to environmental change, through agile organizational structures and flexible manufacturing techniques, will thrive. The others will perish. |
| Emphasis on Tacit Content | The unique skills and knowledge embodied within the employees and structures of a firm are difficult to imitate and can lead to monopoly profits if properly exploited. |
| Absorption of Knowledge | Absorbed knowledge provides support for dynamic decision-making, and can reduce both the costs of non-recurring R&D and the development time for new products. |
| Rapid Innovation | Rapid innovation enables firms to achieve first-mover advantage, and allows them to take advantage of fragmenting markets. |
| Appropriability of Returns | Returns on investment are the fuel of continued innovation. Beyond rigorous intellectual property protection, products should be difficult to imitate, be resistant to retaliation, and incorporate rare tacit knowledge. |

Table 9.1: Summary of the factors that can impart a sustainable competitive advantage to firms. Each factor takes temporal change, technology intensity, and system-level behavior into account. All enable increased productivity through increasing the speed of invention or the appropriability of returns, or by augmenting a firm's ability to absorb and accumulate valuable knowledge.

available information, and then must begin preparing for the next twist or turn. Under these turbulent conditions, *a continuous cycle of decision and execution becomes the strategy of the firm.*

Useful tools for analyzing competitive strengths and weaknesses must be considered in the context of dynamic decision-making. One such framework, referred to as the "resource-based view of the firm," remains valid only if the value of a firm's resources is constantly reevaluated as a function of changing industry and market conditions.[9] In a dynamic competitive environment, weaknesses can become strengths and vice versa, forcing resource-based strategies to become highly amorphous.

The process of strategy formation has been illuminated in recent years by the idealized models of game theory. A brief overview of the contributions of game theory to the understanding of strategic decision-making in firms is presented in the next section. This discussion is followed by the presentation of a simplified decision model that can form a basis for "temporalizing" strategy in a real-world competitive environment. In essence, this model imposes a "date code" on strategic decisions, allowing firms to shift between periods of continuity and chaos.

If we think of decisions as part of a continuous process rather than as isolated events, the interrelationships between factors internal and external to a firm become critical. A firm can no longer be treated as a discrete entity with respect to its external environment, and must come to view its internal structures as a complex, interdependent network. An enterprise should be thought of as an evolving system, driven by system attributes such as boundaries, partitions, feedback, and control. Decisions that impact the future of a firm, therefore, should be made in the context of "global" rather than "local" optimization. The use of systems thinking in the formation and execution of strategy is discussed in a later section.

To survive under rapidly changing conditions, an "enterprise system" must be highly adaptable. Strategic advantage is gained by those firms that can closely match their organizational structure, production operations, and innovation processes to the external competitive environment.[10] It is possible for firms that are subject to the same external influences to respond to these conditions in very different ways. Within a narrow segment of the software industry, for example, one firm might choose to sell a generic product through mail-order catalogs, a second might opt for bundling its product for retail sales, and a third might pursue a strategy of customer collaboration and customization.[11] Depending on the evolving nature of the market, one of these firms will undoubtedly become significantly more successful than its rivals.

Relating this common scenario to ecological systems has become a cottage industry among both strategy scholars and popular writers. Although the similarities are striking, there is one critical distinction between business enterprise and biological ecosystems. Biological organisms have their competitive advantages (and disadvantages) hardwired at birth through their genetic codes. The only way for life to evolve is through the destruction of poorly adapted organisms in favor of more suitable ones. In the evolving business environment, however, it is possible for firms to change their genetic makeup (so to speak), and reinvent themselves through rapid adaptation. The dismal fate of the dinosaurs need not apply to agile firms that can quickly adapt to new environments.

To avoid being consumed by a "Schumpeterian revolution" of creative destruction, technology-intensive firms must choose strategies that enhance the three elements of the innovation cycle discussed in Chapter 4. The dynamic decision-making process described above, for example, depends on a continuous stream of environmental information being absorbed into the firm. These data must go beyond simple "environmental scanning" to include specialized technical knowledge, market research, financial and economic analysis, and forecasts of impending shocks and discontinuities. Chapter 10 is dedicated to this vital aspect of sustainability, including the absorption and accumulation of current knowledge and methods for visualizing and roadmapping possible future conditions.

Rapid innovation is the surest method for firms to resist the forces of "natural selection." New opportunities to fragment existing markets through high levels of customization are continuously arising, offering even the smallest of firms the chance to become a first mover. To take full advantage of these opportunities, organizations must be structured to efficiently utilize absorbed and accumulated knowledge and to rapidly transition new-product concepts to the factory floor. Some important methods for enhancing innovation rates and R&D productivity, including design reuse and mass customization, are discussed in the final section of this chapter.

Last but not least among the three elements of innovation is the appropriability of returns from R&D investment. Studies have shown that it is not uncommon for the market leader in a given technology-intensive industry to expend twice the R&D investment of its competitors. High appropriability is essential to maintaining these leading levels of investment.[12] Naturally, the first line of defense is intellectual property protection, either through legal means or by institutionalized secrecy. Both of these options, however, are of only limited effectiveness, for the reasons discussed in Chapter 6.

A more robust method for increasing the appropriability of returns involves altering the basis of a firm's competitive advantage to be inherently more easily protectable. Products that are resistant to imitation, for example, will provide consistently higher rates of return. Increasing the level of tacit knowledge content of a product makes it less susceptible to imitation, particularly if the skills required are complex or are deeply embedded in the organizational structures and culture of the firm.[13] Such a pedigree renders the sources of competitive advantage ambiguous to outside observers, making their imitation a risky prospect.

Not all of the above factors can be applied to every firm, nor is this an exhaustive list of possibilities. The underlying concepts, however, are broadly applicable: Focus on strategic decision-making rather than strategic planning, adopt a systems approach to all choices, prepare the organization for rapid adaptation, and emphasize the three self-reinforcing elements of profitable innovation.

### Game Theory vs. Practical Reality

There is a somewhat gloomy truism among strategic management scholars, known as the "fundamental theorem of industrial organization," that highlights the practical limitations of game theory.[14] This "theorem" asserts that given the complexity of real-world enterprise, and the ability of theorists to choose simplifying assumptions from a rich set of options, a model can be devised to explain almost any observation. Such is the case with game-theoretic modeling. As a tool for informing studies of market entrance and exit, adoption of standards, and first-mover advantage, it has been reasonably effective. As a framework for the formation of competitive strategies within firms, however, it has been an abysmal failure.[15]

During the 1970s and 1980s, game theory attained a prominent position within the literature, primarily due to its reductionist approach to analyzing competitive strategies among rivals. Early games, such as the Prisoner's Dilemma previously described in Box 5.1, demonstrated the interdependency of competitors' strategies, the potential benefits of cooperation, and the "psychology" of the decision process. Unfortunately, the foundational assumptions of game theory have severely limited the applicability of its predictions.

It is assumed, for example, that players in game-theoretic models behave as *rational individuals*. The term 'rationality' has a very specific meaning in this context: rational behavior is assumed to be *utility maximizing*. This condition was first expressed by the mathematician John Von Neumann and the economist Oskar Morgenstern in their classic work *Theory of Games and Economic Behavior*.[16] They proposed a broader requirement than the profit-maximizing behavior that is often assumed in microeconomics: Utility maximization can apply to any factor that a player considers to be of value. Obviously, this rather vague constraint on rationality leaves open the possibility of unrealistic outcomes. Economists have observed, for example, that when the utility condition is narrowed to consider only *profit maximization*, real-world managers rarely meet the rationality criterion.[17]

Furthermore, for game-theory models to apply to real-world competition among firms, it is necessary to assume that organizations can be treated like rational individuals. Hence, for game theorists the question of how firms should behave in competitive environments becomes: How do rational (perhaps even hyperrational) individuals behave under these conditions?[18] This simplifying assumption stands in sharp contrast to the fact that firms often "speak with many tongues."[19]

The complexity of real-world situations also hinders the utility of game theory for practical applications. Actual competitions are multifaceted, and may display the characteristics of several types of "game" simultaneously.[20] Moreover, as the number of players increases, the ability to predict optimal outcomes declines geometrically. Even when only two major players are involved, the nature of competition often resembles a game of chess, with each player considering long sequences of decisions in the context of his or her rival's current capabilities and past behavior. In such contests, game-theory models are often reduced to an endless stream of "If I believe that you believe that I believe. . . . "[21] In short, game theory provides few usable insights into the complexities of competition in technology-intensive enterprise, and there is little evidence to suggest that it is actually being employed in strategy development.[22]

Occasionally, however, a situation does arise that can be related to the simplified models of game theory. In particular, the Cournot duopoly model is frequently applied to conditions in which two firms dominate their industry (for a survey of some other interesting game-theory models, see Box 9.1). One example of such a scenario is the recent consolidation of the commercial aircraft industry. The actions of the Boeing Company and Europe's Airbus Industrie have many of the elements of the "Developer's Dilemma," in which two competitors must make investment choices under conditions of a limited worldwide market.[23]

In this real-life version of the Developer's Dilemma, each player has the option of investing a sizable amount (estimated at $15 billion) to develop a "super-jumbo" jetliner for high-traffic international routes. The global market has been estimated to be no more than two thousand aircraft, with some projections showing a potential for fewer than one thousand total sales. Under these conditions, if both Airbus and Boeing independently develop a super-jumbo plane, one firm would surely lose its investment, and both rivals could potentially be crippled.

The logical choice for both firms was to engage in a cooperative venture to co-develop the big aircraft. A joint team was set up in 1993, but the alliance soon collapsed, due in part to each competitor's concerns over the other partner "cheating" on the collaborative agreement. Ultimately, Boeing chose to shelve plans for a super-jumbo jet, citing declining market estimates and the damage such a plane would do to its already dominant position in the four-hundred-seat aircraft market segment.

The true benefits of game theory for the practitioners of competitive strategy are conceptual rather than prescriptive. Game theory has brought legitimacy to the idea of cooperation among competitors. Its conclusions demonstrate that it is not necessary for competition to be a zero-sum game; it is often possible to increase the size of markets through mutual cooperation. More important, however, is the emphasis that game theory places on the process of decision-making itself. Although few specific rules can be derived from game-theoretic models, their study has made it abundantly clear that decisions are the building blocks of competitive strategy.

## Box 9.1: A Sampler of Games[24]

A game consists of a set of players whose choices determine their success, as measured by payoffs in some "valuable" commodity. The most familiar game-theoretic model is the *Prisoner's Dilemma,* which can be used to understand the behavior of firms in cartel-like arrangements. Firms in a cartel must decide whether to "defect," by slashing prices or increasing production volumes. If only one firm defects, it will receive considerable benefit. If the remaining cartel members follow suit (as is very likely), everyone loses.

The *Centipede Game* also demonstrates the benefits of trust and cooperation. In this case, players must make a sacrifice to retain a valued relationship, with the expectation that they will be reimbursed for their losses in the future. This is analogous to a customer-supplier arrangement, in which a supplier chooses to absorb a price increase in the hopes that upon the next "play" the customer will reciprocate.

The Centipede Game must be played several times to observe the pattern of outcomes. It is easy to make the choice to sacrifice profits once, in return for an eternal payback period. If the game is allowed to evolve over time, however, a familiar pattern becomes evident. At each decision point, each player must decide whether, upon the next play, the other player will be willing to "sacrifice" to maintain the relationship. Since there will be multiple plays of the game, both players know they will be confronted with this choice on a recurring basis. If the players trust each other, the game can continue indefinitely. If, however, either player comes to believe that the other will not reciprocate in the future, that player's best option is to quit, bringing the relationship to an end.

A related game has come to be known as the *Chain-Store Paradox,* because of its counterintuitive conclusion. In this case, a monopolist (e.g., a large chain store) enters several new markets. In each region, if it is confronted with a new entrant, it must choose to either engage in "predatory pricing" to drive the other out of business or else adopt a more accommodating position. Intuition would suggest that the use of a strong deterrent such as predatory pricing might serve as a warning to future entrants, and therefore be an optimal strategy. According to game-theoretic reasoning, however, the hyperrational monopolist recognizes that it cannot maximize profits through such a deterrent, and therefore accommodates all new entrants. For this paradoxical conclusion to be drawn, some stringent conditions must be met, such as the existence of only a finite number of possible entrants. More important, however, this theoretical game fails to capture the "irrational" effects of fear and intimidation, which have a proven track record as effective deterrents in "mismatched" competitions.

## Strategic Decisions in an Evolving System

Suppose that you are the CEO of a technology-intensive firm that is organized around three functional groups: Marketing, Manufacturing, and Engineering. Your company is currently pursuing an exciting market opportunity, and each group has been given a clear message that the future of the firm depends on a successful product launch. How might each functional group respond to this challenge?

The Marketing organization will recognize that high levels of customization will increase pricing power in the marketplace, and will pressure Engineering to add features and options. Simultaneously, Manufacturing will be given an unrealistically low cost target, as an "incentive" to improve recurring costs. Both groups will be informed of a launch date that is impossible to achieve, since the marketing group recognizes that an early market entry could translate into a first-mover advantage. These "decisions" by Marketing are driven by two primary factors: price and market share.

Manufacturing, on the other hand, sincerely believes that the future of the company is best secured through low production costs and high volumes. Hence, it pushes for fewer options and a simplified product architecture. In addition, it chooses not to expand capacity at this time, since Marketing's demand forecast is "always way off." Manufacturing makes these decisions in the context of two different drivers: recurring cost and capacity utilization.

Last but not least, Engineering recognizes its superior insight into the design of the product, and determines that a breakthrough new technology should be exploited in the upcoming product line, since it may provide better performance and avoid early product obsolescence. The risks associated with the new technology are primarily schedule-related; such risks are an acceptable tradeoff from their perspective, since Marketing always "cries wolf" with respect to product launch dates. Engineering's choices are driven by a third set of factors: technical capabilities and risk.

As the CEO of this fictional (but hauntingly familiar) firm, are you satisfied with the decisions being made by these three groups? Clearly, each function believes that it is acting in the best interests of the organization, yet the choices are conflicting. By allowing the three functional groups to engage in *local optimization*, the strategy of the firm as a whole is compromised. This problem can be exacerbated by internal performance metrics and incentives that are aligned with each group's local drivers, rather than company-wide goals.

Suboptimal decisions of this type can be avoided by considering the entire enterprise as a system. A system is defined as a collection of interdependent elements that work toward a common purpose, typically involving a *transformation process*. In such a system, the performance of the whole is affected by each of its constituent elements, and the contribution of each element is related to that of at least one of its brethren.[25] In the case of firms, one could consider the production function described in Chapter 1 as representing the transformation performed by an *enterprise system.*

Even with this limited exposure to systems thinking, it becomes clear that some of the most cherished "movements" in American industry have failed to treat the enterprise as a system. In the 1980s and early 1990s, the wildly popular Total Quality Management (TQM) initiative emphasized the continuous improvement of each part of an organization, while the dynamic interaction of the various elements was largely overlooked. Likewise, the Business Process Reengineering (BPR) frenzy precipitated a torrent of detailed process maps, in which the vital linkages between processes were often either oversimplified or neglected.[26] In each case, these movements have proved to be of inconsistent value, and when improperly executed, have become productivity sinkholes.

The most basic characteristic of a system is the existence of a *boundary.* The boundary serves to divide the "universe" (which in the case of global enterprise encompasses just about everything) into two regions, the enterprise system and the external environment, as shown in Figure 9.1. System boundaries can be either open or closed. If we treat nations as economic systems, for example, autarky would represent a closed boundary and liberalized trade would exemplify an open boundary.

The various functional elements of an enterprise system are internal to the boundary, and as such, are subject to *feedback and control.* The external competitive environment, on the other hand, is outside the boundary, and hence must be accommodated rather than controlled. To improve the competitiveness of the system, a strategy must utilize internal feedback and control to continually optimize the firm's position relative to its external environment. In this context, the national system of innovation, for example, can provide a significant competitive advantage, provided that a firm's strategies are aligned so as to benefit from it.

A second defining characteristic of a system is that its various constituent elements must share finite resources. These shared resources are referred to as *commons;* for an enterprise system they include discretionary funds, human resources, facilities, technological risk, and even development time. Often these common factors are in fixed supply, forcing a competition among functional elements in the system for their fair share. Furthermore, decisions are often made by individual functions as though the *entire* common resource was available to them. The inevitable result is suboptimal utilization of precious assets.

This common affliction can be mitigated through the use of *partitioning.* Both the responsibility for strategic decisions and the allocation of common resources must be partitioned among the functional groups. To ensure optimal deployment over time, feedback and control should be provided by a central executive function, which is empowered to adjust allocations of commons and verify that local optimization does not occur. The executive function also manages the *interfaces* between functional elements that enable the exchange of critical information and shared resources.

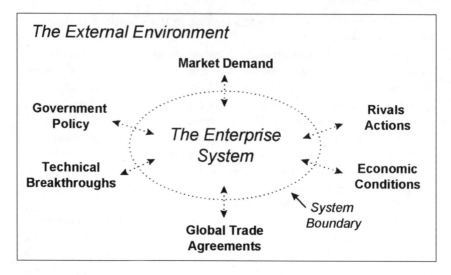

**Figure 9.1:** Conceptual diagram of an enterprise system, demonstrating the role of the system boundary in dividing the "universe" into internal and external environments. In this case, the boundary is permeable, allowing constant interaction between the system and its environment. The external factors shown do not represent a complete ensemble.

In recent years a number of software products have become available that fall under the general heading of "decision support tools." These packages range from the relatively narrow focus of project-management and market-analysis applications to some ambitious attempts at enterprise-wide information management. This latter category includes relational databases for data warehousing, enterprise resource planning software, and even some aggressive attempts at strategy simulation through the use of structured enterprise models.[27] While it is evident that an efficient flow of decision-support information is crucial to successful strategy development, there is a danger in placing too much trust in software tools. Applications that provide tailored feeds of data are relatively low risk, whereas integrated information tools that filter, shape, and analyze data can cause systematic errors in strategic judgment, due to misalignments between the software designer's understanding of "generic enterprise" and the realities of a firm's unique competitive environment.

If we accept that an evolving system is a reasonable metaphor for technology-intensive firms, we can then describe strategy development by answering three basic questions: 1) What are the strategic decisions upon which the performance of the system depends? 2) What methodology (e.g., information, analysis techniques, responsible parties, etc.) should be used to execute these decisions? and 3) How frequently must each of these strategic choices be reevaluated? A decision

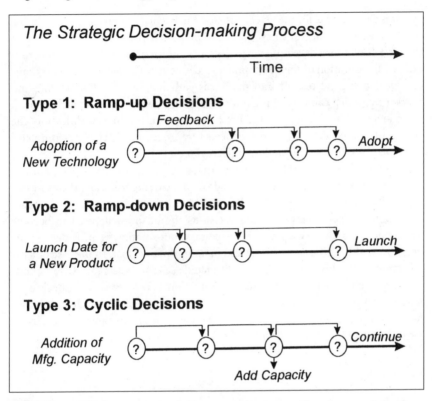

**Figure 9.2:** Examples of how systems thinking can be incorporated into the strategic decision-making process of a firm. In each case, described in the main text, decisions are made at optimal intervals, and are influenced by feedback from both the external environment and the internal executive-control function.

process that addresses these issues will allow for continuous adaptation, while providing periods of relative stability for the implementation and refinement of choices.

The three typical strategic decisions are presented in Figure 9.2, along with a diagram of their possible evolution over time. In the first case, the choice involves the timing for the adoption of a new process technology being developed by an external source. I have classified this as a *ramp-up decision*, implying that as the potential for adoption increases, the decision must be reevaluated more frequently. For example, if a new process technology is still in the "proof of concept" stage, one might need to monitor its progress only every six months or so. Once this innovation reaches the development stage, a more frequent sampling would be justified. Finally, as the new process begins to diffuse throughout the industry, the topic might be elevated to a weekly agenda item for management review.

It is also possible for the reverse situation to occur. In the second example, the choice of a launch date for a new product is presented as a *ramp-down decision*. At the kickoff of a new development project, the choice of launch date (and consequently the duration of the development cycle) is very flexible. During early concept design, a great deal of information is gathered, both by Engineering and Marketing, which can impact the tradeoff between design time and early launch. Ultimately, as the development effort moves toward product qualification, the choice of launch date might be reduced to an occasional "time-to-complete" estimate from Engineering and Manufacturing.

The third category of decision is perhaps the most common. Many choices within an enterprise system must be made on a recurring basis, and therefore can be classified as *cyclic decisions*. The choice of whether to add manufacturing capacity is an excellent example. Cyclic decisions should be structured to become "automatic," in the sense that the necessary information is provided to the decision-maker at the proper time and in an optimal format. In the case of a capacity decision, this information might include sales forecasts for existing products and a schedule for new-product introductions. Each time a cyclic decision is made, the timing of the next choice should be reevaluated, and the adequacy of information should be reassessed. In this way, cyclic decisions can become a living part of strategy development within the enterprise system.

## Decisions Great and Small

No static strategy can succeed for long in a dynamic business environment. The formation of competitive strategies must be a continuously evolving process, accounting both for changes in the external environment and shifts in the internal resources and capabilities of the firm. The essence of strategic management in rapidly changing industries, therefore, is the execution of a continuous stream of decisions, both large and small, that determine a company's long-term success.

In the previous section I noted that most well-developed theories regarding the strategic behavior of firms depend on assumptions of rationality on the part of individual decision-makers and collective organizations. Unfortunately, most experts agree that these assumptions are generally invalid.[28]

There are tremendous inconsistencies in the decision-making processes among firms within a given industry, and even within the hierarchies of the firms themselves. Moreover, there is compelling empirical evidence that individuals and groups alike consistently make *irrational* decisions, due to biases and behaviors inherent in the human animal.

Like it or not, there are distortions in the way individuals and groups make decisions that almost guarantee that suboptimal choices will be made. In the discussion that follows, the reader should consider the examples in the context of his or her own decision history, but be prepared to experience some emotional discomfort.

## Risk Aversion and Irrational Optimism

If you were offered either a guaranteed $100 or a 50/50 chance of winning $200, which would you choose? Even though both options have the same expected value (.50 × $200 = $100), it is roughly three times more likely that you would choose the $100 sure thing. Now consider a slight variation. Suppose you are given $200, but you are required to choose one of two penalties: You must either give back $100 or take a 50/50 chance of losing either $200 or nothing. In this case, even though the value of this choice is the same, studies have shown that you are far more likely to choose the chancier option.[29]

Both the decisions described above have one possible outcome in common: a sure gain of $100. Yet when the decision is stated in terms of a positive gain (the first example), the majority of people will be *risk-averse*, whereas when the choice is structured as a potential loss (the second example), most people will be *risk-seeking*. One of the criteria for rationality in choice is that *only the consequences of the choice are considered*. If that were the case, people would always choose either the gamble or the sure thing. Yet a number of empirical tests of decision-making under risky conditions have shown that the choice is dependent on *framing effects*; the way the offer is worded has a significant effect on the preferred decision.

The most common pattern of irrationality in decision-making involves systematic bias toward risk-aversion when options are attractive, and risk-seeking when options are negative. This is by no means the only way in which framing effects can impact rational choice. Decision-makers are typically more risk-averse when they believe their decisions will be reviewed by others, for example.[30]

Managers often display an excessive aversion to outcomes that may yield a net loss. In one example, a mid-level manager of a large firm said he would accept a 50/50 chance of a $300,000 gain for his company against a risk of losing only $50,000. In this case, the manager was six times more risk-averse than rational decision-making would dictate. The consequences of such behavior are clear: Firms whose managers follow this pattern will systematically choose to pass up highly profitable opportunities.

Another significant area of logical distortion involves the scale of the decision. Big decisions define the strategic trajectory of a firm. The implementation of these big decisions, on the other hand, requires that thousands of small decisions be made at all levels of the company. In a sense, small decisions "operationalize" the big strategic choices of firms.[31] Surprisingly, decision-makers tend to be far more optimistic when making big decisions than small ones.

Big decisions are often framed in terms of past successes and desirable future outcomes, giving them a "feel of believability." Executives who have had several consecutive successes, for example, may come to believe that they have a "hot hand," and are therefore willing to accept higher levels of risk, while spending insufficient time evaluating negative consequences. Their belief in their own invincibility causes them to make irrational choices.

Similarly, in the aftermath of a highly profitable product launch, firms will tend to invest "found money" in projects with much higher levels of risk, despite the fact that they have "nowhere to go but down." On the other hand, companies that are in desperate financial straits will tend to invest conservatively, directing their available funds toward obsolete or mature cash cows, and canceling exploratory ventures.

These behaviors are so common in the domain of enterprise that they appear on the surface to be defensible. Yet if a military general were to attempt a risky foray, despite having already achieved battlefield superiority, we would think him a fool. Similarly, a general in danger of losing his army would be far better off adopting a daring and risky strategy with a chance of victory, rather than sticking to conservative strategies that might lead to ultimate defeat.[32]

This "presumption of success" with respect to broadly framed strategic decisions stands in sharp contrast to the risk-averse behavior that typically applies to small decisions. These everyday choices are usually narrowly framed in terms of the reputation of the manager, internal company politics, local optimization, short-term budget constraints, etc. Thus, a conflict is virtually inevitable between the irrational optimism under which major strategies are formed and the risk-averse conservatism under which they are often implemented.

In technology-intensive industries, decision-makers must face a whirlwind of change and uncertainty. There are no sure things and few predictable outcomes. Survival in such an environment demands high levels of confidence and optimism. Naturally, these same traits can be a barrier to objectivity when evaluating the viability of new technologies. When taken to an extreme, overconfidence can breed contempt for rivals and result in critical errors in judgment.

The majority of people believe that they are above the median in their endowment of positive traits or beliefs.[33] Studies have shown that more than 80 percent of entrepreneurs, for example, believe that their chances for long-term success are 70 percent or better, even though the five-year survival rate for new firms is only around 33 percent.[34] Scientists and engineers are certainly above the median when it comes to overconfidence. They often suffer from delusions of "complete knowledge" of a subject, and harbor unrealistic assessments of their control over future events. Whereas technologists may be risk-averse when presented with small, narrowly framed decisions, it is likely that they will be irrationally risk-seeking when it comes to estimating the potential success of new technologies.

### Faulty Decisions as a Fact of Life

We can define a rational choice as one that meets the following three criteria:[35]

1) It is based on the decision-maker's *current condition,* including financial assets, state of mind, capabilities, etc., 2) it is based only on the possible consequences of the choice, and 3) the basic rules of probability theory are used when the consequences of a choice are uncertain. Despite the reasonableness of these requirements, people violate them on a surprisingly regular basis. When this occurs, due to habits, misconceptions, beliefs, emotions, pride, ego, or other logical distortions, a contradiction results. According to the *law of contradictions,* reasoning processes that reach contradictory conclusions based on the same evidence are irrational.[36]

One of the most disconcerting examples of contradictory decision-making is the "honoring of sunk costs." A *sunk cost* is any investment that is unrecoverable, except as embodied in the activity to which it was directed. Since sunk costs reside in the past, they should have no effect on the decision-making process; each new commitment of funds should be evaluated based on the current status of the project, without undue consideration of previous expenditures. If the potential return on the total investment (including all costs, sunk and otherwise) is less than the required additional funding, the rational choice would be to kill the project. Yet such a sensible decision is rare indeed.

The stock market understands the proper treatment of sunk costs far better than do the typical decision-makers within firms. On the day that Lockheed finally scrapped its money-losing L1011 Tristar product line, its stock value jumped $7^{3}/4$ points, despite a sunk cost of more than $2.5 billion. The most common motivation for honoring sunk costs is pride. Few decision-makers have sufficient strength of character to acknowledge that a new innovation project has reached a dead end, particularly if they had a hand in its inception. This is not just an individual failing; organizations are collectively reluctant to jettison old institutions or structures, even when it is a clear case of "throwing good money after bad."[37]

A variation on the sunk-cost distortion occurs when individuals or firms utilize budgeting to allocate financial resources. The very act of establishing a budget distorts the perceived value of money. Rather than all funds being considered with equal weight, a budget creates an often arbitrary line of judgment. Above the line, money is available to spend without scrutiny. Below the line, each expenditure might elicit a reprimand. Under these conditions, it is difficult to imagine that rational decisions can be made. Vital activities in one functional area, for example, may be truncated because that group is over budget, while nonessential expenditures may be mounting up in a less-strapped department.

Scale can also impact the rationality of decisions. Individuals tend to consider choices in terms of ratios, rather than absolute frames of reference. For example, you might think it reasonable to drive across town to save $25 on food, but would likely not make the same drive to save $25 on the price of a car. Even though the time required to save $25 is the same, the amount seems trivial when compared to the price of a new automobile. In this case the decision process is distorted because the ratios of amount saved to total investment are very different.[38]

When confronted with an unfamiliar problem, our minds search for *anchors* to provide a frame of reference for decisions. Unfortunately, these anchors can skew our perspective dramatically. If individuals were asked to guess the number of jellybeans in a jar, for example, their choices would be very different if they were told to "guess a number between one and ten thousand," rather than if they were told that the allowed range was "between one and fifty thousand." Experiments have shown that individuals will consistently skew their responses toward the greater number, even though that number might seem impossibly large. This effect is present even in non-quantitative situations; extremist rhetoric (i.e., propaganda) serves as a subconscious anchor, subtly distorting the perspectives of those who are continuously subjected to it.

A similar distortion in thinking can be caused by the order in which choices are presented. Experimental subjects with no mathematical background were asked to estimate the value of 8-factorial (the product of all integers from 1 through 8). When they were presented with the problem stated as "What does $1 \times 2 \times 3 \times 4 \times 5 \times 6 \times 7 \times 8$ equal?" the average response was 512. When the question was phrased as "What does $8 \times 7 \times 6 \times 5 \times 4 \times 3 \times 2 \times 1$ equal?" the average response was 2,250. In this case, the order of the numbers significantly altered the perceptions of the subjects. Moreover, the relatively small size of the numbers being multiplied together served as an anchor for their thinking, causing most subjects to guess totals far below the correct value of 40,320.[39]

The source of such disconcerting irrationality is embedded deep within our cognitive processes. Much of our daily problem-solving and decision-making activity is performed without systematic analysis. The human mind accumulates a "look-up table" of reflexive responses to familiar situations, which have been reinforced over years of repetition. Psychologists refer to this cognitive process as *automatic thinking*, in contrast to the formalized structure of *scientific thinking*. When automatic thinking is taken to higher levels of complexity, it is typically thought of as *intuition*. Studies have shown, for example, that chess grandmasters formulate a surprisingly large portion of their game strategies through automatic thinking.[40]

While automatic thinking can be extremely powerful when restricted to a well-defined set of conditions (e.g., the rules of tournament chess), it can be a source of irrational decision-making when the heuristics developed over years of experience conflict with the realities of a new and unfamiliar problem. This is a leading cause of the decline of mature firms under rapidly changing market conditions; they fail to recognize that the complex methodologies and decision processes that had served them well in the past are no longer applicable. New entrants, on the other hand, have no such history, and can therefore evaluate changing market conditions more objectively.

Even when individuals attempt a thorough analysis of an impending decision, the frailties of the human mind limit their chances of success. Our memories, for example, are not reliable when used to estimate the probability of future consequences. We tend to "remember" extreme or unusual circumstances far more vividly than routine occurrences. When asked how long it takes to get from point A to point B on the Los Angeles freeways, for example, most of us will recall our worst experiences, and tend to neglect the more mundane rush-hour traffic. Likewise, we more readily remember great successes or catastrophic failures, and will give them irrational weight when making related decisions.[41]

In general, the past must be used with extreme caution when making decisions regarding the future. There is a natural tendency for individuals to assign causality to unrelated events as we attempt to establish some coherence to our memories. Although we seem to observe "patterns" to our past experiences, these memories are typically not based on a statistically accurate sampling of events. We might, for example, come to believe that a certain golf hat is "lucky," since we seem to always shoot a better round of golf when wearing it.

When taken to an extreme, such thinking can lead to dangerous stereotypes. A bad experience with a new employee from a particular university might lead us to consider that school to be substandard. Failed attempts at implementing new manufacturing processes in the past might carry over into our decisions about adopting new technologies in the future, even if the past failures were completely unrelated. It is important to differentiate coincidence and causality, particularly if there is no obvious causal force connecting the two events.

Memory, opinion, intuition, and experience can all contribute to the making of optimal choices, provided that extreme care is taken to maintain objectivity and ensure applicability. In general, rational decisions are based on objective estimates of future possibilities and probabilities; the past is useful only if it applies directly to the formation of those estimates.

## Organizational Decision-Making

Both the formation of alliances and the development of revolutionary new technologies by firms are subject to gross overconfidence. This irrational optimism is common when the possible consequences of a decision are so monumental as to be impossible to visualize objectively. In such cases, organizations will often "just go for it," without the careful planning and risk-mitigation strategies necessary to guard against failure.

Organizational overconfidence is a symptom of a deeper issue: Technical and functional personnel (e.g., nontechnical administrators, planners, executives, etc.) view the world in fundamentally different ways. When projects are promoted by technical champions within an organization, they are almost always given too high a probability of success. This is an unavoidable situation; few people will choose the thankless task of championing a new technology or strategy if they do not have a strong belief in its potential for success. The aggressive salesmanship of a champion tends to shift the mental anchors of decision-makers within a firm away from more objective measures of success (such as net present value or internal rate of return). We are all susceptible to making irrational choices after being exposed to infectious enthusiasm.

The effects of anchoring on strategic thinking will typically cause firms to search for solutions in the neighborhood of the current alternative. Firms organize their knowledge concentrically, with the most protected beliefs at the center. It is therefore exceedingly difficult to escape the vortex that these central beliefs create in the decision process. This is the reason why visionary leaders exhort decision groups to "think out of the box" or to "abandon old paradigms."

Decisions, by necessity, must be made under conditions of imperfect information. Under such constraints, firms that provide the most efficient and objective flow of information to senior management are the most likely to succeed. An organization structured around product lines, for example, might be more effective at providing balanced information than ones with a functional hierarchy. The informing of strategic decisions will be covered in more detail in Chapter 10.

Organizations will often make decisions based on scanning a set of options and choosing the first "acceptable" alternative. In this case the order and prioritization of choices is critical to effective group decision-making. It is incumbent on the leader of a group discussion to present a fair and objective agenda of options, and to avoid cluttering up the decision process with superfluous information and second-tier alternatives.

Moreover, group dynamics under highly stressful conditions can drive debate into polarized, personality-driven irrationality. In a typical scenario, an executive team is confronted with an uncomfortable dilemma that requires immediate action. With the level of fear already running high, several strong individuals propose their favored solutions, and the remainder of the meeting is spent in an emotional debate over these choices. Frequently such knee-jerk recommendations fail to include the best solution, but the objectivity needed to consider all reasonable possibilities has been lost. All that remains is the narrow (and blatantly irrational) framing of personal pride and stubbornness.

## Maximizing Rationality and Objectivity

The combination of incomplete information, framing effects, and automatic thinking can dramatically distort the ability of individuals and firms to execute optimal decisions. Although these effects are almost unavoidable, they can be minimized. This section describes several techniques that can increase the rationality of choices and compensate for the inherent biases in our perceptions of the world.[42]

Decision theorists have concluded that although it is impossible to achieve a perfectly rational decision process, it might be possible to craft a practical alternative through the use of *bounded rationality.*[43] According to this school of thought, the proper selection of bounds on choice is likely to be the single most important determinant of successful decision-making.

There are two reasons why bounded rational decision-making is a sensible compromise. First, it is unlikely that all the possibilities and consequences that relate to a decision can be gathered in a reasonable period of time. Under such conditions, it is sensible to devise a systematic method for harvesting relevant information over a set period, and then make a decision based on the available options.

The second motivation for accepting the constraints of bounded rational choice is *decision costs.*[44] How much time and money should be spent in gathering information and alternatives before a decision is made? Clearly this depends to a great extent on the magnitude of the decision. For smaller decisions, we are typically comfortable with bounding our choices. Yet even for strategically important decisions, the limitations of time and cost must ultimately prevail.

One of the most effective methods for bounding choice (and simultaneously limiting decision costs) is a procedure known as *elimination by aspect.* Suppose, for example, that you must select a new assistant from hundreds of applicants. After being confronted with an impossibly large stack of résumés, the logic of bounded rational choice becomes vividly clear. How can one make an optimal decision from such a massive set of options without incurring excessive decision costs?

Elimination by aspect involves the selection of screening criteria that enable the rapid (albeit somewhat arbitrary) reduction of choices. Ideally, the screening criteria are ranked in order of diminishing importance. In the example described above, the first criterion for screening candidates might be proficiency in word processing. Once those applicants with insufficient computer skills are eliminated, a second important criterion could be imposed, and so on. Ultimately, you would be left with an acceptably small set of applicants whose résumés can be given a thorough review. All decisions must be informed decisions, but the importance of the choice should determine the depth and breadth of information that must be gathered.

There are some powerful methods now available to assist firms in the gathering and organizing of decision-support information. In a typical firm, the kinds of data needed to inform managers and executives include competitor performance, economic statistics, material prices, consumer demographics, technological risks, process yields, equipment performance, etc. Decision-support systems are designed to harvest the appropriate categories of information and present them in a "predigested" form.

Several types of decision tools are gathered under the collective designation of *data warehousing.*[45] Underlying these information systems are structures such as decision trees, payoff tables, and risk/utility analyses that allow raw data to be categorized, filtered, and prioritized. The form of these tools varies, but typically they are built around a relational database, and utilize query tools, corporate planning models, or expert systems to shape the raw data they acquire. The benefits of automating the decision process are obvious: speed, efficiency, and consistency. The primary risk is that the processes used to eliminate alternatives and prioritize information might not be optimal, and might be poorly understood by the decision-maker.

Even in the pragmatic context of bounded rational decision-making, there will be significant distortions due to the inherent thinking processes of individuals. Is there no place for heuristics and intuition in the decision process? Actually, the making of "intelligent" choices is a high-level tacit skill that can represent a sustainable competitive advantage for firms. The ability to filter alternatives, weigh options, and estimate unknown probabilities takes years of experience to develop. Rather than deny the human aspects of decision-making, firms should seek a balanced mix of analytical (metrics-based) reasoning and informed intuition.[46]

Two familiar categories of tacit decision-making ability play a valuable role in the formation of strategy within firms. Major decisions such as the launching of new product platforms or transitioning to a new technology s-curve require *vision*. Vision-based decisions can be ideally described as broadly framed, long-term, risk-neutral, and objective. The countervailing pressures that often defeat vision-based decisions include short-term incentive programs, inaccurate or incomplete information, and the inability of less visionary individuals to set aside narrowly focused issues and personal prejudices.

Decisions that represent the day-to-day implementation choices required to execute a grand vision are best made through *judgment*. Judgment is based on the accumulated experience gained through years of making related decisions. Judgment-based choices can be ideally described as moderately framed, medium-term, and slightly risk-averse. Unfortunately, the punitive attitude of many firms toward errors in judgment, and the nature of the specialized individuals who often make these decisions, tend to distort judgment into something that is detail-obsessive, short-term, and fear-driven.

The most challenging aspect of decision-making in technology-intensive industries is the estimating of probabilities for various outcomes. Several factors limit our ability to objectively assess such probabilities, beginning with the abysmal understanding that most decision-makers have of basic probability theory. The need for statements such as "the absence of evidence is not evidence of absence" and "because something is possible doesn't mean that it is probable" highlights some of the logical weaknesses that are prevalent in managerial thinking.

The use of past experience as a predictor of future probabilities is a common method used by decision-makers. Provided that the current choice is substantively similar to previous situations, the application of past history can be a useful estimating tool. More commonly, however, past decisions only *appear* to be similar. Often some vital difference in current conditions causes past history to be of little relevance, rendering such information far less valuable in estimating future probabilities. Moreover, much of our past "experience" is derived from sources other than our own perceptions. Vicarious knowledge from colleagues, trade journals, and even the popular media can infiltrate our thinking in insidious ways, thereby corrupting our ability to accurately assess future prospects.

Another source of distortions to our estimates of probabilities stems from what is often called *regressive thinking*.[47] Regression occurs when a future event is not perfectly predictable from past events. Tall parents, for example, will typically have tall children. Yet there is a considerable amount of regression in this relationship. In other words, there is significant statistical variation in the prediction of individual children's characteristics based on parental characteristics. Failure to recognize this type of variability between related phenomena can result in severe errors in judgement.

In general, decision-makers are ill-equipped to evaluate future prospects in an objective manner. As such, a brief survey of probability theory is in order. The probability of a specific event is, by definition, the ratio of its rate of occurrence to the total occurrences of all possible events. A flipped coin has a 50/50 probability of coming up "heads," because the rate of occurrence of heads is one-half the rate of occurrence of all possible events (i.e., heads plus tails). Likewise, the rate of occurrence of snake eyes in craps is 1/36, since snake eyes can be formed in only one way, but there are 36 possible combinations of the numbers on two dice.

Frequently, however, decision-makers will not recognize the total number of possible outcomes, and give too high a probability estimate to events. The example diagrammed in Figure 9.3 illustrates this effect. Suppose that a firm has two groups that manage new development projects: Department A and Department B. During a status review, the firm's CEO announces that 80 percent of projects within the firm that are over budget reside in Department A. Does this mean that the CEO should assign the next new project to Department B, since it appears to have superior budget-management skills?

The answer is a resounding no. As the Venn diagram in Figure 9.3 illustrates, the CEO has not considered the relative number of projects being performed by each group. If Department A was responsible for the vast majority of the firm's projects, it would not be surprising that it would also have the lion's share of over-budget projects. What should be compared is the ratio of over-budget projects in Department A to the *total number of projects in that department.*

In summary, the process of executing high-quality decisions is as important as it is problematic. Although there is no way to completely eliminate the factors that cause irrationality in our choices, we can compensate for them to some degree by considering the methods described above. At the highest level, a

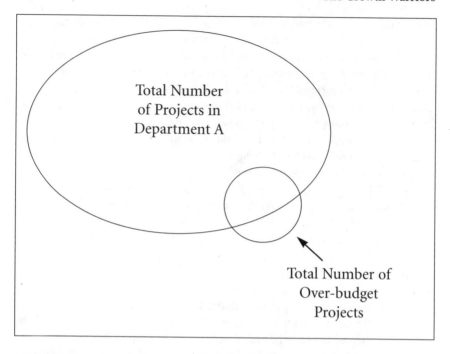

**Figure 9.3:** Venn diagram of the probabilities of a project going over budget in Department A, as compared (incorrectly) to the total number of over-budget projects, and as compared (correctly) to the total number of projects within Department A.

dramatic difference can be achieved by taking just a few simple steps: Consider only relevant alternatives, do not confuse familiarity with probability, and recognize that "reality" as we observe it is a subset of possibility, not vice versa.

## Creative Destruction or Rapid Adaptation?

Is it inevitable that the forces of capitalism must destroy mature industries even as they give birth to new ones? The process of "creative destruction" introduced by Schumpeter suggests an enterprise equivalent of natural selection: As the business environment evolves, firms possessing traits that enhance their chances of survival will flourish, while those that cannot adapt will perish. This biological analogy highlights the most critical aspect of strategy formation in technology-intensive industries: the capacity to sense changing market conditions and rapidly adapt to these new environments. If mature industries can learn to adapt, their destruction is most emphatically *not* inevitable.

Familiar terms such as agility and flexibility have become catchalls for virtually any strategy that survives the test of dynamic markets. Although it is hard to deny the importance of these concepts, they are just superficial descriptors for the attributes of successful firms. Predicting that "agile" firms have a greater chance of survival under evolving market conditions is not a very satisfying prophecy. One can easily predict, for example, that "agile" gymnasts have a greater chance of qualifying for the Olympics than "rigid" ones. Agility is no more than an entrance ticket to rapidly changing markets; adaptability is the key to long-term success.

What does a highly adaptive company carry with it during its market-driven transmutations? Product lines are the first casualty of changing market conditions, due to obsolescence or faltering demand. Intellectual property has a short half-life, and is often rendered valueless long before markets become fully commoditized. Even the identity of firms might be sacrificed during a corporate transformation. International Harvester is now Navistar and Bell Telephone Laboratories has become Lucent Technologies. Is there nothing sacred in the adaptive firm's quest for newness?

The asset that is most likely to follow a firm's evolution into new markets is its pool of specialized human capital. These precious resources are the catalyst for value creation, and embody the tacit experience and accumulated technological competencies that are the essence of an enterprise. Agility allows firms to make rapid adaptive changes to organizational structures, products, and market positions. It is a firm's base of tacit knowledge that translates agility into sustainable market leadership.

## Tacit-Based Strategies in Fragmented Markets

One of the more dramatic ramifications of the recent commercialization of the Internet is the potential for reducing the "stickiness" of prices. In the past, prices have tended to remain stable for extended periods, primarily due to the costs associated with changing them. Often the price of a commodity can vary considerably across geographic distances and among competing suppliers. The power of the Internet to negate distance, reduce price-change costs, and provide virtually instantaneous price comparisons might change all that.

As global competition increases, there will be greater pressure for prices to react rapidly to market demand. The potential for customers to perform rapid price comparisons will eventually drive the last vestiges of monopoly pricing out of commodity industries. What will remain will be an environment similar to today's commodity-futures markets, with bidding and selling prices hovering within a hair's breadth of marginal cost.[48]

The only producers that can escape this low-profit fate are those whose products cannot be "commoditized." If the quality, performance, reliability, customization, or other features of a firm's products make them both unique and difficult to imitate, there remains a potential to derive monopoly profits.

To avail themselves of these excess profits, it may be necessary for firms to intentionally fragment their markets. A fragmented industry is one in which a number of smaller firms compete and no small group of firms controls a dominant share of the market.[49] This is becoming a common scenario in technology-intensive industries, due to the high rates of change in markets and the increased levels of specialization required to create leading-edge products.

Fragmented markets offer excellent opportunities to exploit tacit knowledge, since economies of scale are typically not a dominant factor. Moreover, it is often possible for smaller firms to instigate the fragmentation of a consolidated (and often relatively mature) industry, by exploiting their deep understanding of customer needs and forming "boutique" markets for customized products. An excellent example is the fragmenting of the commodity DRAM and EPROM markets by Silicon Valley custom-chip houses, as discussed in Chapters 5 and 7. Some low-tech examples include the "designer jeans" market in the late 1970s and the proliferation of microbreweries today.

One of the best ways to incorporate tacit knowledge into products is by exploiting relationships with sophisticated customers. Often these customers are "leading indicators" of broader market demand in the future, and can provide a highly collaborative environment to prove out new innovations. It is important, however, that projects be selected carefully to ensure that a commercially viable product can be harvested from the relationship. Otherwise, firms will find themselves in an endless cycle of point-design efforts that cannot be aligned with a well-defined market segment.[50]

One example of a product derived from the combination of tacit expertise and technical skill is the Sentinel software tool developed for Nynex Corporation by two of its operations managers. By transforming their experience, judgment, and creative problem-solving skills into powerful object-based code, these experts were able to create a unique system-management tool that dramatically reduces response times for customer orders and inquiries.[51]

Tacit knowledge need not be directly embodied in a product to provide a competitive advantage. The development of a proprietary project management tool by Black & Veatch LLP, a leading power-plant engineering and design firm, is another example of using software to capture valuable tacit knowledge. Its system is built around a common database that ensures that information is entered only once and can subsequently be shared by all members of the project team. The database incorporates "componentized knowledge" about every aspect of the firm's specialty, including supplies used, shipping times, a schedule history for each phase of the project, cost-estimating data, and construction-management insights. The tool has enabled Black & Veatch to lead its industry with a 9.4 percent market share, despite being only a fraction of the size of such established competitors as Raytheon and Bechtel Group. This is an example of a firm gaining a potentially sustainable advantage by automating its deep understanding of an industry.

Another way for firms to leverage their tacit knowledge base is through the use of simulation tools, both in the design and development of products and in the formation of competitive strategies. The application of computer simulation to product design and development is ubiquitous in today's technology industries. The use of similar techniques to test prospective market strategies is not nearly so common. Yet there is evidence that the use of strategy-simulation tools can provide valuable insight, particularly when they are customized to capture the sophisticated understanding of market conditions within the heads of executive management.

Shell Oil Company recently collaborated with a strategy consulting firm to create a customized "war game" simulator to test customer response in several market segments. This simulation tool gave Shell the opportunity to estimate market acceptance and predict competitors' reactions to a planned introduction of unmanned, self-service gas stations. After several iterations, it became apparent that although the strategy would give Shell an early advantage, its gains would quickly vanish as competitors copied the new distribution concept.[52]

The limitations of simulation tools for strategy formation are the same as those of any automation tool: they are only as accurate as their algorithms and inputs. If a firm has a deep understanding of the competitive environment within its industry, it may gain some strategic benefits through a well-crafted simulation. Firms that lack such competitive insight, however, are destined to have their own inadequacies amplified through such a tool.

## Exploiting the "Turnkey" Interface

Beginning in the 1980s, new entrants into the semiconductor industry were confronted with massive financial barriers. The cost of wafer-fabrication facilities had grown beyond the reach of many small start-ups. Undaunted, entrepreneurs made arrangements to buy "foundry time" from existing factories, thereby launching a new era of "fabless" semiconductor firms. This new business model emphasized the tacit design and marketing skills of these firms, while substantially reducing their front-end capital investment. Not only do these scaled-down companies avoid the commitment of capital, but they also benefit from the economies of scale achieved by these so-called "foundries." In some cases, the firms never touch the physical product; the foundry serves as both a fabrication and distribution center for their chips. This rapidly expanding niche now accounts for nearly 10 percent of the world's semiconductor revenue.[53]

The attractiveness of this "turnkey" production strategy is that it allows new and innovative concepts to be brought to market with remarkable speed. Moreover, firms that adopt this streamlined model can appropriate higher returns on their creations; in addition to controlling the manufacturing of their own designs, they derive additional profits from licensing their design cores to multiple customers. These firms have discarded all pretense of being a bricks-and-mortar enterprise. Instead, they have evolved into a structured pool of tacit skills, selling their non-rival goods on the open market and reaping the benefits of increasing returns to scale. From the standpoint of both operating profits and return on net assets, the results are impressive, as shown in Figure 9.4.[54]

Along similar lines, the contract electronics manufacturing (CEM) industry grew out of a desire by electronics original-equipment manufacturers (OEMs) to outsource the labor-intensive "stuffing" of printed circuit boards. Twenty years ago, total industry revenues were in the tens of millions. In 1997, the worldwide CEM industry accounted for more than $20 billion in sales, with an average annual growth rate of more than 20 percent. For that same year, it is estimated that 16 percent of electronics manufacturing worldwide was outsourced to this industry.[55]

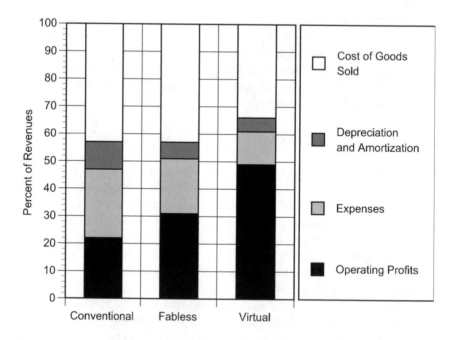

**Figure 9.4:** Percentage of revenue from semiconductor-device manufacturing consumed by cost of goods sold, capital equipment costs, and operating expenses for three different business models. The fabless model assumes that the firm performs the distribution and customer service function, while the virtual model allocates these functions to a contract foundry as well. Source: Intel Corporation.

CEMs are able to capture economies of scale by distributing the costs of capital-intensive manufacturing processes and support systems over a number of customers. The growth of the CEM industry has been especially beneficial to smaller firms. By exploiting the volume purchasing power of these manufacturing specialists, small firms can achieve material costs similar to those of larger firms. This is a critical factor in the electronics industry, since material costs for electronics products can be as high as 90 percent of the cost of goods sold. Furthermore, the rapid amortization of equipment costs by CEMs allows them to upgrade process technologies at a far greater rate than individual firms could justify. Thus, CEM clients benefit from the latest process technologies, without so much as a blip on their balance sheets.

In recent years, CEMs have climbed the production food chain from unskilled assemblers to turnkey manufacturers, and beyond. Today it is not uncommon for a contract manufacturer to perform the majority of product design and development activities, with the client serving as a market-sensitive collaborator. This arrangement is archetypical of the "virtual company" touted in the popular press, wherein the boundaries between firms become highly permeable, allowing the flow of ideas and responsibilities to seek their level.

The presence of CEMs dramatically alters the playing field for new high-technology start-ups. Software developer Palm Computing (now a division of U.S. Robotics), for example, contracted with Flextronics, a leading CEM, to perform the design and development of its highly successful personal digital assistant. The combination of the tacit software skills of Palm and the specialized product development expertise of Flextronics enabled a higher level of productivity than either firm could have achieved independently.

The CEM of tomorrow will not just design and manufacture products for its clients, but it will also manage logistics, sales, customer service, and distribution. In this innovation-friendly environment, technology-intensive enterprise will evolve toward the slender model of the fabless semiconductor firm: a structured pool of creativity and market awareness.

### Design Reuse and Mass Customization

Three closely related development strategies have enormous potential to increase the productivity of firms: design reuse, modular/scaleable design, and mass customization. In a sense, these concepts are just different manifestations of the same underlying principle: by taking a systems approach to the product development process, it is possible to dramatically increase returns on R&D investment.

With the recent adoption by many firms of concurrent design techniques, including design for manufacture and design for testability, there has been an acknowledgment of the co-dependency between products and their associated manufacturing processes. When successfully implemented, this integrated approach to product development enables R&D, marketing, and manufacturing to become a *system for value creation.*

If we extend this system perspective to encompass entire product lines, exciting opportunities to increase productivity become apparent. By considering all models within a product line simultaneously, it is often possible to commonize components, reuse design elements, standardize interfaces, and build in customization features. Although the costs associated with designing such an integrated value-delivery system are somewhat higher than is the case with developing a single "point design," the aggregate productivity of R&D will increase geometrically.

The most obvious opportunity for exploiting systems thinking in product development involves the reuse of existing design elements on future projects. Design reuse is the "low-hanging fruit" of R&D productivity enhancement. By applying existing design elements to new products, firms gain in three ways: The cost of new-product development falls, the time required to introduce new products diminishes, and the return on the initial R&D investment multiplies.

Given this stunning résumé, it is hard to believe that not all firms embrace modular, reusable design principles. Yet the pressures of market timing and tight budgets still render far more "one-shots" than "recyclables." Even knowledge-intensive industries such as software and semiconductors have just begun to tap the potential of design reuse, as discussed in Box 9.2.[56]

Once the desirability of design reuse is recognized, R&D productivity can be further enhanced through design principles such as modularity and scaleability. The goal in each case is to "design once, reuse often." Modular design elements are defined through system partitioning; interfaces, functionality, and form are all established in the context of the entire product line. Electronics-based products can utilize techniques such as standardization of communications protocols and power requirements, sharing of bus addresses, and the use of "building-block" architectures.

Scaleability leverages front-end design efforts along a different axis. In this case, design elements are structured to be scaleable in an appropriate dimension or performance parameter. The printed circuit boards of an industrial motor-control system, for example, might be designed to accommodate several horse-power ratings through simple part changes. The RAM memory within personal computers is scaleable, in that the same power supply, motherboard, etc. can easily be modified to dramatically increase memory capacity. If different amounts of memory required entirely different internal architecture, personal computers would be considerably more expensive.

The concept of mass customization is a synthesis of the above design principles with the marketing and production functions of a firm.[57] A successful mass-customization strategy depends on a multifunctional development process that concurrently designs entire families of products, along with flexible manufactur-

## Box 9.2: The "Logic" of Design Reuse[63]

Nowhere is the need for productivity growth more compelling than in the semiconductor industry. With global competition at a fever pitch and manufacturing-process technology evolving exponentially, it is estimated that for custom-device firms to survive, their design productivity must increase by an average of ten to fifteen times by the year 2001. Much of the needed productivity gains will come through systematic reuse of core circuit design building blocks, often referred to in the industry as "intellectual property elements."

The productivity benefits of design reuse impact the integrated-circuit design process in several ways. There is the obvious benefit of not having to start from scratch on every new design. To manually lay out a typical intellectual property block (consisting of about six thousand transistors) in 0.25 micron technology, a firm might consume twenty-four person-months. By scaling a suitable existing design from obsolete 0.8 micron technology to the more advanced process through automated design-migration tools, the labor required is reduced to roughly two person-months—a 90 percent saving. Moreover, since the previous designs have already been debugged, tested, and proved in the field, additional productivity gains are likely as the product is prepared for launch.

To develop a design-reuse capability, firms must have a means to store, access, sort, and evaluate their legacy of circuit building blocks. This is accomplished through maintaining a design library, using one of several library-development software tools. Even with automation, however, developing a reuse library is often beyond the capabilities of smaller firms. Instead, these companies can purchase design blocks from third-party libraries by paying an affordable licensing fee. Given the obvious benefits, the third-party intellectual property business is expected to explode over the next decade.

Design reuse is not without its limitations. Fundamental shifts in technology, such as the recent migration from 5-volt circuits to a 3.3-volt standard, can render existing designs virtually unusable. Furthermore, if a design building block is not quite suitable for a new application, the cost of modification and "reintegration" can be quite high. For design reuse to yield the highest possible productivity gains, engineers must aim to create circuit elements that use simple, standard components and interconnections. Squeezing the last bit of performance from a new design may render it more difficult to reuse in later development efforts.

ing processes.[58] Mass customization achieves low cost at relatively low volumes through capturing *economies of scope*; a single production line can produce literally thousands of varieties of product without sacrificing cycle time.[59]

The literature identifies at least four distinct marketing strategies that exploit mass customization: collaborative, adaptive, cosmetic, and transparent.[60] Collaborative customizers establish a dialogue with their customers, often with the help of point-of-sale data-collection systems. Examples include footwear and eyewear that can be tailored to a customer's physical requirements and esthetic sensibilities, and computer products that can be configured to specific needs through a rules-based order-entry system.[61]

An adaptive customizer produces a standardized product that can be adjusted to meet unique customer needs. Virtually any product that is programmable or alterable by the user falls in this category. Examples include telecommunications switches that are user-configurable, programmable lighting systems, and adjustable car seats that "remember" the comfort preferences of multiple drivers. This category represents an excellent example of enhancing the value of a physical product by increasing its knowledge content, often through the use of software control.

Cosmetic and transparent customizers offer products that are essentially the same from a technological standpoint, but are packaged and delivered in customer-specific ways. This can be particularly important in global markets, where local tastes and logistical constraints may dictate a preferred method of delivery. Examples of cosmetic customization include multilingual packaging, alteration of package sizes or materials, and the use of returnable containers. Transparent customization exploits advanced knowledge of the customer's needs, and allows the producer to tailor semi-standard products to meet specific requirements. Altering the formula for a popular soft drink to suit regional tastes is an example of transparent customization.

From the standpoint of production, the key concept that enables mass customization is *postponement.*[62] As an order for a product travels through its value chain, there are several points at which choices must be made. At each of these decision points, some degree of flexibility in product configuration is lost. In the fabrication of a personal computer, for example, once the power supply is installed the geographic applicability of the product is determined; different regions of the world have different power standards. By postponing the installation of the power supply until after an order has been received, a PC manufacturer can avoid warehousing dozens of versions of its product for international markets.

Hewlett-Packard has moved beyond the postponement strategy to solve its global power-supply dilemma. Rather than postponing the installation of dedicated 110V and 220V power supplies, it has developed a universal power supply that can accommodate both markets. This "world product" approach has an added benefit: Even after products are shipped to various global markets, they can be reallocated as necessary to correct for imbalances in demand.

The productivity gains that result from the preceding strategies are just part of the story. The tacit skills and proprietary processes needed to develop and manufacture mass-customized products are not easy to imitate. Moreover, these techniques resist reverse engineering, since they span entire product lines. Thus, adopting a systems approach to product development can yield a potentially sustainable advantage in productivity.

# 10

# The Tools
# of the Oracle

## Building an "Omniscient" Enterprise

The use of hyperbole in the title of this section is intentional. Leaders of rapidly evolving technology industries are routinely confronted with a barrage of unrealistic admonitions from business writers, consultants, and pundits. They are told, for example, that firms must thrive on chaos, destroy old organizational structures, engineer new ones, learn how to learn, slash development times, manage cultural diversity, and all the while find a way to pay the bills. Each of these "silver bullet" solutions treats a different symptom, but all of them fail to address the underlying dilemma: rapid, unanticipated, and in some cases catastrophic change.

With respect to enterprise leadership, it seems that "ordinary" strategic vision is no longer adequate. Even the most brilliant visionaries have been rendered laughably wrong by the swirling currents of technological change. Near the end of World War II, Tom Watson, the patriarch of IBM, observed, "I think there is a world market for about five computers." A generation later, Ken Olsen of DEC made a similar prediction with respect to the infant personal computer market by flatly stating, "There is no reason for an individual to have a computer in their home."[1] In such unsettling times, even the development of a *learning organization* seems too weak a response when huge R&D investments are at stake. Given the choice, most executives would prefer to bet their futures on a *knowing organization,* rather than a learning one.

Beyond the myriad challenges faced by traditional industries, technology-intensive firms must contend with an additional risk: Will changes in the competitive environment render a new product valueless before it can reach the marketplace? Such uncertain conditions make the commitment to develop a breakthrough technology a frightening prospect, yet progress would grind to a halt if every firm in an industry adopted a fast-second strategy.

In this section, the ability of firms to *learn from the present* will be explored, as a first step toward analyzing future opportunities. Every available tool for "predicting" the direction and magnitude of future change depends on a deep understanding of current conditions as a starting point. Since no one can actually predict the future, the development of dynamic competitive strategy is reminiscent of the old joke about the two men being chased by a bear; their best strategy for survival is not to outrun the bear, but rather to outrun the other man. Although omniscience should remain a "stretch goal," a persistent advantage can be gained by learning faster and better than your competition.

### The Technology of Learning

The absorption and accumulation of knowledge within a firm can be achieved through a five-step process.[2] First, the groundwork for learning must be laid through the formation of firm-specific human capital. Second, once the employees of a firm are well prepared to absorb external information, channels must be established that provide a high-quality feed of competitive data. In step three, the firm develops methods to filter, condition, and shape incoming knowledge. This is the most critical stage, since the volume of potentially applicable competitive data is mind-boggling. It is vital that the kernels of information that provide the greatest leverage are sorted from the chaff of irrelevant data. In the fourth step, this filtered information is directed toward structures within the firm that allow efficient storage and retrieval of accumulated knowledge. The final step involves the construction of a detailed panorama of the external business environment, as it relates to the firm's competitive future.

This broad perspective, continuously updated and refined over time, serves as a foundation for strategy development. Much as in an impressionistic painting (or a cartoonist's caricature, as the case may be), those aspects that are crucial and relevant must be enhanced, while background features that have little bearing on competition should be suppressed. The art in creating such a landscape lies in knowing which features are distinguishing and which are superfluous.

Companies are currently using some powerful new collaborative tools as well as familiar methods to facilitate the first step of the above process, the development of human capital. A fledgling industry has been launched to meet the growing demand for Web-based training servers, authoring tools, courseware, and training services.[3] President Clinton has taken an active role in mentoring the development of Web-based learning technologies, by recommending that federal-agency training programs become model users of these new software tools.

The use of learning technologies is expected to grow from a 10 percent share of training time in 1995 to capturing more than 35 percent of total training time by the year 2000.[4] In 1997, about one in five companies utilized computer-based training methods, according to a survey of more than two hundred large U.S. companies. Among those surveyed, the number of users was expected to double in the following three years.[5]

Several major firms have collaborated with software developers to create firm-specific educational tools. The giant pharmaceutical firm Eli Lilly & Co. is pilot-testing a custom training application, intended for its research scientists, that runs on either Lotus Notes client or Web browsers. The application, known as the Scientific Performance Improvement Network (SPIN), combines threaded discussion groups, a directory of subject-matter experts, links to databases, and on-line course work. This system is intended not only to impart existing knowledge, but also to create new knowledge by facilitating collaborative interaction among various specialists.[6]

The primary goal of internal training is to enhance the human capital of the firm.[7] This is not an end in itself, however. Human capital is like a carefully prepared vessel: It must be filled with specialized knowledge to become productive. No matter how finely tuned an in-house training program might be, it is only a starting point. Once a firm's employees are properly conditioned, the second step in the process is to connect them to high-quality channels of external information.[8]

In traditional strategic-management curricula, the process of absorbing information from the competitive landscape is referred to as *environmental scanning*. Unfortunately, this is often treated as a passive, almost obligatory activity, without strong connection to tacit skill and strategy development within the firm. Treating absorption as an afterthought is simply not acceptable in today's technology markets. An aggressive, highly integrated program of harvesting and subsuming relevant external data is a more appropriate strategy.

The challenge in acquiring competitive knowledge is to push beyond the boundaries of the routine and mundane. Familiar sources of market and technical information, such as academic literature, conferences, and professional organizations, represent easy pickings. To achieve a competitive advantage in knowledge absorption, firms must also find creative (and even devious) ways to "scoop" their rivals.

If a firm happens to be as financially well endowed as Microsoft, it can simply buy exciting new technologies, along with the human capital that created them. In 1986, Bill Gates initiated the acquisition of a small operating-system company, Dynamical Systems, founded by Nathan Myhrvold. During the course of negotiations, Gates became disenchanted with the technical aspects of Dynamical's product. He continued with the purchase nonetheless, for the express purpose of "acquiring smarts in bulk." Today, Dr. Myhrvold serves as chief technology officer for Microsoft, and several other former Dynamical Systems employees have risen to senior technical positions within the firm.[9]

A somewhat more subtle strategy is used by firms such as Johnson & Johnson (J&J) to acquire proprietary (and therefore rare and appropriable) technologies. J&J has recognized that its own internal R&D staff cannot be expected to bring products to market rapidly while simultaneously developing the next break-through technology. Instead, it has established an organization to seek out promising new technologies from entrepreneurs and universities and arrange to license these innovations into the firm. In this endeavor, J&J has developed tacit skills both in the recognition of new technical trends and in the ability to "seduce" innovative start-ups into allying with J&J, as opposed to the dozens of other major firms competing for the same licensing opportunities.[10]

Other options that can offer a sustainable edge in knowledge absorption include direct investment, technology-sharing agreements, the use of "tech-watching" services, and industry-level cooperative arrangements. Foreign direct investment, in the form of equity participation in foreign research laboratories or firms, can be an effective vehicle for tapping into regional or country-specific tech-nological advantages.[11] Likewise, technology sharing at the industry level, including cross-licensing agreements and co-development consortia, can give members exclusive access to proto-competitive knowledge. These strategies are particularly effective as a foil against international competitors, although in some cases the par-ticipation of foreign firms might offer more advantages than disadvantages.[12]

The expanding need for high-quality business intelligence has been a boon to firms that specialize in "tech watching." In a marketplace where information is paramount, technology researchers are developing considerable clout. These firms often specialize in specific industry segments, preparing thick reports document-ing consumer trends, predicting the probability of success for new products, or identifying the "next big thing." The ability to customize these studies to a firm's unique needs is a powerful advantage of these services. On the downside, however, is the non-exclusivity of the information source; large companies such as Hewlett-Packard spend millions of dollars per year to gain access to every piece of compet-itive research available on the market.[13]

The scope of technology absorption will undoubtedly expand in the new mil-lennium. Improving the effectiveness of industry and government collaboration in this regard represents the next frontier in American technology policy. The econo-mist Paul Romer has proposed taking collaboration among firms to a new level, through the formation of industry investment boards.[14] These self-organizing con-sortia could potentially be funded by a "special tax" on their industry's products, with the proceeds going toward the creation of "industry-specific public goods." Although the primary use of such boards would be to solve technical problems common to all firms within an industry, the concept could easily be extended to include the creation of limited-access data warehouses, or the performance of in-depth market research. Whether Romer's rather ambitious concept is "politically correct" remains unclear. The need for industries to push the envelope of pre-competitive knowledge-sharing, however, will become increasingly evident as global integration accelerates.

The political arena remains the weakest aspect of knowledge absorption by high-technology firms. Industry leaders have been slow to recognize the impor-tance of gaining direct access to policymakers, particularly at the national level. Generally speaking, technologists abhor the rough-and-tumble of legislative pol-itics, and high-technology firms have been slow to form political-action commit-tees (PACs) capable of opening closed doors. On the government side, technology leaders are viewed as arrogant, naive in the ways of politics, and stingy with campaign contributions.

The result is that technology-intensive industries have surprisingly little access to "inside" legislative and policy information. Recently, executives from sev-eral major U.S. firms formed Technology Network (TechNet for short), a political organization with the primary mission of building and maintaining close rela-tionships with key players in national government. Founding members include the CEOs of Oracle, Netscape, Cisco Systems, and National Semiconductor, along with several leaders from the venture-capital community.[15] TechNet is attempting to fill a political void that groups such as the American Electronics Association and the Semiconductor Industry Association have consciously chosen to ignore.

The next two steps toward forming a realistic view of the competitive land-scape are the creation of filters for incoming data and the establishment of flexible structures to store and retrieve accumulated knowledge. From the perspective of an enterprise system, the flow of information might appear as shown in Figure 10.1. At each *gateway*, some level of filtering and shaping is performed to discard useless material and highlight critical insights. This concentrated data stream is then routed to the various knowledge-management structures of the firm. Upon retrieval, additional filtering is performed, along with a reduction of the data into a usable format. Possible retrieval formats include regular management briefings on recent reconnaissance, a set of standardized reports tailored to the needs of each functional manager, or an intranet database with open access to all employ-ees. Some advanced tools that support knowledge management within the firm are described in Box 10.1.[16]

Thus far, we have considered a set of steps that support the absorption and accumulation of competitive knowledge. Although these prescriptions can be daunting to implement, they pale in comparison to the challenges that lie ahead. Building on this foundation of high-quality information, firms can take the final step: establishing an integrated view of the present competitive environment, as a solid basis for developing visions of the future.

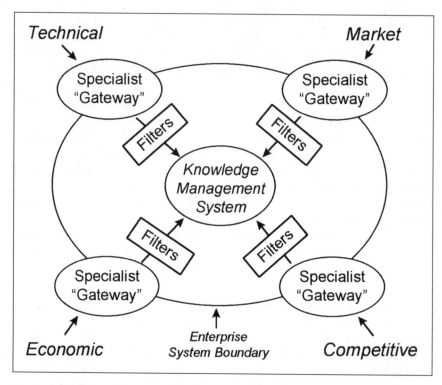

**Figure 10.1:** Conceptual diagram of the flows of information through the boundary of an enterprise system, highlighting the role of filters and gateways to properly concentrate and channel various forms of competitive knowledge.

### A Survey of Forecasting Techniques

It is interesting that the process of anticipating future change in the marketplace is most often referred to as *technology forecasting*.[18] The implication of this term is that all future changes worth considering will be technology-related. In reality, the global competitive environment is also highly sensitive to fluctuating exchange rates, flows of capital investment, the entrance of new competitors, subsidies by governments, regulatory reform, privatization actions, shifts in consumer demand, demographic trends, etc. Narrowing the focus of a forecasting effort to technological change alone would be a fatal mistake.

With that warning in place, however, I must admit that technological change is often the "eight-hundred-pound gorilla" of evolving competition. Not surprisingly, the need for an effective forecasting capability grows in proportion to the rate of change in the competitive environment. While demographic trends such as the "second baby boom" and the "global teenager" may be powerful market forces,

## Box 10.1: Managing What You Know[17]

*Knowledge management* is the process of getting the best available information to the people who need it within a firm in the most efficient manner possible. According to a recent survey of corporate executives, 94 percent believed that they could leverage the knowledge of their firms more effectively if their firms were able to "know what they know." Furthermore, these leaders were acting on this desire; 40 percent of respondents said they had either started or completed the implementation of a knowledge-management system.

Web-based Document Management Systems (DMS) have become an important tool for knowledge management. Although "document library" applications have been available for legal offices and research organizations for some time, the initial cost and ongoing maintenance of these earlier approaches were often prohibitive. The new Web-based tools require far less support, and training costs are minimal; the interface is the same as browsing the Web.

Motorola has recently implemented a browser-based DMS application that covers about 65,000 seats worldwide. System manager Mike Muegel described the knowledge-management system as "a global, company-wide information and knowledge environment." Motorola uses the Open Text Livelink Intranet DMS, one of the few DMS products built from the ground up for the Web. Most other systems use server add-ons to Web-enable their systems. Muegel observed, "We're developing virtual communities where people can discuss and publish information with their peers, no matter where they are."

The market for knowledge-management consulting services and software is expected to reach $5 billion by 2001. This new market is currently being flooded by a broad range of products, some of which embody more hype than knowledge. Even Microsoft and Lotus Development are repositioning their groupware products in light of the knowledge-management trend.

The roots of knowledge management are in the familiar "Big Six" consulting mill. This dubious history has led at least one major analysis firm to predict that, like so many other management fads, the current peak in corporate interest will be closely followed by a "trough of disillusionment." The fickle nature of executives and consultants will not derail the knowledge-management trend, however. The management legend Peter F. Drucker put this fad in context: "If we apply knowledge to tasks we already know how to do, we call it productivity. If we apply knowledge to tasks that are new and different, we call it innovation. Only knowledge allows us to achieve those two goals."

their magnitude and timing can be anticipated with some confidence; after all, the global teenagers who may impact the marketplace five years hence have already been born.

Technologies, on the other hand, evolve both rapidly and unpredictably. Attempts to describe the nature of technological change most often begin with the familiar *technology s-curve*.[19] As an innovation evolves through its commercial life cycle, it follows a reasonably predictable pattern. In its infancy, the value that it can provide is limited. Early adopters with pressing needs will seek out such immature technologies and pay top dollar for them. Beyond this small cadre, however, the commercial viability of the innovation will depend on its ability to "cross the chasm" into mainstream markets.[20] Making this jump often depends on the performance-to-cost ratio. This is where the technology s-curve comes into play.

Virtually every innovation will move through three stages in its life cycle: early development, rapid performance improvement, and maturation. Immature technologies evolve slowly, due to a conspiracy of factors. Initial development funding is often constrained by the risk-averse inclination of firms, and the lack of a revenue stream to justify a stronger commitment. Efficient manufacturing processes may not yet exist, rendering even the most promising new technology too expensive for broad commercial acceptance. The underlying principles of a breakthrough innovation may not be well understood, forcing developers to fight for performance improvements through brute-force experimentation. A final factor is the lack of competition. The knowledge that a rival firm has made a technological breakthrough is often the force that drives a firm into matching that achievement.

The initial period of slow performance growth is usually followed by a rapid, even exponential expansion. At this point in the s-curve, the above factors have been laid to rest, and essential support structures within the marketplace are fully implemented; distribution channels are identified, ancillary technologies have been developed, and customers have been educated. Increasing market demand will draw in new competitors, driving up total R&D expenditures and creating a virtuous circle of improvement, market expansion, and reinvestment.

During the rapid-expansion phase, there are two possible future outcomes: Either the technology will gracefully mature into a long gradual period of commoditization, or it will be struck down in mid-life-cycle by another breakthrough. In the former case, rates of performance improvement will eventually slow as the market becomes desensitized to such enhancements. Ultimately, the focus will shift from performance improvement and differentiation to cost reduction. Industries will consolidate, and technological progress along the s-curve will all but cease.

Such a dignified old age for technologies is becoming increasingly rare, however. At any point along the s-curve, a new innovation may enter the picture, rendering the existing technology base obsolete long before its maturation. This situation is often referred to as "jumping s-curves." For such a jump-shift in technologies to occur, an innovation must be at least a partial substitute for existing commercial technologies, and must enable a higher total value to be delivered to the market. A jump-shift might follow the same technology trajectory but represent a breakthrough in performance, or it may solve the customer's problem in an entirely different way. The shift from 80486 microprocessors to Pentium chips is an example of the former, while the jump from adding machines to electronic calculators is representative of the latter.

The strategic importance of anticipating jump-shifts in technology is obvious. Synopsys' CEO, Aart de Geus, has observed that for companies to achieve sustainable market dominance they must repeatedly identify where the next "10 × technology change" will occur.[21] Moreover, firms must go beyond detecting breakthroughs to developing a picture of how this revolution will affect their competitive environment.

This is where the utility of simple models such as the technology s-curve rapidly diminishes. Traditional forecasting techniques focus on predicting where new innovations fall on the s-curve, the rates of change along that curve, and which substitute technologies are waiting in the wings to cause a catastrophic jump-shift. This information is necessary, but not sufficient, to form a competitive strategy. The final sections of this chapter describe two methods that offer the complex insights necessary to support a comprehensive strategy-development methodology.

Techniques for technology forecasting range from simple trend extrapolation to cumbersome systems that involve hundreds of hours of analysis and debate by dozens of managers.[22] Based on my experience, envisioning the future is not a suitable job for large committees. This approach will tend to yield "lowest common denominator" predictions with little strategic value. In reality, the most powerful insights about the competitive future reside in the heads of a firm's technology leaders. Forecasting techniques should not be used to build a consensus, but rather to prime the heuristic pumps of the most talented visionaries within the firm.

To develop a strategic forecasting capability, a firm must make four choices: How will a foundation for future predictions be constructed? What time frames will be considered? Which methods will be used to analyze the problem? And finally, how will the forecasting activity be integrated into strategy development? Each of these considerations will be addressed in turn.

A foundation for envisioning the future should consist of an objective assessment of current conditions, as described in the previous section, along with a history of change in critical competitive factors.[23] A historical perspective is essential to the extrapolation of trends; one cannot extrapolate from a single point. The creation of integrated, complex perspectives on future prospects may be essential to making big strategic decisions; more routine choices, however, can be founded on a "sketch" rather than a "painting."

In a 1993 interview, Edward R. McCracken, CEO of Silicon Graphics Inc., made the following assessment of strategy formation based on long-range forecasts: "Long-term planning weds companies to approaches and technologies too early, which is deadly in our marketplace and many others. No one can plan the future. Three years is long-term. Even two years may be. Five years is laughable."[24]

Setting the horizon for technology forecasting depends on two primary factors: the anticipated rate of change of a technology and the speed with which a firm can adapt to that change. As the quote above implies, the error in a future prediction increases rapidly as the time horizon moves farther out, rendering long-term forecasts of little value in the strategic decision-making process. If a firm is sufficiently adaptable, however, the time horizon for decision-support data can be drawn in considerably. It makes more sense for firms to work toward reducing adaptation time rather than developing elaborate methods for seeing farther into the future.

A summary of some popular methods for strategic forecasting is provided in Table 10.1. Regardless of the methods selected, the output of a forecasting exercise must meet an important criterion to be useful in strategy development. My own "fundamental rule of forecasting" says that all possible futures must have a probability estimate (including error bars) associated with them, and within a given analysis all probabilities must add up to 100 percent.[25] Despite the common sense of this guideline, it is rarely followed, even by seasoned futurists. It is all too easy to simply state a bold prediction, rather than weigh all possible outcomes and rank-order their probabilities. Without such rigor, however, it is impossible to calculate quantitative investment metrics, such as the net present value of each possible future outcome. It is only from such calculations that a firm's optimal R&D investment portfolio can be derived.

The most obvious approach to forecasting technological change is consultation with subject-matter experts. Unfortunately, this is also the most unreliable method.[26] Experts within a firm are often biased by hidden agendas, or may be too isolated from mainstream thought to render a balanced opinion. This limitation can be mitigated somewhat by the formation of a technology-assessment committee, consisting of thought leaders from various specializations within the firm. The giant Japanese electronics firm NEC has created such a corporate-level committee, which is empowered to plan technical core competencies and anticipate shifts in the competitive landscape.[27]

The most reliable sources of *expert opinion* are external entities that have a proven track record for successful prognostication.[28] The University of Southern California's Information Sciences Institute (ISI) is one such organization. Founded twenty-five years ago by four refugees from RAND Corporation, the institute has

| Forecasting Methods | Advantages | Disadvantages |
|---|---|---|
| Expert Opinion | Inexpensive Convenient | Biases of experts can render predictions useless |
| Trend Extrapolation | Easy to understand Software available | Inaccurate even for short term Does not take causation into account |
| Time-series Estimation | Inexpensive More accurate than trend extrapolation | Good only for short term Does not take causation into account |
| Regression Analysis | Good accuracy-to-expense ratio | Must know causal variables and predict future values |
| Delphi Method | Forces a consensus prediction | Expensive, time-consuming, subject to expert biases |
| Scenario Building | Forces consideration of multiple possible futures | Too qualitative for some decision support applications |
| Strategic Roadmapping | Connects forecasting insights to strategy development activities | Can revert to "strategic planning" obsolescence if not rigorously updated |

**Table 10.1:** Comparison of several important methods for strategic forecasting.

been turning out a steady stream of practical innovations, including the first portable computer and the Internet domain name system currently in use worldwide. Not all the projects at ISI have obvious commercial applications; current activities range from the development of soccer-playing robots to immersive virtual-reality systems. These high-profile demonstration projects almost always have a kernel of practicality in them; the robo-soccer project is an ideal vehicle for refining computer vision, artificial intelligence, and autonomous-agent technologies.[29]

The *extrapolation of current trends* is still the most popular method of technology forecasting. The assumptions upon which trend analysis is based are restrictive: the future is assumed to be a continuation of the recent past, and is predictable from an understanding of the underlying historical data. Extrapolation suffers from the time-horizon limitation in the extreme; as the time of interest moves farther from the most recent data point, the accuracy of the prediction rapidly decays. On rare occasions, such as the famous Moore's Law of semiconductor technology, a trend line will be valid over extended periods.[30] More commonly, the underlying assumptions of the extrapolated trend are violated almost immediately, rendering the forecast strategically useless.

A number of more sophisticated forecasting methods are currently in use, with varying degrees of success. *Time-series estimation* is an improvement over trend extrapolation, in that it utilizes statistical techniques to separate the systematic behavior of data from random variations over time. *Regression analysis,* a generalized form of time-series forecasting, may use several variables other than time to assist in the prediction of future trends. As with most common forecasting methods, inexpensive software is available to assist in time-series and regression data analysis.[31]

The so-called *Delphi method* was developed by RAND Corporation in the 1950s as a means to predict future trends in military and defense technologies. This cumbersome technique involves sending questionnaires to a select group of "experts," and iterating the responses until a consensus is reached. Comparisons of Delphi projects performed over several decades have shown that although a consensus can often be achieved, the group's agreed-upon prediction is no more likely to occur than those derived from far simpler methods. As with all forecasting approaches based on expert opinion, the biases of the individuals tend to corrupt the accuracy of their predictions.[32]

All the above methods can be useful under some circumstances, but I believe that the final two techniques, *scenario-building* and *strategic roadmapping,* offer the best combined support of the decision-making process. These methods will be expanded on in the sections to follow. Whichever approach is used, it is important to recognize that there is no such thing as a "fact" about the future. Predictions and forecasts are useful as inputs to strategic decision-making only when evaluated objectively and comparatively. Unfortunately, it is the nature of human reasoning to distort perceptions of the future, in much the same way that we distort the decision process itself.

If there is one area in which human thought is inherently irrational, it is in dealing with future prospects.[33] Entrepreneurs entering a new market are almost always biased toward optimism. Leaders of mature industries threatened with jump-shift technology change are typically pessimistic about the commercial potential of the new innovation, at times to the point of denial (witness the statement by Ken Olsen quoted at the beginning of this chapter). Even the most capable executives may lack the imagination necessary to envision alternative futures. Those leaders with adequate imagination may lack the ability to synthesize multiple technology trends into a set of coherent possibilities. Threading through all of this is the "official future" embedded in the culture of the firm, often based on unrealistic bravado about competitiveness, capabilities, and prospects for industry dominance.

As with the strategic decision-making process, there is no cure for the irrationality that plagues our assessments of the future. There is, however, a method that can force the senior management of a firm to confront their biases, and consider even the most unattractive possibilities. The technique of scenario-building can be used to create vast panoramas of the future, or to envision the consequences of an isolated decision. Regardless of the scale, its strength lies in the consideration of several contrasting outcomes, a process that can alter the perceptions of even the most single-minded leaders.

## Envisioning a Scenario-Based Future

Every firm has a process by which strategies are developed, investment choices are made, and portfolios of projects are optimized. These methods may be highly structured in one firm and ad hoc in another, based on the firm's unique culture and leadership and the market environment it must serve. This diversity of process is matched in the literature; journals and technology-management texts are rife with complex prescriptions for the formation and implementation of strategy, most of which are represented as a source of competitive advantage.[34]

Is there a single strategy-development methodology that can elevate the competitiveness of all technology-intensive firms? Undoubtedly, the answer is no. Among successful firms, the *mechanics* of strategy development vary widely. What is common among all successful firms (almost by definition) is their ability to take prompt and decisive actions that consistently yield positive outcomes. While it is vitally important that the chosen methodology for strategy development be highly efficient and well suited to the firm, it is the underlying factors of vision and decision that are at the core of sustainable market leadership.

Competitive advantage is gained and retained by making effective decisions that enhance the three elements of the innovation cycle. As described in the previous chapter, strategy development, by whatever path, is a continuous process of decision-making, based on both an objective picture of current conditions and acute perceptions about impending change. The remainder of this chapter will focus on honing these central abilities, by introducing some conceptual tools that can sharpen strategic intuition, perception, and vision.

*General Theory of Scenario-Building*

The ability to dream about the future is innate in all of us. Our daily lives consist of a continuous cycle of visualizations about upcoming events, comparisons to current conditions, and references to past histories. We routinely develop "scenarios" to get to work on time, manage our workloads, or provide for our entertainment. In this context, scenario-building is as familiar as breathing.

Neurobiologists have suggested that the ability to tell ourselves stories about the future is "hardwired" into our brains, in a process closely linked to the formation of speech and language.[35] Animal behavior is dominated by such abilities; herds migrate based on seasonal "triggers," and pack animals spontaneously establish hunting patterns based on images of how their prey might respond to attack. But in the human animal, such visualization has evolved far beyond an instinctive ability. We are able to imagine our lives unfolding in a hundred ways, and even place ourselves in futures that we can never actually experience.

In *The Art of the Long View,* Peter Schwartz defines scenarios as "a tool for ordering one's perceptions about alternative future environments in which one's decisions might be played out."[36] In other words, scenarios are essentially structured "dreams" about the future, in which plot lines are constructed and possible outcomes evaluated. The formation of scenarios is an art rather than a science, and is heavily dependent on heuristic reasoning and creativity.

The use of scenarios as a means to evaluate possible future environments has its roots in the military planning exercises of World War II. Herman Kahn of RAND Corporation became well respected in defense circles for his use of cause-and-effect sequential scenarios to anticipate conditions leading up to a nuclear confrontation with the Soviet Union.[37]

The first important industrial application of scenario-building was performed by General Electric in the late 1960s to visualize the social and economic conditions that might exist in America in the 1980s. Rather than attempting to create valid cause-and-effect pathways, GE was concerned only with evaluating plausible future conditions. This path-independent approach is necessary when outcomes are dependent on a complex range of factors beyond the control of the scenario-builder.

One of the most famous and strategically important applications of scenario-building by a commercial enterprise was the development of an "oil crisis" scenario by members of the Group Planning team of Royal Dutch/Shell in the early 1970s. The essential point in this case is that the scenario planners, including Pierre Wack and Ted Newland, did not *predict* the OPEC oil crisis. They did, however, generate an oil shortage scenario that was realistic enough to galvanize Shell's executives into performing contingency planning. The result was that Royal Dutch/Shell fared far better than its competitors when something akin to this dire scenario came to pass.[38]

There is no single "right" way to utilize scenarios in strategy development. There are, however, some guidelines that will improve the utility of the process. As I suggested in the previous section, a scenario should be treated as a rational, explicit description of the future, and should be assigned an estimated probability of occurrence relative to all other possible outcomes. If the scenarios are based on extrapolated trend data or other quantitative metrics, then these probabilities may be directly applicable to the development of R&D investment strategies. More often, the assignment of probabilities serves to rank-order possible outcomes, and ensure that a single "most likely future" does not become a de facto prediction.

Another important consideration in constructing scenarios is that they must be internally self-consistent and comparatively distinct. The former condition simply implies that scenarios must make sense. It is not particularly useful, for example, to consider a scenario in which energy prices are on the rise but demand for energy is nonetheless rapidly increasing. Likewise, the invention of a microprocessor that can crunch data beyond the "light-speed limitation" should not be part of a rational set of scenarios. Physical laws, economic principles, and even human nature should not be violated without a compelling reason.

The condition that scenarios should be comparatively distinct is a bit more subtle. One of the greatest dangers in constructing scenarios is that the results will be boring. Considering three scenarios in which the only difference among them is a few percentage points of market share is a waste of the technique. As any good physicist knows, the way to gain insight into an intractable problem is to consider the extremes. How does a system behave at its limits? What happens if a variable goes to zero? What happens if a different variable becomes infinite? Similarly, an effective suite of scenarios should span a significant range of possibilities, and might even be intentionally polarized to highlight a critical strategic weakness.

Finally, it is important to recognize that the purpose of scenarios is to alter perceptions, motivate contingency planning, uncover future opportunities and pitfalls, and generally support the strategic decision-making process. For scenarios to have value in this role, they must be continuously updated or even replaced as events unfold over time. The future will belie even the most ambitious scenarios; much like dreams, they should be experienced and then rapidly discarded.

Beyond these basic "rules," the literature on scenario-building diverges. The actual process of creating scenarios can be as complex as the BASICS system developed by Battelle Memorial Institute or as simple as an informal brainstorming session.[39] The process that I propose in Table 10.2 is straightforward, cost-effective, and can be tailored to a firm's individual needs and market environment.

The first step in the process is to define the question to be addressed through scenario-building. In principle, scenarios can be used to inform virtually any decision, but the wording of the problem can make a difference in how effective the technique will be. It is appropriate, for example, to ask the question, "Will market conditions in two years be favorable for the launch of a new product?" It would not be appropriate, however, to ask, "Should I launch a new product?" The former query focuses the scenario exercise on possible market conditions in two years, whereas the latter directs attention toward the decision itself, something entirely within your control. Scenarios should be performed only when a decision is directly impacted by forces and uncertainties *outside of your control.*

Before the scenario-building process can begin, the critical variables that impact the decision of interest must be determined. These variables can be divided into four closely related categories: leading indicators, driving forces, predetermined elements, and critical uncertainties.[40]

The idea of a *leading indicator* that points to an important event in the future has been around since the days of the Pharaohs.[41] In those years in which the waters of the Nile rose up to flood the fields of ancient Egypt, a bountiful harvest could be anticipated. If, however, the waters were weak and failed to overflow their banks early in the growing season, crops might fail and famine could spread. Clearly, it would have been advantageous for the Pharaoh to know in advance the future behavior of the Nile, so that he could make contingency plans.

---

## A Four-Step Scenario-Building Process

| | |
|---|---|
| **Step 1:** Define the Question | Wording is important. Be sure that the question is clearly phrased in terms of the decision that must be made and the driving forces that impact it. |
| **Step 2:** Choose the Variables | These include leading indicators, driving forces, predetermined elements, and critical uncertainties. |
| **Step 3:** Develop Scenario Plots | Plots must be self-consistent, comparatively distinct, and relatively few in number. |
| **Step 4:** Estimate Probabilities | Avoid mindsets and other biases. Be sure that the probabilities for all scenarios add up to 100 percent, even if it is necessary to define a "dummy" scenario that captures "everything else." |

---

**Table 10.2:** A simplified method for constructing scenarios in support of the strategic decision-making process.

Fortunately for the Pharaohs, the Nile offered a leading indicator of its future behavior. Far upstream, in what was then known as Nubia, two major tributaries of the Nile join together. The White Nile, whose clear waters have their source in Lake Victoria, joins with the darker waters of the Blue Nile. Early in the spring, priests at a temple strategically located at this confluence would examine the color of the waters there. If the combined flow appeared clear, then the weaker White Nile would dominate the downstream behavior of the river, and flooding would be minimal. If, however, the dark waters of the Blue Nile dominated, then a strong flow could be expected, and a bumper crop was in the offing. In this way, a causal indicator of future flooding provided protection from famine for Egypt's people, and enhanced the job security of the Pharaohs.

Leading indicators are not in themselves a factor in the future; instead, they foreshadow changes in important factors known as *driving forces*. Every scenario-building exercise must incorporate one or more of these driving forces as its primary variables. If you were considering the launch of a new computer-game product, for example, the driving forces in your scenarios might include the state of multimedia technology, the demographics of the regions you intend to target, and the platform standards to which your game must adhere. In general, driving forces are the factors that should receive essentially all of your scenario-building attention.

For a scenario to be valid under realistically complex conditions, however, two other building blocks must be considered: *predetermined elements* and *critical uncertainties*. A predetermined element is something that one can be reasonably sure will not change in the future, and therefore behaves as a constant among various scenarios. If, for example, we were using scenarios to analyze an entrance strategy for the Chinese domestic market, it would be reasonable to assume that a steadily increasing population and a growing "middle class" are predetermined elements. Of course, catastrophic political upheaval could invalidate this assumption, but it is more fruitful to hold constant this socioeconomic background and focus on more subtle driving forces, such as consumer preferences and trade liberalization.

If the risk of political turmoil in this example is so great that it cannot be ignored, then it becomes a critical uncertainty. Critical uncertainties lurk in the background, with the potential to invalidate an entire suite of scenarios. Depending on the level of risk, it may be necessary to include scenarios on both sides of the uncertain factor. As with all aspects of the scenario-building process, the identification of driving forces, predetermined elements, and critical uncertainties is an art that can be mastered only through experience.

Once the factors relating to a strategic decision have been determined, a set of stories about the future can be constructed that include reasonable variations in these factors. *Plots* are used to tie together the elements of a scenario into a complete picture of future conditions. There are infinitely many plots that might be constructed, but only a few are typically necessary to illuminate strategic issues. A minimalist approach to scenario-building might utilize the following three generic plots: The future is like today but better, the future is fundamentally different but offers opportunities, and the future is a nightmare.

In the "future is like today but better" scenario, the plot line should follow a logical extrapolation of recent trends. If your firm's market share has been steadily rising, for example, you might construct a scenario in which you achieve market dominance. How will your firm protect itself against new entrants? Can you ramp up capacity and expand core competencies to enable this outcome? Perhaps most important, is this leadership position sustainable, or is it built on a "one-shot" product success?

The "future is fundamentally different but offers opportunities" scenario is where technological change can best be considered. A typical plot might involve a new innovation coming along that is a substitute for your current products. Can you position your firm to hop onto this bandwagon? How can you best appropriate returns from your existing technologies? Will the innovation capture the entire market, or fragment it in ways that can offer advantages to your company?

Finally, a negative scenario should always be considered. It is not particularly useful to consider a completely dismal future. Instead, I recommend selecting a plot that is frighteningly believable. Catastrophic predictions, even when plausible, tend to be shut out by decision-makers. An insidious plot, however, may grab attention and alter overly optimistic perceptions.

There are many alternatives to these simplistic choices, but in any case it is important that the number of plots be kept to a reasonable minimum. The consideration of too many futures will dilute insight, and often can result in unnecessary confusion and expense. Remember to take your own biases into account when selecting plots and fleshing out stories; if you are building scenarios about a pet project, you might begin by considering several negative plots to compensate for your inherently optimistic mindset.

### Probabilities and Delusions

There is a common lament among strategists that establishing probabilities for complex and uncertain future events is futile. Anyone who has visited the racetrack or a sports book in Las Vegas should have little sympathy for this argument. From a novice's perspective, it seems virtually impossible to handicap a set of nearly identical horses, or to choose point spreads for highly competitive sports teams. Yet rational individuals with sufficient knowledge of the "causes" of victory and the history of the participants are willing to wager good money on their probability estimates.

The choice of a gambling example here is revealing. All too often, entrepreneurs convince themselves of the certainty of their success, rather than confronting the realities of chance and uncertainty. It is often difficult for us to consider futures that fail to match our preconceptions or follow our carefully conceived plans. By forcing the explicit estimates of probabilities for scenarios, this common mental trap can be avoided. The business of technology is a gamble, even for the most powerful industry titans. Those who doubt this are fooling themselves. The only way to win at such a probabilistic game is to know the odds and attempt to work them to your advantage.

Often scenarios are derived from a sequence of events that the scenario-builder believes to be highly probable. This tends to restrict insights to only those that match a preexisting mindset. The outcomes of the most plausible scenarios should never be assumed to be the most probable, nor should one assume that an outcome is impossible if no plausible scenario comes to mind (absence of evidence is not evidence of absence). Moreover, it is not necessarily true that a series of probable events will combine to yield a probable scenario. For the purposes of probability estimation, it is best to decompose each scenario into the important factors from which it was derived, and estimate probabilities based on these factors alone.

It is a mathematical truism that the combined probability of two events occurring together can never be more probable than the less likely of the two events. For example, it is impossible that the combined probability of a flipped coin coming up heads on two consecutive occasions (which equals 0.25) could be greater than the probability of the coin coming up heads on a single occasion (which equals 0.50). Our own preconceptions, however, can cause a violation of this fundamental law of probability, known as the "compound probability fallacy."[42] The chances of meeting a person who both jogs *and* works out at a gym, for example, *must* be lower than the chances of meeting someone who either jogs *or* works out at a gym. Yet because these activities have a close logical association, we tend to give the compound probability an unrealistically high estimate. Without careful consideration, it is easy to falsely assume that the chances of someone jogging and someone both jogging and working out at a gym are *essentially the same.*

To construct strategically useful scenarios, the builder must first become consciously aware of his or her preconceptions about the world. This mindset includes attitudes about all aspects of the external environment: prejudices based on personal experience, poorly informed opinions, conventional wisdoms, and other nonobjective reasoning.[43] In fact, one of the most powerful benefits of scenario-building is that the process often focuses a harshly objective light on our most cherished, but often ill-founded, beliefs about the future. It is better to be confronted with our irrationality than to be undermined by it.

Microsoft founder Bill Gates recently observed, "We usually overestimate what we can do in two years and underestimate what we can do in ten."[44] The rate of technology growth is often not accurately predicted, because innovators presume that the proof-of-concept activity is the hard part. Actually, the learning necessary to convert an invention into a commercial reality can take years to develop. It was several decades after the first demonstration of coherent spontaneous emissions that the laser found broad commercial applicability. Similarly, the discovery of high-temperature superconducting compounds in the early 1980s was thought to signal a revolution in high-efficiency electric motors, lossless power transmission, and frictionless magnetically levitated (mag-lev) trains. We are still awaiting the first significant shots to be fired in that commercial revolution.

As a final warning, one should never assume that a suite of scenarios represents all possible futures. If a firm concentrates its strategies on a limited set of scenarios, it may come to believe that action need be taken only in respect to those possible futures.[45] This false sense of security might leave the firm even more vulnerable to unexpected events. Scenarios are a means to illuminate future possibilities; they should not become blinders that narrow the perceptions of the firm.

### Constructing Stories About the Future

When should scenario-building be applied? In principle, the technique can be used to support any decision, from determining manufacturing capacity requirements to evaluating the benefits of a major acquisition. There are two broad classes of decision to which scenarios are particularly well suited: consideration of new-product introductions, and explorations into major shifts in the competitive environment. These two important categories of scenario-building will be considered below through some conceptual examples. The reader should note that these cases have been oversimplified from a technical standpoint to illustrate the scenario-building process more clearly.

Suppose that your firm is a leading producer of computer data storage devices. Your current product line includes a full spectrum of PC-compatible disk drives and hard drives that utilize traditional magnetic storage media. Projections show relatively slow growth in this saturated market, prompting your firm to consider entering the rapidly expanding optical data storage arena. How could scenarios be used to help determine the best strategy for this major new thrust?

The first step is to clearly define the goal of the scenario-building exercise. Your management team has determined that it can be ready to enter this new market in approximately two years, thereby establishing the time frame for your scenarios. Moreover, a review of your firm's operations indicates that the primary concerns with respect to this expansion are technical. Hence, an appropriate phrasing for your strategic question might be: "What technological conditions will exist two years hence that would enable a successful entry into the optical-data-storage market?"

With this clear goal in mind, a set of variables must be selected. A good starting point is to identify the *enabling technologies* that will determine the performance, cost, and features of optical-storage products. One of these key enablers is the semiconductor diode laser, which is used to read and write data onto optical media. Disk capacity and data density are fundamentally limited by the wavelength of light produced by these lasers; shorter wavelengths enable higher densities. Currently, the shortest available wavelengths are in the yellow-green region of the spectrum.

Fortunately for your firm's future, the engineering group has been absorbing information on applied research in electro-optics for several years in anticipation of just such a product-line expansion. In particular, you uncover a press release from 1995 announcing the first demonstration of a blue-wavelength light-emitting diode (LED) by a small Japanese firm, Nichia Inc. Although LEDs themselves are not useful in the production of optical storage devices, their underlying technology is closely related to the production of diode lasers, and hence this breakthrough is a *leading indicator* of the availability of shorter-wavelength lasers in the very near future.

A brainstorming meeting with your engineering team highlights several other *driving forces*, including the rate of evolution of current optical storage technology and the size and format of the media "disks." *Predetermined elements* include a steadily increasing demand for high-density data storage and a growing base of PCs throughout the world. Finally, your team notes at least one *critical uncertainty:* Will the PC architecture be replaced by network PCs, which might require little or no permanent data storage capability?

From this limited set of variables, a suite of three plots can easily be developed. If we follow my generic recipe from the previous section, we might begin with a "future is like today but better" scenario. The plot in this case would involve a steady evolution of current optical storage technologies, based on extrapolations of historical rates of performance improvement and cost reduction. This is essentially an "s-curve scenario," in which the primary challenges lie in determining where the industry currently falls on the curve, the rates of change, and the points of inflection. To add believability and depth, additional details should be added to this skeletal plot, including competitors' anticipated actions, major market segments, positioning strategies, etc.

The second generic plot describes a more interesting future. In this scenario, the blue LED leading indicator proves to be a harbinger of a "jump-shift" in the industry: blue-violet laser diodes become commercially viable in two years, causing the almost immediate obsolescence of current platforms.[46] This plot offers both change and opportunity. By anticipating this fundamental shift in enabling technologies, your firm can be in on the ground floor of a new wave of high-performance products.

Finally, it is important to include a negative scenario. In this case, the best choice might be one that assumes that the critical uncertainty described above actually comes to pass. The plot would involve a tectonic shift in the computer industry away from autonomous desktop computers to the network-computer architecture being promoted by Sun Microsystems and others. With high-bandwidth data transmission a reality in two years (or so goes the scenario), it will no longer be necessary to permanently store gigabytes-worth of software at the user's site. Instead, applications will be downloaded when they are needed from remote servers, and third-party "storage depots" will be formed to manage personal files and other permanent records. Local data-storage requirements will be minimal, driving the market for your new product line into severe contraction.

The last step in the scenario-building process is to estimate the probabilities of each scenario. Given the powerful market pull for increased data-storage densities, and the knowledge of a "proof of concept" for blue LED technology, it would be reasonable to give the jump-shift scenario much higher probability than the status quo future. But what about the negative plot? Critical uncertainties are troublesome to handicap, but with the relatively short time horizon for your decision, and the huge installed base of PCs, it seems safe to assume that remote high-capacity data storage will have a growing market for at least the next six to ten years. This would likely give your firm sufficient return on investment to warrant the new-product launch, even if the negative scenario proves to be a portent of long-run events.

A word of warning is in order, however. If the probability of the status quo scenario is estimated at 10 percent, the jump-shift at 85 percent, and the negative scenario at 5 percent, your firm will have made a major strategic error. The error is not in the relative probabilities, nor in the fact that they sum to 100 percent, as they rightly should. The flaw is that the three scenarios *do not capture all possible futures.* By normalizing the probabilities to reach the desired total, you would have neglected the possibility that something outside of your three plots might occur. Suppose that holographic (i.e., three-dimensional) optical data storage becomes rapidly commercialized? Or perhaps an innovative neural-network architecture gains market acceptance, requiring data storage to be juxtaposed with processing functions, much like in the human brain. As Andrew Grove of Intel Corp. has observed, paranoia about technological change is not just a cliché, it is a survival trait.[47]

So goes the process of scenario-building when the motivation is internal strategy development. This is not the end of the story, however. It is not sufficient for firms to envision the future by looking from the inside out; it is also vital that they evaluate their competitive positions from the outside in. Dramatic changes in the external environment can throw even the best strategies into chaos. Firms that anticipate an impending tsunami well in advance can best anchor themselves for the impact. Hence, it is important that firms monitor the developments of enabling technologies in all fields, even those developments that appear to be unrelated to a firm's core competencies.

In early 1998, IBM announced a breakthrough in the processes associated with integrated-circuit fabrication.[48] For several decades, the vast majority of semi-conductor devices have been produced using silicon substrates, silicon-dioxide insulating layers, and aluminum conductive interconnects. This ubiquitous archi-tecture has a fundamental weakness, however: aluminum is not a particularly good conductor of electricity, at least when compared to copper or gold. Hence, the use of aluminum interconnects has placed a fundamental limitation on the speed, density, and power demands of integrated circuits.

In this context, IBM's announcement of a commercially viable copper-inter-connect process was earthshaking. Within a matter of weeks, several major semi-conductor-equipment manufacturers had announced their intention to focus future development efforts on this new process, and Intel presented its own R&D results on an alternative copper-deposition process. If this breakthrough follows the historical trendline for transition of proof of concept to production, copper-based integrated circuits will have already entered the marketplace by the time this book is in print, promising gigahertz speeds and dramatically lower power con-sumption.

From the standpoint of our earlier fictional firm, this breakthrough could place a premium on data-access times for all permanent memory and storage media. Read/write times were not even addressed in our first set of scenarios (al-though they probably should have been, in a real-life exercise). With processing speeds set to multiply, however, data-storage speeds could become a primary dif-ferentiator in this new technological environment. Competitive advantage in this fictional case would be gained by those firms that anticipate such a shift in critical design parameters and position their products accordingly.

Dramatic breakthroughs in enabling technologies should immediately prompt a round of scenario-building in virtually every technology-intensive firm. As the tree diagram in Figure 10.2 demonstrates, a revolution in such a critical enabling technology as copper interconnects will propagate throughout virtually every industrial sector. Increasingly, firms must look outside their specialties to detect the next wave of change. Developing a vision of the future is not sufficient; sustainable advantage depends on understanding the interconnectedness of things as well.

## The Technology / Strategy Roadmap

The scenario-building process illuminates a specific decision or action. Strategy development in a diversified firm, however, requires the construction and mainte-nance of hundreds of such scenarios on an ongoing basis. How can these relatively narrow insights be organized so that their inherent linkages and interdependen-cies are clearly evident?

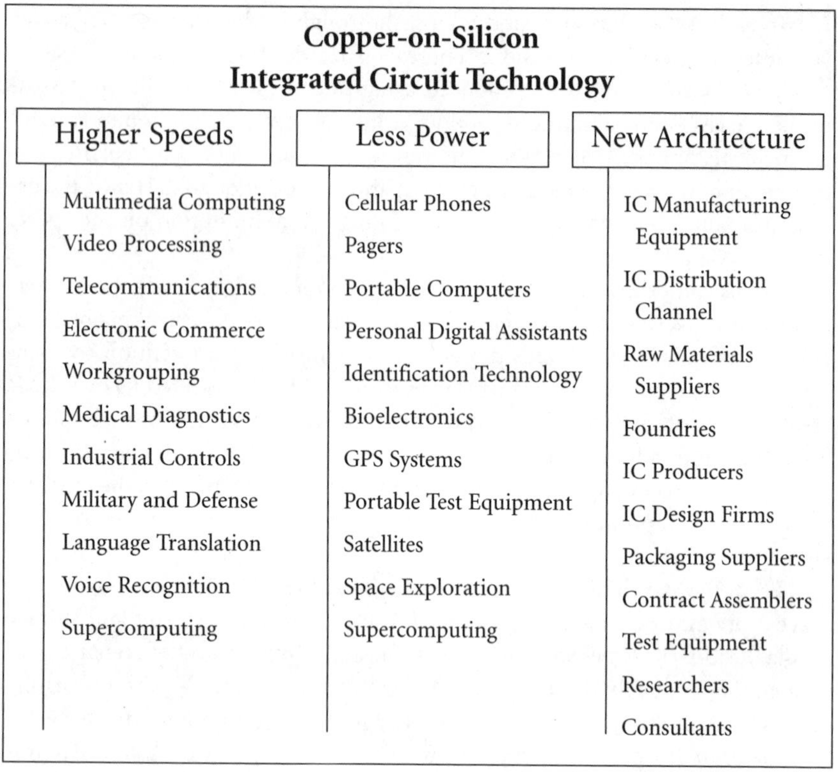

| Copper-on-Silicon Integrated Circuit Technology | | |
|---|---|---|
| **Higher Speeds** | **Less Power** | **New Architecture** |
| Multimedia Computing | Cellular Phones | IC Manufacturing Equipment |
| Video Processing | Pagers | |
| Telecommunications | Portable Computers | IC Distribution Channel |
| Electronic Commerce | Personal Digital Assistants | Raw Materials Suppliers |
| Workgrouping | Identification Technology | |
| Medical Diagnostics | Bioelectronics | Foundries |
| Industrial Controls | GPS Systems | IC Producers |
| Military and Defense | Portable Test Equipment | IC Design Firms |
| Language Translation | Satellites | Packaging Suppliers |
| Voice Recognition | Space Exploration | Contract Assemblers |
| Supercomputing | Supercomputing | Test Equipment |
| | | Researchers |
| | | Consultants |

**Figure 10.2:** Diagram showing the cascading effect of a fundamental breakthrough in an enabling semiconductor-fabrication technology. The listed industries and products are just a sampling of the breadth of impact such an innovation can have.

Over the last several years, a technique has emerged that offers the structural advantages of outdated strategic-planning methods, but in a lean and highly adaptable format. The process of *technology roadmapping* is rapidly becoming the standard for the communication of forecasts and strategies in dynamic industries. Rather than expounding on static conditions, these terse documents focus on change over time and the interrelationships between events. Roadmaps provide a predictive "motion picture" of the future competitive environment, whereas scenarios shine a tightly focused spotlight on specific decision points within this broad panorama.[49]

### Plotting a Course for Technology Enterprise

There are no rules for constructing technology roadmaps, but some guidelines can nonetheless be defined. Simplicity is at the heart of this technique: the top-level forecast for an entire industry can be captured on a single page. Details are typically added in a tiered structure that closely matches the natural decomposition of technology-intensive products and markets: the top-level roadmap establishes a time base and fixes critical milestones, while more detailed lower-tier maps display the needs and requirements that must be satisfied to achieve those milestones.

From a mechanical standpoint, technology roadmaps most often take the form of either a Gantt or a fishbone chart, with time driving the horizontal axis and various technological factors displayed along the vertical axis. Whatever the format, the roadmap presents an integrated picture of likely future events, alternative paths, potential obstacles, and anticipated breakthroughs. To be of value, these documents must be continuously updated based on changes in current conditions and the outputs of ongoing scenario-building and technology-forecasting exercises.

The emergence of this decision-support tool in the early 1990s represented an important milestone in the rise of intra-industry cooperation in America. Some of the earliest examples of technology roadmaps were developed by industry associations to encourage the focusing of limited R&D resources onto a consensus list of needs. By increasing the efficiency with which funds and human capital are allocated to specialized tasks, such a tool can dramatically improve the productivity of an entire sector. The building of a consensus-based, industry-specific technology roadmap has proved to be one of the most valuable roles for high-technology industry groups, and has become a powerful justification for their existence.

The Semiconductor Industry Association (SIA) is a pioneer in the development of industry-level technology roadmaps. Throughout the 1990s, the SIA has sponsored the construction and maintenance of the National Technology Roadmap for Semiconductors (NTRS) to capture the breakneck pace of technical change within that industry. The 1997 version of the NTRS embodies the work of more than six hundred scientists and engineers, whose contributions chart the future course for one of America's most productive industries.[50]

The SIA's organizational structure for creating the roadmap is based on several Focus Technology Working Groups (FTWGs), which are coordinated by a central Roadmap Coordinating Group, as shown in generic form in Figure 10.3. The FTWGs are formed around critical technology areas within the semiconductor industry, including interconnects, design and test, lithography, and assembly and packaging. In addition, the oversight group has established Cross-Cut Technology Working Groups, which address disciplines spanning all seven focus areas. These infrastructural technologies include environmental protection, defect reduction, metrology, and computer modeling and simulation.

The SIA's roadmap forecasts the evolution of the industry over the next fifteen years, by displaying the "paved roads" of proven technologies and the "gravel tracks" of promising alternatives, along with guideposts that identify innovative trails yet to be blazed. The drumbeat for progress along these pathways is set by the two-decades-old criterion of Moore's Law, which continues to serve as a valid predictor of progress for the global integrated-circuit industry. The primary purpose of the SIA roadmap is to provide a high-quality database of impending needs and requirements, and to establish a framework for the enormous R&D investments required to meet those needs. The NTRS is not intended to be a prescription or a strategic plan. According to the SIA, "It is not the purpose of the Roadmap to dictate which options should be pursued by the research community . . . the intent is to enhance communication and encourage innovation to meet the needs expressed in the Roadmap." [51]

The development of industry roadmaps has been working its way up the food chain in recent years. Three important industry groups, representing large segments of the electronics sector, came together for the first time in 1994 to discuss the interlinking of their roadmapping activities. Following the well-established format of the NTRS, the Institute for Interconnection and Packaging of Electronic Circuits (IPC) first issued its technology roadmap for printed circuit boards and other interconnection technologies in 1993. The National Electronics Manufacturing Initiative (NEMI) came out with a similar document in 1994, covering the fabrication, assembly, and testing of electronics products. The NEMI is a subgroup of the Clinton administration's National Science and Technology Council, and represents a collaborative effort by industry, government, and academia. [52]

Under the sponsorship of the Advanced Research Projects Agency, several trade associations, including the SIA, NEMI, and IPC, met in 1994 to compare their technology roadmaps and consider ways to integrate them. Agreement was reached between these three important groups to exchange current and future roadmaps, and to begin the process of interlinking key elements among them. Furthermore, it was agreed that the publication dates would be synchronized on a two-year cycle, beginning in 1996. Ultimately, if such interlinked roadmaps evolve upward to the highest levels of technological interdependency, the productivity and return on investment for all technology industries will be greatly enhanced.

This raises a vexing question. Should industry-wide technology roadmapping activities be national or global? The SIA has confronted this question recently by considering the addition of inputs from Europe, Japan, and South Korea into its highly respected biennial document. For several years, the international semiconductor community has been requesting the development of an international version of the NTRS. On the negative side, observers note that the United States now has a substantial lead in many segments of the industry, and has little to gain from foreign participation. In addition, there would be considerable added cost and delays associated with the global coordination of such a complex endeavor. [53]

On the positive side, however, are the benefits of achieving a broader and in some cases more accurate picture of the current state and future prospects of the

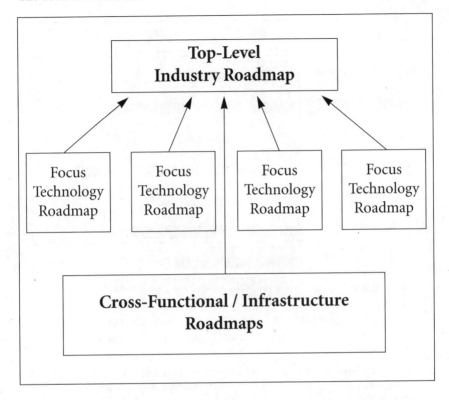

**Figure 10.3:** Conceptual diagram of a multilevel technology roadmap, as developed by an industry association such as the SIA. Note that each focus-technology roadmap could be expanded into a more detailed third tier, and so on. This hierarchical structure can be adapted for use within individual firms.

industry. Although the United States is the most powerful force in the global semiconductor industry, Japan has a long-standing lead in such critical enabling technologies as photolithographic equipment. Moreover, the Japanese technical community has a fundamentally different picture of the way global technology will evolve over the next decade, as shown in Figure 10.4. By excluding the cooperative participation of such major technological powers as Japan and the European Union, we may be setting ourselves up for competitive disaster if our version of the future is less accurate than the ones envisioned overseas.

### Mapping the Competitive High Road

Technology roadmaps for entire industries are formed by building a consensus among leading firms, university experts, government advisors, and the customer community. These efforts are, of necessity, restricted to pre-competitive collabo-

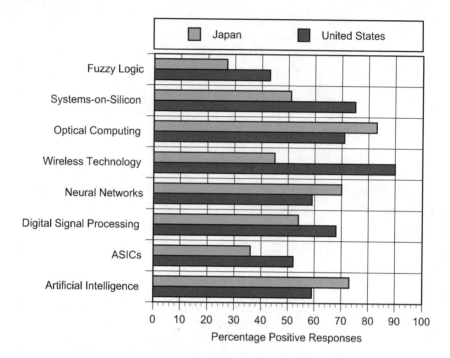

**Figure 10.4:** Comparison of survey results for engineers in the United States and Japan. Respondents were asked to rate the growth prospects of several advanced technologies over the next few years. The sizable differences in results indicate a fundamentally different picture of the future competitive environment, and highlight the need for international cooperation in technology roadmapping activities. Source: *Electronic Engineering Times*, September 1, 1997.

ration, since the resulting document will be widely distributed at both the national and international levels. How then can this framework be adapted as a tool for strategy development within the firm?

Competition in technology-intensive industries is driven primarily by two time-based factors: markets and technologies. In the absence of a clear vision of the future, a firm's product strategies are limited to those formed in *reaction* to these two fundamental forces. The availability of a reasonably accurate roadmap of evolving market needs and technological capabilities, however, would enable the development of *proactive strategies* that anticipate the firm's future competitive environment. High rates of change within an industry make such foresight essential; few of us would consider rafting down a swiftly flowing river without a guide to warn us of impending dangers.

A simplified approach to firm-specific *strategic roadmapping* is shown in Figure 10.5. In this model, the familiar technology roadmap is paired with a similar projection of market conditions. A market-demand roadmap might include predictions of shifts in consumer preference, important demographic trends, and promising opportunities in emerging markets. The anticipated responses of competitors could also be mapped, thereby providing insight into which segments might offer high rates of return or first-mover advantages.

The market-demand roadmap must be complemented by a technology map similar to the industry-level versions described above. Indeed, a firm-specific technology roadmap should be closely aligned with such broad consensus documents. Applicable industry milestones can be augmented by the anticipated progress of internal R&D efforts and the availability of absorbed and accumulated knowledge, derived either from tacit learning or through formal licensing agreements and strategic alliances. The innovative efforts of both rivals and key suppliers should be given a prominent place in a firm's technology roadmap.

From the intersection of these two documents, a product-line strategy can be derived that anticipates future opportunities and exploits economies of scope and scale within the company. This strategic roadmap should feature at least three categories of innovation: platforms, derivatives, and breakthroughs. *Platforms* represent a modular, scaleable base that can be expanded to become a diverse and customizable product line. The resulting tailored products, the *derivatives* of this architecture, address profitable niche markets. Finally, even as a product platform is expanding into a constellation of derivatives, new technological *breakthroughs* should be incubating in the R&D laboratory, ready to burst forth when the time is right.

Motorola began the use of firm-specific strategic roadmaps in the mid-1980s as a hedge against the increasing complexity and diversity of its product lines. The result was a technology-roadmapping process (using Motorola's terminology) that spanned all operations of the corporation and highlighted the linkages and interdependencies therein. Each business unit created its own roadmap, which consisted of technology forecasts, a cross-competency matrix, resource-allocation plans, intellectual property issues, and product descriptions. Observers both within and outside the firm have given this pioneering approach to dynamic strategy development the credit for Motorola's dramatic turnaround in the early 1990s.[54]

Philips Electronics also uses an integrated strategy-development process that yields roadmaps of products, technologies, and their mutual interdependencies.[55] The firm's approach is characterized by combining future forecasts of technology evolution and market-demand projections to create a product-line strategy that is continuously updated and refined. Each line of business is described by its own map, with short-life-cycle products such as portable audio equipment assigned a three-to-four-year time horizon, and underlying core technologies such as optical data storage given as much as a decade of visibility. Roadmaps are produced at several levels of aggregation, from top-level overviews that support corporate strategy decisions to more detailed product-specific maps that guide daily strategy implementation.

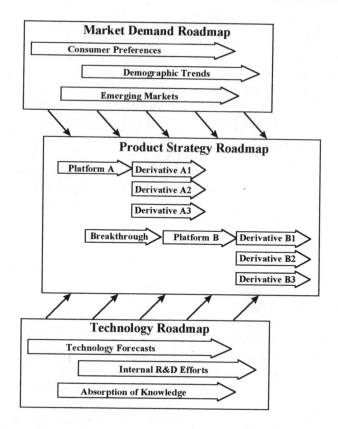

**Figure 10.5:** Concept diagram of an integrated process for the formation of a strategic roadmap within a firm. Note that both the market-demand and technology roadmaps must be continuously updated and expanded to support optimal strategic decision-making.

Beyond Philips' primary goal of achieving an interlinked product/ technology/market approach to long-term planning and vision-building, it has benefited in more subtle ways from the development of strategic roadmaps. The process of developing the cross-functional maps has stimulated learning and communication across all of the firm's core disciplines. Moreover, the highlighting of links between market research and engineering capability has yielded significant improvements in the quality and timeliness of new-product specifications.

As the preceding examples demonstrate, firms at the leading edge of innovation are beginning to recognize that effective strategies must be time-based, adaptable, cross-linked, and decision-driven. The concept of strategic roadmapping offers an efficient framework in which to capture these essential elements. Without being too clever with the roadmap metaphor, I suppose one could conceive of a multilevel document that organizes all the cascaded roadmaps within a firm into one "strategic mapbook." Whatever its title might be, a taut and integrated framework based on scenarios and roadmaps might be the ideal strategy-development tool for rapidly changing technology markets.

# 11

# Models for Sustained Market Leadership

We have traveled a circuitous route to arrive at the final and most obvious source of sustainable advantage: the market itself. Ultimately, all firms must withstand the acid test of customer preference, whether manifested in the mercurial taste of consumers or the pragmatic demands of the industrial supply chain. Persistent market leadership cannot be attained by capitalizing only on near-term opportunities; firms must establish long-term, mutually beneficial relationships with their customers if they are to be granted tenure in the marketplace.

As has been the case throughout this book, I will propose several simplified models that shed light on this most vital element of sustainable advantage. On this topic, however, I must caveat my remarks. To suggest that the concepts described herein represent fundamental principles that drive customer behavior in technology markets would be an overstatement. Unfortunately, no such coherence exists in what has accurately been described as "the inside of a tornado."[1] At best, the models that follow provide insight into the formation of persistent bonds between firms and their customers.

Case studies of technology firms are given a great deal of attention in both the popular and academic literature, but I believe that this emphasis might be misplaced. The selection of firms for study is often biased toward the extremes of either success or failure (otherwise they would not be of popular interest), resulting in case studies that offer little value in detecting broad trends or patterns. Moreover, these isolated anecdotes are often oversimplified, having been filtered first at the source (by the employees of the firms being studied), and subsequently by the author. With respect to technology enterprise, it seems unlikely that the subtle essence of competitive advantage can survive such a crude distillation process.

I believe that the true value of case examples lies in their ability to breathe life into sterile theories by providing vivid evidence of their validity.[2] They play the role of "laboratory experiments" in a regime in which controlled experiments cannot be accurately performed. It is in this spirit that I provide "empirical" examples, both to enhance the credibility of the models I propose and to encourage their refinement in the future as the behavior of technology markets is further revealed.

## Model 1–The Rosebush Analogy

It seems at first glance that success in today's technology markets demands that a firm be both big and small at the same time. There is little doubt, for example, that the fastest growth and the highest rates of innovation are achieved by small, entrepreneurial start-ups. At a time when large, established firms are struggling to maintain profit margins in the low double digits, tiny firms are creating value at explosive rates, often multiplying their investors' equity by factors of tens or hundreds in just a few short years.

Yet in technology markets, size has its advantages as well. Few of the hundreds of high-tech start-ups that are founded each month will survive beyond their first round of funding. Even those that achieve monumental market success risk losing control of their innovations to industry giants, or may be forced to pay homage by sharing their newfound wealth through alliances with powerful industry forces. Big business carries the clout of market access, brand recognition, financial muscle, production and distribution capacity, and perhaps most important, executive management acumen. Small firms are notoriously poor at executing the "end game" of new-product development, in which production and marketing expertise dominate over engineering talent.[3]

Therein lies the paradox: It appears that a firm must have both the hunger and adaptability of a small start-up and the power and sophistication of a major player to achieve long-term leadership in technology markets. There is currently a factor of one hundred thousand between the revenues of the smallest and largest technology-intensive firms in the United States, and that gap is steadily widening.[4] Can there be a single model that captures the advantages of both ends of this spectrum within a single enterprise?

Not only is there an appropriate analogy, but it is likely that you can ponder its validity in the comfort of your own backyard. A rosebush has the ability to retain its beauty month after month, despite the relatively short life cycle of its individual flowers. In a continuous process of "creative destruction," the rosebush sends forth new buds, nurtures flowers in full bloom, and sheds dying carcasses.[5] Although each blossom might last only a few days, by generating many buds with overlapping life cycles the bush sustains its aggregate beauty for an extended period.

To create a new bud, the rosebush must expend energy. As the bud matures and begins to open, this energy expenditure is at its maximum. Once the petals begin to die, the energy stream that had been temporarily diverted to the bloom is drawn back into the center of the plant, to be redirected to the creation of the next beautiful flower. The core of the bush remains in a dynamic equilibrium, buffered from the cycle of life and death that is taking place at the end of its stems.

The parallel with the evolving nature of high-technology enterprise is evident. The core capabilities and resources of a firm are analogous to the rosebush, providing a flow of "energy" to its various product lines. Each line of business is provided with a stream of resources (e.g., R&D funding, human capital, facilities space), but only for as long as this allocation is warranted. As older products reach maturity and their rates of return begin to decline, they are shed by the firm, and the precious resources they consume are redirected to other, more promising innovations. Over time, such a strategy would appear as shown in Figure 11.1.

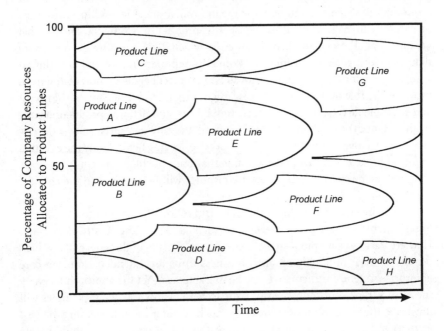

**Figure 11.1:** Conceptual diagram of the rosebush model for allocation of a firm's resources to multiple lines of business. Several product lines are active at any given time, but each is at a different stage in its life cycle. Human and physical capital are consumed by the various product lines in proportion to their potential to generate a future profit stream. Once a product line begins to wane, it is "trimmed" from the portfolio of the firm, and the "energy" being consumed by it is reallocated to other, more promising projects.

Two key elements of the rosebush model can have considerable strategic importance when extended to firms. The first is the need to generate a continuous, overlapping, diversified stream of products, with no one product being dominant. Many firms still follow the "pipeline" approach to innovation, in which one new-product introduction follows another in a sequential pattern. Both resource utilization and cash flow are forced into cycles of boom and bust by this process, with a single market failure often representing a significant risk to the firm.

To avoid such a fate, multiple projects should be maintained at all times, in various stages of development. As market conditions fluctuate, this portfolio of R&D investments should be continuously reevaluated, using tough criteria such as internal rates of return and opportunity costs. The vast majority of projects will not be worthy of a firm's resources throughout a complete life cycle, and should be terminated promptly to avoid undernourishing other, more promising opportunities.

The second insight that the rosebush analogy provides is the need for careful partitioning between the "bush" and the "buds." To capture the full benefits of economies of scope, the autonomy of the innovative activities within a firm must be carefully managed. In the early stages of product development, an "internal venture-capital" model might be followed, in which R&D teams are given no more than a general goal, a deadline, and some discretionary funds, and are allowed complete freedom to create and innovate. As a project begins to show commercial promise, the role of the core enterprise might expand to include establishing cost and schedule targets, providing marketing inputs, and defining production constraints. Once the product has been launched, the core enterprise may withdraw again from active involvement in the enterprise, leaving behind only the essential financial linkages, and an understanding that once the bloom begins to fade, its stem will be cut and the resources it consumes will be returned to the core for reallocation.

Virtually every technology-intensive firm follows this model to a greater or lesser extent. Innovation powerhouses such as 3M, Texas Instruments, and Hewlett-Packard have pioneered a "mainstream" version of this concept, in which dozens or even hundreds of small business units are extended from the core enterprise to fend for themselves in the marketplace.[6] Yet the approach taken by these firms represents only the tip of the iceberg. As the following examples will illustrate, the rosebush model provides much broader insight. Merging the best features of big and small enterprise in creative new ways has come to dominate the strategic behavior of firms in virtually every high-technology sector.

### A Walk Through the Garden

The scope of the rosebush analogy is best illustrated by exploring the full range of possible applications. On one end of the spectrum is the "perennial start-up," in which a small group of entrepreneurs form company after company in a continuous cycle of innovate, cash out, and repeat. In highly volatile markets such as Internet applications, a cadre of cutting-edge engineers with marketing savvy can execute this cycle as frequently as every two years. In this extreme case, the entire energy of the "core team" is directed into each new venture until sufficient value is created (or until the start-up fails). At the appropriate time, equity in the venture is sold to the highest bidder, through either a public offering or purchase by another firm. The founding individuals then regroup, add or subtract talent as necessary, and begin the process once more. This model is becoming increasingly common in technology clusters such as Silicon Valley, due to relatively low entry barriers for new start-ups and high concentrations of experienced risk-takers. One such entrepreneur, Scott McLoughlin, founder of the (revealingly named) Adrenaline Group, observed, "Some people play tennis. Some people play golf. I do start-ups."[7]

In reality, this high-risk business model violates both of the fundamental principles derived from the rosebush analogy. Innovations are produced in series rather than in parallel, resulting in unacceptably high risk. Furthermore, the partitioning between core and peripheral activities is essentially nonexistent. All that is retained at the core of these perennial start-ups is the tacit knowledge and experience of thrill-seeking entrepreneurs, who carry their hard-won wisdom from venture to venture.

A less extreme example is provided by Idealab, a self-proclaimed "start-up factory" established by successful innovator Bill Gross. Gross previously founded Knowledge Adventure, a developer of educational software, and has spent most of his life initiating innovative ventures. After several entrepreneurial successes, Gross recognized that his personal "core competency" was bringing new commercial ideas to life. He established Idealab as a means to multiply this talent, and to offer opportunities to other creative entrepreneurs as well.

Even the physical layout of his planned Idealab "factory" is strikingly similar to the rosebush model. Gross envisions a hub-and-spoke arrangement, with each spoke intended to house an entire fledgling company. New innovators are carefully screened, and subsequently given a "start-up in a box": a template for company structure, financial arrangements, management methods, and even the development of logos and stationery. The goal is to eliminate virtually all barriers to innovation by exploiting economies of scope to the greatest extent possible.

Using proceeds from his previous ventures, Gross funds initial efforts with seed money (typically $250,000 per company), and continues to support the venture with administrative assistance and consultation. Once a start-up demonstrates success, it is spun off into an independent firm, with Idealab retaining a minority equity stake. By releasing start-ups to grow at their own pace, Gross believes that all parties benefit. In one recent example, Worlds Inc., a 3-D graphics start-up, was spun off as a $5 million entity, and within one year had grown to a thriving $77 million business.

In addition to exploiting operational economies of scale, Idealab also funds research into "infrastructural technologies" that benefit all of its hatchlings. One example is the development of data-compression technology which allows Websites to load at twice the normal speed. Such an innovation adds value to all of Idealab's Internet start-ups, but would be far too expensive for a single small firm to develop. This focus on creating a core of facilities, expertise, and infrastructure that is shared among dozens of start-ups permeates every aspect of the organization. As Bill Gross puts it, "I'm trying to factor out all the common business problems and put them in Idealab."[8]

The concept of creating small innovative ventures and evolving them into independent companies is becoming an accepted strategy in technology-intensive industries. Thermo Electron Corporation, a broad-based engineering firm with interests in biomedical, environmental, and energy technologies, has passed over the spinoff method of releasing promising lines of business in favor of a "spinout" approach. The traditional spinoff model typically involves selling a poorly performing unit to shareholders, with the proceeds being returned to the parent firm. In Thermo Electron's spinout strategy, the cash derived from the sale of a minority stake in a promising new venture is provided to the start-up as growth capital. The parent firm retains a majority equity stake, while executive control is transferred to the new firm's management team. In this way, Thermo Electron is free to pursue upcoming opportunities, while receiving a steady stream of profits from its erstwhile "flowers."[9]

Perhaps the most exciting opportunities for the synergistic joining of big and small enterprise are in the biotechnology industry. Alliances between major pharmaceutical firms and small innovative start-ups have become commonplace in this sector, highlighting the importance of both extremes of scale to successful technology development. Small, specialized biotech firms are best able to pursue promising market niches and perform "outside-the-box," cutting-edge research, while the industry giants have the financial depth, production capacity, and market power needed to extract the full value from these innovations.

One of the legends of biotechnology, Genentech, has been reluctant in the past to pursue alliances with smaller firms, perhaps because it still considers itself to be "small." Unfortunately, this philosophy has earned Genentech a reputation for "developing only one product at a time." To dispel this image, it has recently adopted a new research strategy that exploits elements of the rosebush model. For years the firm has pursued promising leads for new drugs by using small, relatively isolated teams that were constrained by a relatively short time horizon. This approach was limited, however, by the rate at which Genentech's small research teams could comb through its massive gene-data warehouses.

Recently, Genentech unveiled the results of a long-range project to improve the underlying efficiency of its product development process. The Secreted Protein Discovery Initiative, referred to as "Speedy" by company insiders, allows researchers to generate many times the number of new-product leads than was possible with previous methods. As much as one-quarter of Genentech's research staff is now working in what amounts to an assembly line for innovations, with impressive results: The new approach generated five hot product leads in a matter of months instead of years.[10]

This dramatic shift in strategy by Genentech brings the firm more in line with the two principles of sustainability highlighted in the rosebush model. The Speedy system promises to fill Genentech's product development portfolio with a number of parallel projects, while capitalizing on the synergy of performing data-warehouse searches en masse. Similarly, by relocating the partition between the activities of individual development projects and the core enterprise, a far more efficient and diverse innovation process has been achieved. Incidentally, Genentech has recently become more open to forming alliances with start-up firms, a further acknowledgment of the growing interdependency between big and small enterprise in dynamic technology industries.

At the opposite end of the spectrum from the "perennial start-up" described earlier are giant innovators such as Microsoft and Intel. Microsoft is notorious for acquiring droves of start-up firms in key industry segments, often preferring to buy hot new technologies rather than develop them internally. Microsoft's penchant in this regard is so notorious, in fact, that the business model for many recent West Coast start-ups has been based on creating an attractive acquisition target for the giant firm.

A more subtle strategy is being pursued by Intel Corporation. Rather than buying start-ups to extend its internal technology base, Intel provides substantial venture capital to entrepreneurs in a wide range of fields, as a way to "pave the road" for the growth of their core lines of business. Intel's industry leadership depends to a great extent on a steadily increasing demand for computing power. Hence, the need to create high-value justifications for faster and more complex microprocessors is deemed a matter of survival.

With so much at stake, growth in the demand for computing power cannot be left to chance. Intel currently invests hundreds of millions of dollars in venture capital, supporting research spanning every aspect of the processing, transmission, storage, and display of data, and even the generation of information content. Although Intel has the resources to develop much of the needed technology internally, it seems to prefer an arm's-length relationship with small innovative firms, providing seed money, technical information, and the credibility associated with the Intel name.[11] Again, the synergy of large and small is apparent; even the massive force of the Roman legions depended for its progress on teams of engineers who prepared a smooth road ahead.

From the high-risk gambles of cyclic start-ups to the subtle, forward-looking maneuvers of industry giants, the applicability of the rosebush model persists. In each instance, a core of valuable competencies remains in a dynamic but stable equilibrium, while evolving markets are serviced through extensions of this core into specific innovative activities. Evidently, the ability of life to continuously renew itself can provide inspiration to those who struggle for survival amid a whirlwind of change.

## Model 2–Three Variants on the Loyalty Game

Among the factors that can impart a sustained competitive advantage to firms, customer loyalty is king. In the days of American preeminence in mass production, all that mattered was brand recognition: With markets expanding rapidly, customer loyalty was treated as secondary to making the initial big sale. Today, with increasing global competition and rapid product obsolescence, the first sale is just a starting point. The survival of firms depends on their ability to attract and retain customers even though the marketplace is continuously jumping s-curves and shifting standards. To gain such allegiance, something more than brand recognition is needed; customers today expect tangible benefits from their participation in an enduring buyer/supplier relationship.

In the sections that follow I will offer a general model for how market share evolves over time in technology-intensive industries, followed by three variations on the theme of customer loyalty. In each instance, a well-defined benefit is received by the consumer in return for a tacit commitment to provide future revenue to that firm. Whether it be lower life-cycle cost, reduced risk, or greater convenience, the incentive must be both explicit and significant for a firm to be embraced by an increasingly demanding technology market.

### Diverting the Profit Stream

In the salad days of the mainframe computer, data-processing systems were considered a long-term capital investment. In the 1960s, the justification for the purchase of a big IBM 360 might have been based on a decade-long payback period. In relative terms, the pace of technological change was leisurely; each major system purchase was an event in itself, with decision-makers having ample time to gather data and evaluate their options.

Today, purchases of information technology (IT) are no longer discrete events. With a two-year obsolescence cycle for most IT hardware (and only slightly longer for software), the selection of suppliers has evolved into a continuous process. Buying decisions are no longer based on methodical considerations. Instead, much of the detailed decision process is delegated to the supplier, based on its reputation and brand identity. In a sense, buyers are "subscribing" to the innovative product streams of trusted firms.

As I scan my home office, it is apparent that I have chosen to commit my stream of purchases to a handful of familiar brands. For example, it is unlikely that I will ever buy a printer that fails to carry the Hewlett-Packard logo, nor will I cast aside my Microsoft Office Suite anytime soon. Although a number of factors impact my buying decisions, I am strongly motivated by severe risk-aversion: I cannot stand dealing with dysfunctional IT products. This emotional bias is no doubt the driving force behind my choices of brands. Once I have established a satisfying, risk-free relationship with a supplier, it would take a major jolt for me to alter my buying habits.

It is ironic that rapidly changing technologies may actually enhance the importance of customer loyalty. From the standpoint of risk-aversion, however, this is a reasonable outcome. If consumers can no longer trust high-technology products to remain at the leading edge for extended periods, where can they turn for stability? With new products and technologies swirling around us, we are forced to place our trust in the innovators themselves. Often the tradeoffs among products are too complex and the options too vast for the typical buyer to execute an informed purchase decision. Instead, we commit to purchasing Sony entertainment products, Motorola communications equipment, H-P test instruments, and Oracle databases. In a sense, we are delegating our decision-making responsibilities to these proven innovators. As long as these firms continue to be worthy of our trust, they will be granted a steady stream of sales revenue.

The most common basis for trust in a buyer/supplier relationship is the minimizing of product life-cycle cost. Although purchase price is still a major differentiator, most buyers today are looking beyond their initial investment in a product to consider aggregate cost over time. The life-cycle cost of a telecommunications system, for example, would include purchase price, installation, training, establishing interfaces to existing infrastructure, operating expense, system reliability (downtime for mission-critical technologies is incredibly expensive), functional use life, maintenance, service, and the cost to upgrade hardware and software in the future. Any supplier that can create a value proposition which minimizes these aggregate costs will attract the favorable attention of long-term customers.

In addition to these practical considerations, several more subtle factors can affect the buying decisions of cost-sensitive consumers. *Switching costs*[12] are often cited as determining the long-term behavior of customers. Presumably, once a user is "locked in" to a given product architecture, the costs associated with shifting to an entirely new platform might be a strong deterrent to switching suppliers. This was the logic behind the "proprietary system" philosophy adopted by DEC for its minicomputers and by Apple for its Macintosh platform. In these cases (and many others in the 1970s and 1980s) the goal was to inoculate customers with a proprietary architecture that forced them into returning to the source for maintenance, upgrades, options, and accessories. Today, firms that use such blatant means to force customers into "shotgun" relationships are often punished in the marketplace. Open system architectures and non-proprietary standards have become the norm. Today, firms that attempt to manipulate buyers into involuntary sole-source arrangements are viewed with suspicion.

Switching costs have become a pivotal issue in technology markets, due to two closely related factors: *network externalities* and *intercompatibility*. In the telecom sector, for example, the value of a cellular phone service is largely determined by the network externalities associated with the geographic coverage of various transmission standards. Performance is clearly a secondary consideration: Improved transmission quality is of little consequence if there is no one out there to talk to. Traveling up the food chain within the telecom industry, the intercompatibility of base stations, switching gear, repeaters, and transmission lines has become a vital necessity. With respect to networked technologies, the cost of adopting an incompatible product is "infinite," in the sense that the desired functions simply cannot be performed.

Similarly, the market for personal computer operating systems has been driven into a virtual monopoly by the benefits of interoperability and network externalities. Windows-based PCs communicate with each other effortlessly, can be loaded with a rich variety of compatible software, have a wealth of available peripherals, and perform properly most of the time. Each of these factors weighs heavily on the buying decisions of both new and returning customers. Most important, however, is the learning curve associated with adopting a new operating system. Each time a new technology is substituted for a familiar one, there will be a significant waste of time, money, and efficiency. After nearly two decades of industry dominance, almost every American office worker has some competence in the use of Windows-based software. Under these conditions, the cost for a firm to adopt an alternative solution (assuming that a viable one existed) would likely be prohibitive.

In mission-critical industries such as medical instrumentation, telecommunications, and information processing, switching costs must be considered in terms of the buyer's exposure to unacceptable risk. In these sectors, product reliability must surpass "four nines," meaning that full system operation must be achieved at least 99.99 percent of the time. This requirement translates into less than one hour of downtime per year. Under such constraints, adopting an unproven technology or transitioning to an unfamiliar supplier would be foolhardy at best. Moreover, a poor choice in such a critical arena can be career-limiting, elevating the naturally risk-averse tendencies of decision-makers to the level of paranoia. It is not surprising that firms with a reputation for superior reliability and quality retain leading market shares in mission-critical technologies.

Furthermore, such firms are often drawn into a positive feedback cycle that will tend to increase their dominance over time. It is common for risk-averse decision-makers to "poll" other buyers of technology to gain confidence in their decisions. As a result, there is a well-established relationship between the market share of a firm and its probability of adoption by new buyers.[13] Moreover, firms that have been anointed as worthy of trust in a given market segment can exploit this reputation by "pulling through" related products or services. The word-of-mouth testimonials of satisfied customers and the ability to expand relationships with loyal patrons tend to reinforce the already powerful advantage of trusted firms.

A final strategy for gaining customer loyalty involves improving the convenience of the buying process itself. Firms that are known to be cutting-edge innovators, for example, can become allies to customers who are unfamiliar with advanced technologies. Through educational outreach programs and demonstration projects, such firms can gain the respect and trust of lay buyers, and may become valued partners in the decision process. Likewise, the ability of firms to provide "one-stop shopping" for complete system solutions, rather than serving as suppliers of specialized components, can reduce both the risk and the cost of a major technology purchase.

As the rate of technological change increases, the traditional concept of "market share" begins to lose its meaning. After all, static pie charts can capture only a single moment in time, whereas strategies in technology markets should be based on both the *direction* and the *rate of change* of market behavior. A better way to evaluate a firm's competitive position within a given market segment might be to envision a stream of revenue flowing from customers to producers. As time progresses, this stream divides and reconstitutes, with the volume of flow to various firms representing their evolving relative positions in the marketplace.[14]

If markets can be thought of as streams of revenue, then the strategic goal of every firm should be to divert that flow into its own coffers. This can be accomplished in several ways, which include becoming an accepted industry standard, gaining a first-mover advantage, giving away something to capture market share, or fragmenting a consolidated market. To illustrate the concept of revenue streams and the strategies that can divert them, I will describe a familiar case example, presented graphically in Figure 11.2.

By now the story of how an American firm, Ampex Corporation, demonstrated the first commercially successful videotape recorder in the mid-1950s, only to have the mass market for this technology captured by Japanese rivals, has become painfully familiar. Likewise, the epic standards battle between Sony Corporation's Betamax format and the poorer-performing VHS format has become a twice-told tale in every MBA program. I will, therefore, begin my description of the evolving market shares of various home-video technologies at the cusp of the Betamax/VHS standards battle, and extend the story into the digital age.

Sony delivered the first Betamax videocassette recorders (VCRs) to the U.S. market in 1976, followed closely by the introduction of VHS-format video recorders by JVC (a subsidiary of Matsushita). Initially, the Betamax standard gained a significant market share, due to a slight first-mover advantage, the market power of the Sony name (which was already associated with quality innovations in the mid-1970s), and the acknowledged performance advantages of the Betamax system. Within a few years, however, the VHS format began to capture the flow of revenues from the home-video market, due primarily to JVC's liberal licensing of its VHS standard to competing firms. High levels of competition resulted from this open-standard strategy, yielding significant cost reductions for VHS-format recorders and the eventual acceptance of this standard by the global marketplace.

By the late 1970s, the home-video market was expanding at an exponential rate. Philips Corporation recognized an opportunity to divert some of this growing revenue stream by fragmenting the market. It could not hope to topple the now-established VHS standard, but it could potentially grab a share of the high-end video-enthusiast market by introducing a system that offered vastly superior picture and sound quality. In 1978, Philips began commercial production of video compact-disk (also known as laser-disk) systems. Unfortunately, a combination of factors caused the laser-disk product line to perform poorly in the marketplace: the cost of the systems was beyond the reach of most consumers, the number of available titles was limited, and the system could only play back prerecorded programs. Hence, the laser-disk format diverted only a trickle of revenue from the established home-video standard.[15]

Another attack on the ubiquitous VHS standard came from an entirely different direction. Cable TV had infiltrated the homes of a large percentage of urban and suburban households by the early 1980s, and firms within that industry recognized that they also had an opportunity to divert some of the revenue stream away from VCR technology. In this case, the value proposition was more attractive to the average consumer: First-run movies and, more important, popular sporting events were offered on a "pay-per-view" basis. This new technology had the advantage of essentially zero switching costs; in fact, since no special hardware was needed, viewers could opt to use their VCRs for recording *General Hospital* while watching their favorite sports teams and movies using the pay-per-view cable delivery system.

Thus, for nearly a decade, three competing technologies (VCRs, laser disks, and pay-per-view) established relatively stable shares of the home-video revenue stream. Yet even in the mid-1980s, a careful listener could hear the distant rumblings of a cataract. Philips had skillfully realigned its laser-disk technology to service the audio market, and through alliances with Sony and others, had established its compact-disk (CD) format as the new standard in the music industry. It was only a matter of time before an optical-storage technology would burst upon the video scene as well.

As of this writing, the standards battles between Betamax and VHS are being played out again in the digital videodisk (DVD) arena. Even as the DVD market is beginning to flourish, rival formats have emerged that might divert revenues from this fledgling standard and potentially confuse buyers of the next generation of home-video appliances. The Divx standard was developed by a partnership backed by Circuit City Stores Inc., and is aimed at capturing a large slice of the lucrative video-rental market as it transitions from VHS cassettes to the five-inch digital CDs.[16]

Divx technology enables a low-cost disposable digital videodisk, designed for customer convenience; there is no need to return the rental to a neighborhood specialty store. The disposable disks could, at least in principle, be purchased at any retail outlet for a nominal fee (estimates are currently set at $5 per view), and after viewing can be discarded. Thus, the Divx technology not only threatens to divert the revenue stream of DVD producers, but it also attacks the flow of revenue received by thirty thousand video-rental stores in the U.S. domestic market.[17]

It is not clear as of this writing whether the Divx standard can find a place in the delta forming around digital home-video revenues. The conceptual diagram shown in Figure 11.2 suggests that the market will be segmented based on consumer convenience preferences, with pay-per-view retaining a small share of sports devotees and the DVD and Divx standards dividing the main tributary of the revenue stream. Of course, nothing in technology markets is permanent; all these current players are in a race to appropriate sufficient returns from their nonrecurring investments (Divx technology is reported to have cost roughly $100 million to develop) before the next tidal wave hits. Permanent storage media for music and videos may be rendered obsolete early in the next millennium by the transmission of high-bandwidth entertainment products directly into residential homes through cable or digital wireless technologies.

In relative terms, the revenue stream from home-video technology is easy to visualize. The markets for Internet technologies, on the other hand, are far more complex. There are literally dozens of opportunities for the diversion of revenue streams just at the entry point to the World Wide Web. Hardware suppliers contest the physical interface, software producers vie for allegiance to their operating systems, the Web-browser wars have become latter-day legends, while various "portal" providers are clamoring to be the first site viewed by Web surfers.[18] At each step in this Internet gauntlet, fierce competition is being waged for the attention and loyalty of customers, with each firm knowing that it must position itself quickly at the estuary of what will become the next mighty river of commerce and growth.

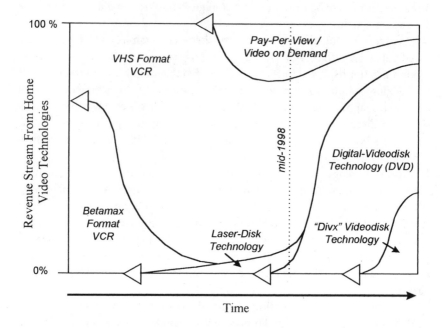

**Figure 11.2:** Conceptual diagram of the share of total industry revenue captured by various home-video technologies, as a function of time. Each innovation causes a diversion of the profit stream away from incumbent products, and might even render existing technologies obsolete. Note that this figure is not intended to be quantitatively accurate.

### The One-Stop Shop

In recent years, several market strategies have emerged for gaining customer loyalty in technology-intensive industries. The three variations described in this section combine the advantages of lower life-cycle cost, reduced risk, and greater convenience, in distinct ways. Each strategy is oriented toward building long-term supplier/buyer relationships, rather than closing one-time sales. Case examples were selected both to illuminate the unique aspects of these strategies and to demonstrate that they are often employed as potent combinations.

The *one-stop shop* strategy emphasizes customer convenience and risk reduction, and has proved to be highly successful among lay buyers of complex system solutions. The concept is quite simple: A single firm provides a complete, risk-free "solution" to its customers' "problems," along with the expertise to implement, maintain, and upgrade the system's capabilities over time. A one-stop shop must possess expertise in all aspects of a given market segment, including a deep understanding of its customers' needs and expectations, familiarity with all major suppliers of system components, problem-solving skills, system-integration skills, proficiency in customer education and training, and much more. In a sense, this strategy mandates a "vertical" structure for the delivery of value to end users, in contrast to the horizontal value chains preferred for production operations.

Examples of the one-stop shop strategy abound, particularly in the information and telecommunications sectors. Throughout much of the 1990s, customers planning to procure complex system solutions were forced to engage outside consultants to support their purchase decisions. It is virtually impossible for even the largest firms to maintain sufficient in-house expertise to make informed choices in every technological arena. The need for assistance in selecting information systems in particular was so desperate that companies were willing to hire their accounting firms to provide decision support. In this way, the Big Six (now Big Four, or is it Big Three?) accounting firms captured a significant slice of the revenue stream flowing to the information technology sector during the 1990s.

Somewhat belatedly, computer hardware and software firms recognized this pilfering of their rightful revenues, and have recently adopted a total-solution sales strategy. Most of the major IT manufacturers have established in-house consulting services to support their customers' product selection, implementation, training, maintenance, and upgrade needs. The recent $10 billion purchase of DEC by Compaq was motivated in large part by its desire to absorb a highly trained, fifty-thousand-person army of direct sales and service personnel. Likewise, Oracle and Cisco Systems have developed respectable in-house consultancy divisions, and are positioning themselves as "solution providers." Even Microsoft has realigned itself as a supplier of enterprise-level solutions.

The most dramatic example of the power of the one-stop shop strategy was the rebirth of IBM in the latter half of the 1990s. After decades of faltering hardware and software sales, and repeated misinterpretations of market needs, IBM's new CEO Lou Gerstner reshaped the company into a provider of non-proprietary hardware and software solutions. Rather than limiting IBM's sales consultants to recommending only IBM products, Gerstner gave the Global Services (GS) division license to draw the "best-in-class" hardware and software from all suppliers, both domestic and international. The results were impressive. With a focus on developing "e-business" Internet commerce strategies for its clients, IBM's GS division increased its revenues by a factor of nine in the period from 1990 to 1997, and continues to achieve annual growth rates in excess of 20 percent as of this writing. With 1997 revenues of $19.3 billion, the GS division represented more than a quarter of IBM's total earnings, a proportion that was expected to grow to more than half of all sales by 2003.[19]

Somewhat ironically, the source of IBM's advantage as a one-stop shop is a legacy pool of multidisciplinary talent accumulated during its vertical, command-

and-control days.[20] Such a deep and rich base of tacit knowledge is more than capable of solving even the most challenging customer problems. The globally recognized IBM name inspires customer confidence, while its expanding suite of market-leading internal products, such as the recently acquired Lotus Notes groupware application, provides a pull-through for the sale of additional products and services. Lou Gerstner's ultimate goal is not to resuscitate the dominant role of the old IBM, but rather to position Big Blue as a "thought leader" in the global information-technology industry.

It is easy to recognize the advantages of the IBM strategy. Rather than face a confusing maze of products and the daunting prospect of integrating hardware and software from multiple suppliers, customers are willing to place their trust in IBM's reputation and share responsibility for buying decisions with a vastly more knowledgeable partner.

This first example captures the key characteristics of the one-stop shop strategy: Customers enjoy a convenient and relatively "safe" buying experience, in return for diverting their stream of purchases through the balance sheets of the one-stop shop. Unfortunately, at the highest levels of the enterprise food chain, implementation of such a strategy requires the inimitable resources of an IBM or an Oracle. The salient aspects of this strategy, however, can be exploited at virtually any level of the supply chain, and with almost any product.

The marketplace for mobile phones, for example, has become a jumbled mass of incompatible standards and shifting technologies. Despite a robust demand for greater geographic coverage, there has been slow movement toward convergence within the U.S. market, and even less consensus globally. Under these conditions, an executive whose career depends on being accessible anywhere in the world would be forced to carry half a dozen mobile phones, and would still experience significant gaps in coverage.

In this situation, both elements of the one-stop shop strategy are strongly represented; the customer's risk of being out of touch, combined with the inconvenience of juggling several geographically challenged handsets, leaves a gaping hole in the mobile phone market awaiting an integrated worldwide solution. The Iridium international consortium was formed to meet this need, by providing a truly global mobile-communications system accessible through a single (albeit bulky) handset. Based on a constellation of sixty-six overlapping communications satellites, and a network of two hundred land-based cellular phone providers in ninety countries, the Iridium system promises clear reception and transmission at any point on the globe.[21]

Motorola and other consortium members hope to carve out a slice of the revenue stream from existing cell-phone providers, while expanding the global market for mobile phones to regions with poor or nonexistent coverage. Admittedly, the bulky handsets and ubiquitous coverage carry a hefty price tag:

Service charges may run as high as $7 per minute for some calls. Yet this is a small price to pay for itinerant executives whose careers depend on reliable global communication.

One of the most common applications of the one-stop shop strategy is as a means to divert monopoly profits from one layer of a supply chain to another. Generally, a value chain cannot efficiently support more than one layer of monopoly profits. The "dominant" link in the chain can typically force all other participants into commodity pricing, by leveraging their unique technological advantages. If one considers the electronic components that constitute a computer motherboard, for example, the supplier of the microprocessor clearly garners the lion's share of the profits. Under these conditions, the other levels of the PC motherboard supply chain should be developing plans to usurp the dominant supplier and capture their fair share of the revenue stream.

In the semiconductor industry, there is an emerging trend toward "systems on a chip" (SOCs). As the density and sophistication of integrated circuits have increased, functionality that previously required dozens of specialized integrated circuits can now be gathered onto a single powerful chip. Motorola, for example, has virtually abandoned the microprocessor market (currently dominated by Intel) to focus on SOCs for telecommunications, digital video, and a host of embedded-processing applications. The SOC approach offers unsurpassed performance in a smaller package with reduced power consumption, ultimately enabling a one-chip cellular phone, set-top box, or digital television.[22] Although reduced customer risk and increased convenience play a minor role in this rendition of the one-stop shop, the primary driver here is performance.

The underlying goal of the firms that developed SOC technology is to divert revenues from both ends of the value chain. By gathering functionality under a single umbrella, the SOC manufacturer is essentially eliminating the specialty-chip supplier from the value chain. More important, however, the SOC developer can often capture the monopoly profits of the end-product manufacturer (i.e., the OEM) as well. The recent explosion in demand for compact videodisk players, driven in large part by the karaoke craze in Asia, provides an excellent example. Unlike previous introductions of new consumer electronics products, industry giants such as Sony and Panasonic were forced to compete with upstart Chinese videodisk firms almost immediately.

The difference in this case was the availability of a low-cost video-compression integrated circuit that performed most of the critical functions of the system. The availability of such an SOC component enabled price-slashing Chinese producers to immediately compete with the technological giants, with the majority of the profits from this market flowing to the maker of the video-compression SOC. As one executive of a major U.S. electronics firm put it, "When all the intellectual property is trapped inside the silicon, there's very little value you can add."[23]

A final example of the one-stop shop model is worth noting. For several decades, pundits of advanced technology have touted the coming of the "age of digital convergence." In this grand vision, all forms of information, including text, music, video, voice communications, and data transmission, will be subsumed into a single delivery system for homes and offices. The scope of this vision is staggering: A single delivery system for all forms of digital communication, linked to every home and business on the planet.

As recently as the mid-1990s, this grand scheme seemed to fade into the wistful daydreams of marketing executives. Delays in the development of digital-television technology, the lack of convergence on standards, and most important, a disappointing lack of progress in extending high-speed data transmission the "last mile" into homes and offices, conspired to postpone digital convergence indefinitely.

All that may have changed with the 1998 merger of AT&T and Tele-Communications Inc. (TCI), a major cable-network provider. For the first time since the adoption of the Telecommunications Reform Act of 1996 there has been a significant breakthrough in the "last-mile" infrastructure blockage that has stifled competition in residential markets. The combination of high-bandwidth cable access, and an enormous customer base in cable television, Internet services, and long-distance telephone, may allow this new conglomerate to finally reach critical mass. As Van Baker, director of consumer research at Dataquest observed, "We could have a universal pipe in the home that handles all flavors of what will then be digital, including video, voice, and data."[24] Indeed, the owners of such a vital link would control what may be the most important one-stop shopping franchise of the new millennium.

### Bundling, Non-rival Goods, and Antitrust

The one-stop shop model suggests that there is more to creating value in high-tech markets than being the best within a narrow specialty, particularly when the intended customers are technically unsophisticated. In fact, this convenience-oriented model is a special case of an increasingly important marketing strategy known as *bundling*. Firms that offer their customers one-stop shopping are essentially "bundling" their multidisciplinary knowledge to provide enterprise-level solutions.

Bundling can be defined as the selling of distinct but complementary products as a single package, at a single price.[25] This strategy can enable firms to increase their revenues by capturing synergies among products and by exploiting their market strength in one segment to expand into related segments. Whereas the advantages of one-stop shopping (i.e., risk mitigation and convenience) are intended to directly benefit the buyer, the broader application of bundling strategies is not always so customer-friendly. Too often this technique is used to coerce and manipulate consumers into excessive expenditures for unwanted products.

The concept of bundling is so ubiquitous that it is often not recognized as an explicit strategy. The inclusion of in-flight meals during air travel, the provision of an extended product warranty, or the "standard features" of new cars are all examples of *transparent bundling*. In these situations, customers come to expect a certain grouping of products and services, and would react negatively to offerings that deviate from this norm. Most of us would be shocked, for example, if a new software product failed to provide an automated "help" feature.

In its simplest form, a bundling strategy might involve the inclusion of detailed product documentation, instruction videos, after-sales consulting, or on-site training, to facilitate the customer's utilization of a new product. The goal of an *information bundling* strategy is to eliminate barriers to the adoption of a product by reducing the "skills gap" associated with implementing an unfamiliar technology. Moreover, information bundling provides an excellent opportunity for firms to highlight their products' unique advantages and differentiating features. Finally, the information that is bundled with a product helps position that product within the value chain: Detailed design documentation and engineering support are typically bundled with the purchase of industrial equipment, for example, while little technical information is included in the sale of a consumer product.

Bundling can offer economies of both scope and scale, thereby providing a cost saving to customers. Compared to separate components, a bundled product might be less costly to manufacture, will require less inventory transactions, can be sold by the same sales team, and can be shipped in the same trucks. This type of bundling, however, tends to reduce the level of customization provided to customers. An over-the-counter cold medicine, for example, might bundle several symptom-relief ingredients into a single tablet, allowing a significant cost saving when compared with the purchase of separate remedies. Unfortunately, cold sufferers who lack the appropriate set of symptoms are forced to ingest (and pay for) unnecessary medication.

For system-level products, a bundled solution can often achieve better performance than a more modular architecture that is burdened with standardized component interfaces. The performance of a personal computer motherboard, for example, is dependent on the interplay of the microprocessor, on-board memory, graphics chips, printed-circuit-board layout, and data-bus design. A "generic" motherboard architecture, capable of accommodating a variety of microprocessor chips (potentially from different manufacturers), could not achieve the same computing performance as a highly integrated "purpose-built" product. Likewise, the system-on-a-chip concept described in the previous section effectively bundles the functionality of a handful of specialized integrated circuits into a single device, thereby achieving significant performance and cost advantages.

Bundling may become necessary when interfaces among related products are not standardized. This is particularly true in the early stages of a new technology market, but will tend to diminish as an industry matures. As interface standards become established, industries will typically flatten into a horizontal structure of specialized suppliers. In this more mature market scenario, bundling can be a detriment; users are not as free to select best-in-class solutions to their specific problems. In the early days of home-stereo systems, for example, all components (i.e., turntable, amplifier, tuner, speakers) were bundled into a single console. As audio technology improved, discrete (i.e., unbundled) components began to dominate the market.

Bundled and unbundled strategies are natural adversaries, and industries can shift rapidly from one paradigm to the other. In the early stages of a new industry, buyers are risk-averse and will tend to seek out bundled solutions that inspire the confidence needed to adopt a new innovation. In the past, there has been a general tendency for industries to unbundle their products over time, as markets become more highly segmented and customers become educated in the detailed tradeoffs among components.[26] I believe this trend is not likely to be pervasive in future technology markets, however. As discussed in more detail below, the effects of network externalities and interoperability in high-tech industries may strongly favor the bundling of products and services.

It is not uncommon for seemingly unrelated products (from different manufacturers) to be bundled, often to create an enhanced brand image for all the products involved. Such *alliance-based bundling* can benefit customers, provided that there is some legitimate synergy among the bundled products. The bundling of a "free" mountain bike with the purchase of a sporty automobile is intended to complement the lifestyle of the consumer. Less trivial examples include alliances that enable the bundling of customized motor controllers with industrial motors, or the inclusion of consulting services from an unrelated firm with the purchase of an enterprise-software package.

From the standpoint of technology-intensive industries, the final two bundling strategy variations listed in Table 11.1 are the most important. Among the greatest challenges for high-tech firms is the continuing need to "create" markets for their most recent innovations. Bundling offers opportunities for firms to leverage their current market position to build an early market for new products.

The more customer-friendly of these new-market strategies involves tying together an existing product with a new item in a manner that aggressively promotes the benefits and value of the recent innovation. Netscape Communications, for example, borrowed a page from the strategy book of its archrival, Microsoft, to market an expanded version of its Netcenter Website in 1998. By tightly integrating the functionality of its popular Navigator Web

| Variation on Bundling Strategy | Description |
|---|---|
| Transparent bundling | Combinations of products or services that have become standard (and expected) in the marketplace. |
| Information bundling | The inclusion of drawings, manuals, training material, etc. to position product and reduce barriers to adoption. |
| Bundling to capture economies of scope or scale | Combinations of products or services that enable cost reduction through synergies in production, sales, distribution, storage, etc. |
| Bundling to improve performance | Superior system-level performance is achieved through a higher degree of integration. |
| Bundling to reduce risks of new technology | Bundled products can provide a "turnkey" solution for customers who are wary of new innovations. |
| Alliance-based bundling | Combining products or services (often from more than one firm) that have a natural (or contrived) relationship in the marketplace. |
| Bundling to promote new products | Use of bundling as a "pull-through" strategy to build a market for new product introductions. |
| Bundling to exploit proprietary products | Firms that possess a proprietary technology in high demand can leverage their market power to sell a much larger "package." |

**Table 11.1:** Several variations of a bundling market strategy that are applicable to technology-intensive firms.

browser with the interactive features of its Website, Netscape asserted that it could offer a richer and more productive Internet experience. In this case, the products were technically not being sold as an inseparable package; each item could function without the other. Moreover, as has been Netscape's habit in recent years, the new Website-integrated browser was to be given away free, placing the burden of profit generation on advertising fees from the Netcenter Website and on pull-through sales of more sophisticated Netscape products.[27]

The above example underlines several emerging characteristics of high-technology markets. The first and most obvious feature is the importance of global brand recognition. It is not sufficient for high-tech firms to adopt the "rosebush" model for innovation described previously; they must also have a means to leverage their existing customer base and recognizable brand to create markets for each new "bud." The power of brand identity in high technology has dramatically increased in recent years, as discussed in Box 11.1. In an era of over-whelming technical complexity, a trusted global brand name is a vital "icebreaker" for new-product introductions.

The seemingly desperate "giveaway" strategy repeatedly used by Netscape points to a second unique characteristic of high-technology markets: the growing impact of non-rival goods and increasing returns to scale. Information-based products are non-rival, meaning that they can be sold again and again at virtually no cost to the owner (see Chapter 3 for additional discussion). Netscape virtually created the Web-browser market in the early 1990s by giving away its innovative Navigator Internet application to all takers. A similar strategy has been adopted by Adobe Systems to increase the market share of its portable-document-format (PDF) products. By giving away a free "reader" at any Website that uses its format-ting standard, it has generated broad awareness of the Adobe brand and the benefits of PDF-based document creation.

The key element in these giveaway strategies is increasing returns to scale. Once a non-rival good (e.g., software, entertainment media, literature, com-puter games, technical information, and most other knowledge-based products) has been created, it can be used in any way that generates revenues, with little regard for marginal production cost.[28] Essentially all that matters is that firms' non-recurring R&D investments are recouped (along with an acceptable profit) over the life of the product. It is not unreasonable, therefore, for firms to give away a non-rival product to achieve any of the following goals: 1) receipt of ad-vertising revenue, 2) sale of other bundled products, 3) sale of upgrades to, or future versions of, the same product, 4) increased brand recognition, 5) genera-tion of goodwill, 6) capture of network externalities, 7) positioning of the product as an industry standard, 8) building credibility for an entirely new market, and so on.

A final characteristic of high-technology markets highlighted by the Netscape example is the impact of network externalities and interoperability. Through its initial giveaway strategy, Netscape captured a large early share of the Web-browser market by the mid-1990s. With many Internet users embracing its initial offering (certainly a value at twice the price), Netscape was well positioned to exploit this nascent loyalty to sell its network software products, and to promote the smooth interoperability of its commercial Website. Since the majority of early Web surfers were using the Netscape Navigator browser, Website designers from around the globe began tailoring their creations to "look good" on this emerging standard. It appeared as though the network externalities gained from a common Internet doorway might drive Netscape into a virtual monopoly position in the Web-browser market.

Readers who are even dimly aware of recent happenings on the Internet know, of course, that this fairytale outcome was not to be. The slumbering giant, Microsoft, finally awoke in 1996 to discover that it had snoozed through the formative stages of the World Wide Web. In a manner typical of a startled giant, Microsoft launched an all-out attack on the Web-browser market, specifically targeting Netscape as the enemy. Throughout the late 1990s this drama played out, as described in the final section of Chapter 6. As of this writing, Netscape has been forced to give away the source code to its latest Web browser in a last-ditch attempt to stave off the beast, while Microsoft has stomped its way into a virtual dead heat with Netscape through aggressive promotion of its Explorer Web browser.

What does this example have to do with bundling strategy and customer loyalty? The pivotal assertion in the Department of Justice's 1997 antitrust case against Microsoft was that the firm was using its monopoly power in PC operating systems to build another monopoly in Internet software. By bundling the Explorer Web browser into its Windows 98 product, it was claimed that Microsoft could exploit customer loyalty to essentially lock out all other entrants. Moreover, it was asserted that the software giant had coerced major PC manufacturers to bundle its earlier Explorer versions with their preloaded products, thereby excluding Netscape from this preferred position.

It is likely that any analysis of this ongoing tempest will be outdated before this book is in print. Given the importance of this case, however, I will risk making a few premature observations. First, it should be acknowledged that there is another side to this story. Microsoft has repeatedly asserted that it did not violate a 1995 consent decree barring the firm from taking undue advantage of its near-monopoly in PC operating systems. To the contrary, it has correctly noted that the consumer is better off for its actions, since integrating the functionality of a Web browser into its Windows operating system will indeed provide smoother interoperability and enhanced performance.

With respect to the coercing of PC manufacturers, the jury is still out (both figuratively and literally). Although fear of Microsoft's market clout has rendered most of the involved parties "speechless," several allegations have surfaced that appear to support the government's case.[29] In any event, it is likely that such brute-

## Box 11.1: Building Brand Identity in High-Tech Markets[32]

The Iridium consortium's market strategy demanded that it establish a global brand identity virtually overnight. With two powerful competitors, Globalstar and ICO, racing to market with alternative worldwide mobile phone services, Iridium hoped to exploit its first-mover status to become a "reputation leader" before its rivals could leave the starting gate. Unfortunately, the brick-sized Iridium phone has proved to be a hard sell. As the consortium's director of marketing put it, "If we had a campaign that featured our product, we'd lose."

Building brand recognition is becoming increasingly important to high-technology firms, given the blinding rates of change and contentious rivalries that characterize these markets. Rather than relying on traditional advertising campaigns that create *interest* but convey little *information,* technology leaders are adopting a more integrated approach, in which brands become a symbol for the company itself. Such trademarked slogans as "Intel Inside" and "100% Pure Java" are intended to position their respective owners (Intel Corporation and Sun Microsystems) as *thought leaders.* In this new marketing arena, trademarks are transformed into "trustmarks," intended to build customer loyalty over an extended period.

With brand identity becoming an integral part of corporate strategy, traditional advertising firms are finding themselves upstaged by marketing and management consultancies. Technology firms are avoiding the excessive hype of Madison Avenue in favor of creating subtle but powerful images of quality, reliability, and innovation leadership. With the battle shifting from market share to "mind share," firms require enterprise-level guidance. Trite promotions such as Saatchi & Saatchi's latest brand-management tool, entitled the Brand, Resources, Advertising, Information, and Network (BRAIN), highlight the relatively narrow focus of even the most prominent ad agencies.

As technology products permeate every corner of the planet, the responsibility for buying decisions is shifting from technical specialists, such as information-technology managers, to nontechnical managers, executives, and even boards of directors. Brand identity can have a disproportionate influence on such lay consumers, given their natural tendency to be intimidated by complex and unfamiliar new technologies.

force tactics will not be tolerated in the future, particularly if public exposure of such unseemly behavior begins to impact customer perceptions and loyalty. As of this writing, Gateway 2000 Inc. and several other major PC manufacturers have announced that they will bundle both Microsoft Explorer *and* the Netscape Navigator browser with their preloaded machines.[30]

Underlying this case is the question of fair practices under conditions of "natural" monopoly power. An arcane principle of antitrust law known as the *essential-facilities doctrine* bars companies that control a large or irreplaceable (i.e., essential) part of the economy from arbitrarily banning others from its use.[31] The essential-facilities doctrine has been cited as applying to the case of Microsoft versus the U.S. Department of Justice, particularly if the software firm ultimately wins the Web-browser market battle and ends up controlling virtually the entire front end of the Internet.

While the entrepôt to the Internet seems to meet the criteria of an essential facility, I believe that such an antiquated doctrine (the essential-facilities doctrine dates to the turn of the twentieth century) may be irrelevant in a sector characterized by blinding change. Today, Microsoft's hegemony is challenged by Java-based operating systems running on PC platforms, "set-top boxes" that run several other brands of software, and the "network computer" architecture promoted by powerful forces within the software and computer-hardware industries. Adjudicating antitrust action based on static (albeit monstrous) market share is an outdated concept. What matters in dynamic technology markets is *how tight a grip is held on the flow of sector revenues over time.* Today, Microsoft holds the throttle on much of the software industry. The reader should not be surprised, however, if a few years hence this software giant discovers itself marginalized by unexpected market forces that are beyond its ability to control.

### The Technology Club

The importance of customer loyalty to sustaining leadership in high technology has caused a shift in the focus of competitive strategy toward building long-term supplier/buyer relationships. There has been a proliferation in recent years of so-called *soft strategies* that bypass the traditional head-to-head rivalry among firms in favor of expanding the size and scope of markets and creating steady streams of revenue from satisfied customers.[33]

From a relationship-building standpoint, the most powerful soft strategies being employed by high-tech firms are those that draw customers into a *technology club.* The general principle is straightforward: Create an incentive program for customers that makes recurring patronage economically and psychologically appealing. The funding for such an incentive program can be easily justified. The cost of capturing a new customer is far greater than the relatively minor investment required to maintain an existing relationship. Moreover, a well-defined base of returning buyers can substantially reduce the risks of future innovations, and can enable far more accurate projections of market demand over time.

Three general methods can be used to form a technology club. The most subtle involves shifting the buying process from an emphasis on one-time purchases of discrete products to a "subscription" structure that bundles both products and services into an attractive recurring package. The goal here is to satisfy a customer's ongoing need for a specific capability. Gateway 2000 Inc., for example, has begun offering customers a novel program in which a personal computer, software and hardware maintenance, and Internet-access services are provided for a single monthly fee.[34] By converting the purchase of a PC into a subscription for a continuously upgraded "desktop computing service," Gateway hopes to retain a larger share of new and returning customers.

The most familiar example of the club strategy is the frequent-flyer programs of major airlines. In this instance, returning passengers are given a financial incentive in the form of free air travel and other perks. Variants of this marketing approach are emerging in virtually every commercial sector. A new chain of Wolfgang Puck restaurants offers a "Puck's Perks" program to recurring customers. Likewise, several supermarket chains are using bar-coded identification cards to offer returning customers special discounts. The advent of so-called smart cards, which exchange customer information and transaction data with merchants' on-site card readers, further enhances the benefits of this strategy.

Can the "frequent-flyer" concept be applied to technology markets? Most software producers already provide incentives for returning customers, by offering steep discounts on upgrades to their earlier products. Firms producing technology-intensive hardware can offer attractive "trade-in allowances" to customers that can be applied to the purchase of their latest model. More subtle arrangements might include providing access to "developers' kits" and proprietary product documentation to valued long-term customers. In each case, a dividing line is established between new and returning customers, so that the recurring group is rewarded for its continuing loyalty.

The final method for forming a technology club follows the "giveaway" approach described in the previous Netscape example. By offering the source code for its Navigator Web browser to programmers around the world, Netscape had hoped to benefit from a flood of Navigator-specific applications written by third parties.[35] The initial giveaway was intended to create a "club" of developers whose loyalty to Netscape would be based on the unprecedented inside information they were granted. In this case, the members of the technology club were a part of the value chain itself, rather than being end users of the product.

The technology club strategy can be used to build valuable and sustainable relationships beyond the confines of the commercial marketplace. With a little imagination, one could envision strategies for employee retention fitting within this model. Savings plans, profit-sharing programs, bonuses, and stock options are all means to retain valued workers for extended periods. Perhaps human-resource managers should consider applying loyalty strategies such as the technology club to the human-capital marketplace. It would seem that protecting a firm's vital pool of tacit knowledge is surely worth the creation of a "frequent-innovator" program.

## Model 3 – Technology as Art and Craft

During the first half of the twentieth century, astronomers and cosmologists were faced with a troubling dilemma. On the one hand, most people (scientists and otherwise) felt sure that the universe was a stable, static, and eternal place. It was well known that the planets moved about the sun, but it had always been assumed that the stars in the heavens were comfortably immobile. On the other hand, scientists were confronted with compelling new evidence (i.e., the "red shift") that the universe was actually expanding at an alarming rate. How could these two opposing views be reconciled?

Today, science has reached a consensus that the universe as we know it is actually the vestige of a cosmic explosion, known as the Big Bang, that occurred tens of billions of years ago. In less-enlightened times, however, some creative solutions were proposed to resolve the "expansion paradox." One of these proposals suggested that the universe might be cyclic, pulsing outward, then inward, at some natural rhythm. Although astronomers perceived the universe to be expanding at that time, it was suggested that in the future the system would reverse itself, so that later astronomers might observe the universe to be contracting.

How does this slice of scientific history relate to sustainable advantage? Markets for high-tech products behave in much the same way as the cyclic model of the universe described above. New technologies enable industries to expand rapidly: major firms and start-ups alike burst forth to generate a rich variety of products. As the novelty of the new technology wears off, however, these same industries will begin to "collapse" into a commodity structure, only to be fragmented again when the next enabling wave of technology hits.

If a company's strategy is based solely on extending the technological frontier, it will be exposed to the roller-coaster ride described above. Firms that utilize technology as a *medium,* however, rather than as an end in itself, can insulate themselves from these disconcerting oscillations. This does not imply that these companies are less innovative: Their creative talent extends beyond the research laboratory to address the more human aspects of market demand. Firms that generate value through "artistic" use of technology can ride the ebbs and flows of technical progress far more easily.

If one considers our current technological capabilities objectively, it seems apparent that if the human race ceased to extend the scientific frontier tomorrow, it would take decades for us to fully exploit our existing knowledge base in the marketplace. For every firm that makes its fortune at the leading edge, there are dozens of firms that create value through clever utilization of existing technologies. Synthetic fabrics enable the fashion industry to break new ground, high-tech materials benefit the sporting-goods sector, electronic musical instruments expand the depth and range of artistic expression, and embedded microprocessors instill intelligence into even the most mundane household appliances. In all these instances, it is the creative use of existing technologies that provides value.

The innovative use of existing technologies can be divided into three loosely defined categories: fad, craft, and art. The idea of a *high-tech fad* is a relatively new concept, made possible by the ubiquitous presence of technology in our current society. In general, a fad can result from any new creation that appeals to people's sense of fun, curiosity, self-image, or aesthetic sensibilities.

Given the typically short duration of fads, it would seem that they could not form a basis for sustainable market leadership. Yet this is clearly not the case. Automobile companies have exploited fads for years to accelerate the sales of new car models. In the 1950s, nearly every full-sized car sported "wings." In the 1960s, the convertible sports car came into its own. The 1970s car market was dominated by high-power muscle cars with racing stripes. And so it goes to this day: The vaguely European luxury sedan is the "in thing" in the late 1990s. Such fads are often cyclic: Volkswagen introduced a "new Beetle" in 1998 that merged advanced automotive engineering with a nostalgic appearance.

Fads are becoming increasingly common in high-technology products as well, facilitated by the rapid flow of global information. The "Gigapet" craze of 1997 provides an excellent example. From its origin in Japan, this relatively low-tech electronic toy spread throughout the world in little more than six months. The demand for Gigapets did not result from their technological sophistication; the fact that it was an information technology product was incidental. Their appeal came from the desire of children the world over to own and care for a cute little pet. Evidently the human aspects of innovation can become a powerful engine of value creation when coupled with the exciting potential of new technologies.

In 1996, a three-person graphics firm called Unreal Pictures teamed with the multimedia division of Autodesk, Inc. to create some interesting computer-generated animation characters. One of those characters was a gyrating infant that came to be known as the "Dancing Baby." From its origin as a tutorial for

Autodesk's customers, this "freeware" animation spread throughout the Internet, often carried as an attachment to e-mails between friends. Within a year, this weird little character became a guest star on a television series, and was featured in commercials aired during the Academy Awards. Although little income was generated by this fad, Unreal and Autodesk benefited from almost instantaneous brand-building publicity.[36]

What is important to recognize in these two examples is that it was not the underlying technology that made global stars of these creations. Application-specific integrated circuits (ASICs) allowed the construction of a small, low-cost toy that seemed to have a mind of its own (i.e., the Gigapet), while advanced three-dimensional video graphics formed the basis for the boogying baby character. Yet it was the humor and imagination embodied in these innovations that accounted for their appeal.

The disconcertingly rapid diffusion of technological fads often works in favor of their creators, since there is little time for competitors to imitate the recipe, or for pirates to undermine market demand. Borne on the "buzz" that permeates chatrooms and e-mail exchanges (along with more traditional radio and television exposure), knowledge of intriguing new products can proliferate around the globe in a matter of weeks. Although it may be tough for firms to get a fad started, once the effects of word-of-mouth begin to snowball the innovator will be in for a wild ride.

The impact of technology on *craft* is somewhat more enduring, but also goes through frequent cycles. Music, art, clothing, architecture, sports, entertainment, and most other forms of human expression are continuously reinvented by creative minds. As new media become available (e.g., three-dimensional video graphics, laser light shows, interactive software, electronic music), new opportunities emerge for creative talents to express themselves. Often, old themes will be reborn in entirely new forms. Traditional household appliances can become trendy items through creative use of form, color, and automation features.

"What was old is new again" applies to technology-based creations. As technology progresses, time-honored stories or characters (e.g., the Titanic, Dr. Dolittle) are brought to life first in movies, then on television, and perhaps again in movies enhanced by modern special effects. Action-hero characters might also be featured in multimedia video games and high-tech toys. The tremendous value generated by Disney characters, for example, has been derived from the sensitivity and imagination of their creators: Modern technology has simply enabled these familiar characters to become an even greater part of our lives.

*Artistic creativity* is becoming almost as powerful a source of productivity growth as traditional R&D is in many industries. The entertainment sector has experienced explosive growth in the 1990s, propelled by the innovative largess of the information technology industry. George Lucas' Industrial Light and Magic

(ILM), for example, has formed a mutually beneficial long-term relationship with Silicon Graphics Inc. (SGI) that has enabled both firms to advance their respective crafts. In 1993, ILM and SGI formed the Joint Environment for Digital Imaging (JEDI) alliance to establish the world's most advanced production facility for the creation of digital imagery. Since that time, many of the advances in video graphics achieved by SGI have been in response to the obsessive demands of ILM. Likewise, ILM has retained its industry-leading position by combining technological savvy with cinematic artistry. The formation of such non-exclusive alliances, merging the complementary creative abilities of two or more firms, represents a largely untapped source of productivity growth.[37]

Educational and entertainment software firms have long been at the vanguard of technological artistry. Since the early days of Pong and Pac-Man, this industry has evolved in lock step with its chosen high-tech media. In 1993, Broderbund Software Inc., a highly successful pioneer in PC-based multimedia products, transformed the video-game industry with its introduction of *Myst*. This gentle and haunting portrayal of a mythical island world, created in partnership with Cyan Inc., sold more than four million copies, making it the most popular software game in history. Again, the success of this unlikely hit was derived not from clever technical gimmicks, but rather from the powerful sense of mystery, loneliness, and stark beauty that players of the *Myst* game experienced.[38]

Fortunately for those of us who are not gifted with artistic ability, there is still a wealth of opportunities to create value through technology-based craft. As is the case with the more traditional arts, science and technology often spawn the development of craft industries that flourish in the shadow of mainstream production. In the early stages of an emerging technology, clever "tinkerers" can often find ways to combine, enhance, and modify basic design elements to serve niche markets. As the noted physicist and writer Freeman Dyson observed, "In the future, many of the small enterprises of today will be consolidated, and new small enterprises will have to find different niches to fill. There will always be niche markets to keep small software companies alive. The craft of software will not become obsolete."[39]

One such craft company is Viewpoint Data-Labs International, a small firm with an unusual specialty. Viewpoint has set about building the world's largest library of three-dimensional images. Whether the subject is the human body, Godzilla, or the Eiffel Tower, Viewpoint creates detailed 3-D models and scans them into its database. The demand for such images is booming: Medical schools require accurate renderings of organs, universities use images of "atoms" to explain the workings of matter, and movie studios employ images of anything and everything to create realistic special effects. By combining expertise in the crafts of solids modeling and 3-D imaging with a penchant for detail and realism, Viewpoint essentially founded the industry segment that it now dominates.[40]

At the other end of the spectrum from "technology as craft" is technology as a medium for pure artistic expression. Since the first cave paintings were rendered in celebration of animist beliefs, the human animal has sought new and better ways to express itself artistically. The advent of video graphics, lasers, mechatronics, and computer automation has allowed artists to explore new creative horizons. Technology has become such an important part of modern society that a new school of art has formed over the last thirty years, known as the Art and Technology movement. Contributors to this movement include Andy Warhol, Jasper Johns, John Cage, Robert Rauschenberg, and many others. The boundaries of contemporary art have been dramatically expanded through the work of these expressive talents, and in so doing, the role of both art and artist in modern society has been redefined.[41]

In a fascinating example of art/technology fusion, the artist Joel Slayton presents his audience with a small Plexiglas cube that has been outfitted with motion sensors. As the cube is manipulated by the viewer, its movements are displayed graphically on a color computer monitor. In this way, his work, entitled *To Not See a Thing,* challenges the viewer to interact with, and become part of, the visual experience.

What is most intriguing about this artwork is the manner in which people react to the unfamiliar situation. Regardless of their background or technical sophistication, viewers' behavior seems to follow a common pattern. At first they handle the cube delicately, as though it were quite fragile. After a time, they begin to explore the motion/graphics interface with more elaborate movements. Finally, they will almost always try to compete with the computer that is rendering the animated graphics, using ever more violent movements in an attempt to fool the system.

Embedded in this unique artwork is a powerful metaphor for the impact of technology on our culture. At first, an unfamiliar technology is treated with caution by modern society. As our level of comfort increases, we begin to explore the possibilities of the new technology in creative and exciting ways. Ultimately, our competitive spirit dominates our behavior, driving the formation of new technology-intensive industries.

The question of whether technology is a medium for artistic creation, or is in fact a form of art in itself, is best left to philosophers. What matters most to society is the capacity that science and technology provide for growth, learning, communication, and creative expression.

# Conclusion

## An Economic Argument for Environmental Sustainability

The most frustrating aspect of writing a book of this scope is that it must, of necessity, be incomplete. The practical limitations of time and space have forced tough choices to be made on almost every page, and despite my sincere efforts, several important topics have not been adequately addressed. Unfortunately, this is an unavoidable situation, made all the more difficult when a neglected subject is of great personal significance.

Since the industrial revolution, economic development has been based on two implicit assumptions: 1) that the resources of the Earth are available in unlimited supply as inputs to production, and 2) that the environmental damage caused by firms, industries, and nations in their pursuit of wealth has no economic significance. Clearly, in these enlightened times the falsity of such assumptions can no longer be ignored. The goal of this final essay is to mitigate my sin of omission with respect to environmental issues by providing some insight into how the self-serving assumptions of the past can be supplanted by pro-growth/pro-environment strategies for sustainable economic development.

The World Commission on Environment and Development has defined *sustainable development* as economic growth that "meets the needs of the present without compromising the ability of future generations to meet their own needs."[1] From a less human-centered perspective, sustainable development implies that the rates at which we consume resources and create pollutants must be moderated in such a way that economic growth can continue indefinitely without causing irreversible environmental damage.[2] Fortunately, such a balance is achievable: The natural environment has a powerful capacity to cleanse itself. Even such devastating calamities as the 1989 *Exxon Valdez* petroleum spill have healed themselves over time. Although the ecosystem of Prince William Sound has been irrevocably altered, life has returned, and its former beauty has largely been restored.

When the levels of contaminants and pollutants exceed the planet's ability to process these wastes, the delicate balance of the ecosystem begins to shift. To achieve sustainable economic development, it is vital that we know at what levels of pollution the line is crossed from self-regeneration to a nonsustainable or even an irreversible condition. As of this writing, no such quantitative knowledge exists, leaving open the possibility that we have already done irreparable harm.

Economically speaking, environmental damage from industrial activity is a negative externality that, in cases such as global warming, can create undesirable spillovers to every corner of the planet. These negative spillovers are perpetuated by the fact that environmental costs are not accounted for in economic transactions. This market failure causes producers to have no financial incentive to reduce waste or eliminate emissions. In the absence of either effective government regulation or countervailing market forces, shared resources such as lakes and rivers, the atmosphere, rain forests, and fossil fuels will be subjected to unrestrained use by profit-seeking individuals. The inevitable result is environmental tragedy.

Political factors exacerbate the problem: Governments seem unable to override special interests on environmental issues and establish policies that enable market forces to act in the proper direction.[3] Although policymakers and industry leaders have begun to recognize the importance of sustainability issues, the ability to "price" environmental resources (and more important, to determine the "cost" of their degradation) is still relatively primitive.[4] As the Nobel economist Kenneth Arrow recently noted, "The market is no good when it comes to factoring greenhouse control into energy prices, because there is no way anyone can buy the benefits of preventing climate change."[5] Despite this limitation, however, the environmental debate has progressed over the last decade from the antiindustry, antiprofit, and antigrowth rhetoric of radical environmentalism to the realization that business must play a central role in achieving the goal of sustainable development.[6]

The most reasonable concept for allocating the costs of environmental damage is the "polluter pays" principle. The challenge in executing such a policy ideal lies in avoiding the imposition of undue hardship on narrow industry segments. Moreover, if improperly implemented, such a plan could have a significant negative impact on economic growth.

Over the very long term (and with some pessimistic assumptions), one analyst has suggested that the price of environmental damage from greenhouse-gas emissions alone could reach as high as 20 percent of world economic output.[7] Clearly, something must be done to prevent such an economic and environmental catastrophe. Unfortunately, with the costs of protection measures also being high, it is hard to find a politician who would willingly impose this steep bill on domestic industries. Yet without some reasonable safeguards, there is no guarantee that increased wealth will make people better off in the future. Unrestrained development can make society richer, but at what point does the degradation of our surroundings and the risks to our health render growth in per capita income sadly irrelevant?

The most problematic environmental concerns are those that impact common resources, with global warming being the most prominent among these issues.[8] Both the atmosphere and the greenhouse gases being emitted into it are common to everyone on the planet, yet it is in no country's best interest to take unilateral action. Indeed, such action would be fruitless: Even if the United States reduced its substantial greenhouse-gas emissions to near zero, the rate of atmospheric degradation over the long term would hardly be affected. It is only through multilateral action by both advanced and developing nations that further destabilization of the Earth's climate can be avoided.[9]

At best, the science of global warming is ambiguous. Simulation models, the most valuable tools for predicting complicated system behavior, are as yet incapable of capturing the enormous complexity of the Earth's climate. What is well established is that the chemistry of the atmosphere is changing. Airborne carbon dioxide, one of the primary contributors to global warming, has increased by one-third since the industrial revolution, and appears to be well on its way to doubling.[10] If such a change in atmospheric chemistry occurs, the Earth's average temperature may increase by several degrees, with potentially catastrophic results.

The release of carbon dioxide ($CO_2$) accounts for 86 percent of U.S. greenhouse-gas emissions. Hence, for the purposes of this simplified discussion, we can think of the global warming problem as essentially a carbon dioxide emissions problem. Furthermore, since up to 98 percent of all $CO_2$ emissions are caused by the burning of coal, oil, and gas to generate electric power, achieving significant reductions in $CO_2$ emissions can be accomplished only through the reduction of energy waste in all aspects of industrial production. Energy represents a critical input to production for both the advanced and developing nations, as was made painfully clear during recent oil shortages.

The range of attitudes toward environmental restrictions across industries is striking, with energy-intensive sectors typically giving the cold shoulder to the implementation of multilateral environmental agreements. This position has been taken to an extreme by some self-interested industry leaders, whose antiregulation rhetoric has been disconcertingly similar to the "scientific" propaganda of major cigarette manufacturers during the 1970s and 1980s. Capitalizing on the indeterminate results of scientific investigations into global warming, influential leaders such as Exxon Corporation's chairman Lee Raymond have campaigned both domestically and in developing nations for a twenty-year moratorium on greenhouse-gas controls, ostensibly to give science sufficient time to "thoroughly understand the problem." A sensible counterpoint to this cynical perspective has been offered by John Browne, the chief executive of British Petroleum: "The time to consider policy dimensions of climate change is not when the link between greenhouse gases and climate change is conclusively proven, but when the possibility cannot be discounted and is taken seriously by the society of which we are a part."[11]

A similar lack of unanimity exists among nation-states. Developing countries feel entitled to compensation in return for environmental regulation, based not only on economic equity and ability to pay, but also on their belief that the industrialized world bears the moral responsibility for the current degraded state of the environment. Moreover, developing nations mistrust the motives of the advanced economies, referring to environmental measures as "eco-colonialism." They further object to the industrial North's obsessing over long-term issues such as global warming, while ignoring the more urgent environmental priorities of the South, such as polluted water, unsanitary sewers, and rampant deforestation.[12]

In December 1997, a global-warming treaty was forged in Kyoto, Japan, by negotiators representing 159 nations. According to the so-called Kyoto Protocol, the United States is required to reduce its emissions of greenhouse gases by the year 2010 to a level that is 7 percent below domestic emissions in 1990. European nations are assigned a target of 8 percent below their emissions in that same year, while Japan somewhat reluctantly agreed to a target reduction of 6 percent. To meet the goals of the Kyoto Protocol, the United States will have to reduce emissions of carbon dioxide, carbon monoxide, methane, and other carbon-based gases by roughly one-third of the current projections for U.S. output of these gases in 2010.[13]

One of the most promising results of the Kyoto meeting was a recommendation that an international market be established in *emissions credits*. The concept is simple: Nations whose firms reduce their emissions below a specified level can sell their excess reductions to firms in nations that are over their limits. In this way, market forces can be engaged in the battle to control global warming. In a recent speech to Congress, Senator Robert Byrd stated that "reducing projected emissions by a national figure of one-third does not seem plausible without a robust emissions-trading and joint-implementation framework."[14]

The international trading of emissions credits allows firms and nations alike to reduce their output of carbon-based gases in the most economically efficient way. Under such a regime, firms would have the flexibility to select the most efficient methods for achieving their emissions targets, either through immediate action or by purchasing credits from other firms or nations until an optimal improvement plan could be implemented. Command-and-control policy measures that force firms to adopt "quick-fix" solutions can be suboptimal in terms of the cost to firms, and more important, in terms of overall effectiveness. If firms are given time to implement the most efficient and cost-effective changes, the economic impact of emissions reductions can be dramatically reduced.[15]

The trading of emissions credits between nations can result in far greater economic efficiency, since the cost of reducing the emissions of advanced manufacturing processes in the North can be as much as ten times higher than the cost of the same improvement in the South. In a credit-trading system, firms in the United States that buy emissions credits from China, for example, would essentially be subsidizing the modernization of China's energy-related industries. Since all emissions are equivalent from a global perspective, this could be a far more cost-effective solution than attempting to make incremental improvements at home. An international "commodity market" in emissions credits would allow the market price for these credits to approach the marginal cost of emissions reductions worldwide. Assuming that the transaction costs are kept low, trading would lead to the highest possible efficiency in solving this global problem.

There is an excellent recent precedent for the concept of credit-trading to reduce environmental damage. In 1990, the U.S. Acid Rain Program was created with the goal of halving the emissions of sulfur dioxide by domestic utilities. According to a study by the Government Accounting Office, this credit-trading system has decreased the cost of pollution reduction to half of what was expected under the previous rate-based measures, and well below industry and government estimates. Moreover, by 1995, one-third of all utilities that complied with this measure did so at a net profit, due to unforeseen savings associated with the changeover to low-sulfur coal.[16]

What effects will environmental regulations have at the level of the firm? It is easy to make dire predictions of energy price increases and costly process modifications. Yet the above example suggests that there may be new opportunities as well. Since energy is an important input to the production functions of most high-technology firms, energy-saving measures are essentially the same as cost-saving measures. Moreover, the marketplace offers a value premium for many environmentally safe products, potentially offsetting much of the recurring costs associated with these socially responsible strategies.

The development of so-called *green products* addresses both the product designs themselves and the manufacturing processes required to produce them. The two complementary goals of green design are: 1) to prevent waste, by reducing the weight, toxicity, and energy consumption of products, and by extending their service life, and 2) to improve the management of energy and materials, through techniques such as remanufacturing, recycling, composting, and energy recovery.[17]

Firms that wish to adopt green design as a long-term strategy should consider the following questions: What are the relative waste levels of alternative manufacturing processes? How are products to be managed after disposal? How can product designs incorporate recyclability? What is the environmental impact of component materials? And most important, how can the total life-cycle energy consumption of products, from raw materials through production and ultimately to disposal, be minimized?

Many companies have already adopted corporate-level environmental strategies. Techniques such as voluntary environmental audits and product life-cycle impact analyses are the first steps toward integrating environmental costs into the core strategies of firms. Guidance for enterprise-wide management of environmental issues is provided by global standards such as the International Standards Organization's ISO-14000 series. Those firms requiring outside expertise will find that there is no shortage of available consultants, spanning all aspects of environmental protection.

There has been a surprising shift in the preferences of consumers in recent years toward environmentally friendly products, a movement that is often referred to as the "greening of the marketplace." Ultimately, such ethical consumers may have the final word in global environmental protection, by insisting on high standards of corporate citizenship. Programs such as eco-labeling provide an opportunity for firms to advertise their social responsibility, while imparting a "green tinge" to their corporate brands.

The economic and political power of environmentalists has expanded dramatically in the 1990s. Currently, many European nations host active and influential "green parties," and several countries offer voluntary eco-labeling programs that can have surprising leverage in the marketplace. Germany, for example, has a Green party that consistently captures a 10 percent share in political polls, and is expected to become a partner in government at the federal level within the next several years.[18] Germany has embraced environmental activism more enthusiastically than any other European country, and has maintained an active eco-labeling system for classifying products, known as the Blue Angel program, since 1978.[19]

Furthermore, restoration of the Earth's environment will offer tremendous innovation opportunities well into the next century. The market for environmental products and services is expected to reach $300 billion by 2000.[20] According to one estimate, 40 percent of global economic output in the first half of the twenty-first century will be derived from environmental or energy-linked products and technologies.[21] U.S. firms currently hold an edge in important nonpolluting energy technologies, including reliable solar power, gasoline alternatives based on agriculture, zero-emissions fuel cells, and so on.[22]

Once market forces begin to act on problems such as global warming, the profit incentive will fuel the creative fires of entrepreneurs and innovators, potentially yielding faster-than-expected emission reductions. A decade ago, an international negotiating team met in Montreal to establish a protocol for eliminating the use of chlorofluorocarbons (CFCs), the compounds linked to depletion of the Earth's ozone layer. At that time, both government and industry predicted an economic catastrophe. Instead, CFC emissions have declined so rapidly that replenishment of the ozone layer now is expected to occur early in the twenty-first century. Much of this tremendous progress was the result of a skillful realignment by the manufacturers of air conditioners and refrigerators to non-ozone-depleting refrigerants, a transition that went virtually unnoticed by consumers.[23] Thus, the human capacity to invent, adapt, and solve intractable problems represents our greatest hope in the battle to save the natural world.

On a business trip to Malaysia in 1997, I experienced firsthand the futility of economic development in the face of environmental disaster. From the window of my room at the Regent Hotel in downtown Kuala Lumpur, the Petronas Towers, symbols of the emergence of Malaysia as a global economic force, should have been an impressive sight. Despite my being virtually at the foot of these world-class skyscrapers, however, my view was completely obscured by thick brown smoke carried by the winds from intentionally set wildfires in the rain forests of Indonesia. On the streets below, businesspeople covered their faces with handkerchiefs as they waited for cabs, while pedestrians scurried to escape the noxious air.

At what level of environmental degradation does increasing wealth reach a point of diminishing returns? If quality of life is the true measure of economic gains, then surely unbreathable air renders all of us poor indeed. Likewise, the climatic turmoil that may result from global warming will not discriminate on the basis of per capita income. Thus, environmental responsibility is an inevitable mandate, otherwise the benefits of economic growth will ultimately be negated by a steadily deteriorating quality of life.

The ethical use of technology, coupled with responsible action on the part of industries and governments, can simultaneously raise both the quality of life *and* the quality of the environment throughout the world. If we can all learn to cooperate toward such a goal, the effects could be significant within our lifetimes. To achieve this happy outcome, all that is needed is for each of us to accept our responsibilities as caretakers of the beautiful planet that abides us.

# Notes

## Introduction

1. Barney, C., 1997, "Bewildered New World," *Upside*, November Issue, pg. 110. I highly recommend the three-volume work by Manuel Castells, collectively titled *The Information Age: Economy, Society and Culture.* See 1) Castells (1996), 2) Castells (1997), and 3) Castells (1998).
2. Sawyer, R. D. (translator), 1994, *Sun Tzu: The Art of War,* Westview Press, pg. 179.

## Chapter 1: Productivity and Economic Growth

1. "What makes economies grow?" *Chief Executive,* April 1998, pp. 36–41.
2. For those who have missed this elegant euphemism, the term refers to those centrally planned economies of Eastern Europe that are "transitioning" from the ravages of communism to the harsh realities of capitalism.
3. Unless otherwise noted, I will use the term 'growth' to mean an increase in aggregate economic output, as measured by real (i.e., inflation-adjusted) gross domestic product.
4. The Balinese culture achieved remarkable agricultural abundance and a deep and rich artistic tradition in an essentially closed economy. They have lived in harmony with their island environment for the last several thousand years, jeopardized in recent times by growing tourism. On a more ominous note, the vision of notorious Cambodian revolutionary Pol Pot was to revert his nation to the stable agrarian culture that had existed for centuries before the intrusion of the West.
5. The World Bank (1997, Selected World Development Indicators, Table I, pg. 214). These numbers reflect 1995 exchange rates and have not been corrected using purchasing power parity (PPP) factors.
6. The rate of growth of the world's population has been gradually slowing. Between 1980 and 1990, the average growth rate was 1.7 percent per year, but declined to 1.5 percent per year from 1990 to 1995. This slowdown is observed across all income levels, although the rate of growth in low-income countries is still more than double that of the advanced economies. Source: The World Bank (1997, Selected World Development Indicators, Table 4, pg. 220).
7. Abelson, P. H. (1996, pg. 34).
8. Doremus, P. N. et al. (1998, pg. 25).
9. Many Japanese firms are associated through "enterprise groups" known as *keiretsu*, which encourage mutual cooperation and inter-firm commerce. The *keiretsu*,

much like their pre-World War II predecessors, the *zaibatsu*, typically have a large bank at their center which administers much of the debt and equity of the group's members. See Doremus, P. N. et al. (1998, pg. 43).

10.  If it is not obvious to the reader that investments with long-term payback can play a vital role in the economic survival of firms, consider the decisions that individuals make when planning their own economic future. A college education is an obvious example of a long-term investment that has substantial payback potential, but would have far too long a time horizon to be funded by the typical American firm.

11.  "China's school of business," *The World In 1998*, The Economist Publications, pg. 92. This blithe neglect of returns to shareholders by Japanese firms is likely to change in the near future. With a rapidly aging population and a woefully inadequate social security system, returns on equity by publicly held Japanese companies that average about 4 percent per year can no longer be tolerated.

12.  Doremus, P. N. et al. (1998, Table 3.8, pg. 52).

13.  Although there is typically an initial boost in the stock values of merged firms, in the long run their performance has been less than stellar. In the 1990s, nearly half of all mergers have created firms whose stocks have not performed as well as those of their industry peers. See "Global swarming," *The Los Angeles Times*, December 14, 1997.

14.  "SGI's plight reflects rapid growth, strategy discord," *The Wall Street Journal*, October 31, 1997.

15.  "Break out with a break-up," *The World in 1998*, The Economist Publications, pg. 91.

16.  Rifkin, G. (1997). For a sobering perspective from one of the firms which Cisco acquired, see Wiegner, K. (1997).

17.  Krugman, P. R. (1994b, pg. 62).

18.  This is the desired result. It turns out that some factories in Soviet Russia were achieving *negative* productivity; they actually subtracted value from the inputs they consumed.

19.  Here I am assuming that the production function has constant returns to scale, meaning that a growth in inputs, in the proper ratio, will yield a proportional growth in outputs. For an accessible discussion of this subject, and the broader topic of long-term growth, see Dornbusch, R. and S. Fischer (1994, Chapter 10).

20.  Although new capital equipment eventually wears out, these costs can be accounted for in the capital-input term. The new technology still provides a sustainable increase in output.

21.  See, for example, 1) Young, A. (1992), and 2) Young, A. (1995).

22.  This subject has become quite controversial, as several contradictory interpretations of the data have been proposed. For an overview of this debate, see International Monetary Fund (1997, Box 9, pg. 82). For some contradictory data that imply a factor-of-two increase in Singapore's productivity from 1962 to 1986, see Dollar, D. and E. N. Wolff (1993, pg. 159) and Chen, E. K. Y. (1997).

23.  Whether Singapore has shown impressive productivity growth over the last two decades or has been virtually stagnant is not relevant to the near-term productivity of capital. At any instant in time, a severe shortage of labor can slash the marginal returns to capital, regardless of long-run trends.

24.  Greenwood, J. (1997).

25.  There is empirical evidence that larger factories are more likely to employ newer technologies than smaller plants. See Dunne, T. (1994).

26.  See 1) Hall, B. H. (1993a), and 2) Hall, B. H. (1993b).

27.  "High-tech industries, led by Internet, boost U.S. growth and rein in inflation,"

*The Wall Street Journal*, April 16, 1998.

28.  See 1) "The zero inflation economy," *Business Week*, January 19, 1998, pp. 28–30, and 2) "Patents tell an upbeat story," *Business Week*, December 1, 1997, pg. 28.

29.  See 1) Solow, R. M. (1957), and 2) Solow, R. M. (1956).

30.  I will follow the convention of the literature in this field and use the terms 'productivity growth' and 'technical progress' somewhat interchangeably.

31.  Economists at the time considered technical progress to be *exogenous*, meaning that it was assumed to be independent of controllable factors within the economy. As I will demonstrate in Chapter 3, this somewhat unrealistic assumption has been abandoned in favor of more recent models of *endogenous growth*.

32.  Denison, E. F. (1985).

33.  Generally, total-factor productivity is used in the context of technological progress, and labor productivity is often used as a reflection of capital-to-labor ratios and the relative skill level of the workforce.

34.  Arrow, K. J. (1962a).

35.  Sources: 1) "Singapore eases rules on immigration," *International Herald Tribune - Internet Edition*, August 26, 1997, 2) "The fight to keep making something," *Business Week*, December 30, 1996, and 3) "Towards a learning nation," *Singapore Business Times - Internet Edition*, June 4, 1997.

36.  Grossman, G. M. and E. Helpman (1991, pg. 6).

37.  Regarding the limitations of empirical data on productivity, one of the foremost researchers in the field notes that "We are caught up in a mixture of unmeasurement, mismeasurement, and unrealistic expectations." See Griliches, Z. (1994).

38.  This is the well-known "Solow residual method," which has been subjected to many tweaks and reinterpretations in the literature.

39.  This potentially critical weakness in the measurement of postwar productivity is discussed in detail in Nordhaus, W. D. (1997). Nordhaus cites an example of the relative price of domestic lighting, in which the traditional price-index method of output measurement understates the actual "value" of modern lighting products by a factor of one hundred over the last century.

40.  These tidbits are drawn from a fuller exposition of this topic, provided in Baumol, W. J. et al. (1989, Chapter 3).

41.  Baumol, W. J. et al. (1989, pg.12).

42.  The two databases that support this discussion are found in 1) Kendrick, J. W. (1973), spanning the period from 1884 to 1969, and 2) Maddison, A. (1982), spanning the period from 1870 to 1979, and can be extended to 1984 with 3) Maddison, A. (1987).

43.  For a detailed discussion, see Baumol, W. J. et al. (1989, Chapter 4).

44.  Denison, E. F. (1989, pg. 5). This alternative data set, based on recent Labor Department statistics, was selected because it offered the clearest insights into the productivity slump during the postwar period.

45.  See, for example, 1) Baumol, W. J. et al. (1989, pg. 70), and 2) Romer, P. M. (1989b).

46.  Griliches, Z. (1988). Perhaps more significantly, there is evidence that *innovation* suffered a sharp decline in the 1970s. See, for example, Englander et al. (1988).

47.  National Academy of Engineering (1993, pg. 24).

48.  "Productivity increased a sharp 1.7% last year following strong gain in 1996," *The Wall Street Journal*, February 11, 1998.

49.  Except where noted, this section follows the treatment in Dollar, D. and E. N. Wolff (1993).

50.  Barro, R. J. (1991).

51.  Baumol, W. J. et al. (1989, pg. 91). It is interesting to note that as of 1985, the United States was no longer the world productivity leader. Norway was foremost in all three important productivity metrics. The U.S. was in second place behind Norway in labor productivity, third in total-factor productivity behind Norway and Australia (and tied with Canada), and fourth in capital intensity, following Norway, the Netherlands, and Finland. See Dollar, D. and E. N. Wolff (1993, pg. 103).

52.  Tilton, J. E. (1971).

53.  In the world of auto racing, there is a strategy known as *drafting*, in which drivers position themselves directly behind the lead car for several laps to take advantage of the lack of wind resistance. A country that originates new technologies faces the "resistance" of high investment and risk. Once innovations have proven to be productive, however, the risk of adoption drops precipitously.

54.  See, for example, 1) Jones, C. I. (1995), and 2) Pack, H. (1994). By way of rebuttal, see Romer, P. M. (1994).

55.  See Dollar, D. and E. N. Wolff (1993, pg. 63), or Grossman, G. M. and E. Helpman (1991, pg. 7).

56.  This is often referred to as the *vintage effect*, in which new capital has a higher productive potential than old capital.

57.  The Organization for Economic Cooperation and Development (OECD) counts most of the advanced nations of the world as members, including the United States, much of Western Europe, and Japan.

58.  For perspectives on cross-national and cross-industry productivity growth, see 1) Costello, D. M. (1993), 2) Fagerberg, J. (1994), and 3) Dowrick, S. and N. Gemmell (1991).

59.  Gerschenkron, A. (1952).

60.  Dollar, D. (1992).

61.  Dean, E. (1984).

62.  The literature often refers to the separation between advanced and developing economies as being caused by a *technology gap*. See, for example, Fagerberg, J. (1987). Another interpretation suggests the combined effects of an *object gap*, which requires significant capital investment to bridge, and an *idea gap*, which may be closed more easily, resulting in rapid benefits to the world's poorest countries. This alternative view does not obviate the need for a minimum educational level necessary to internalize new ideas. See Romer, P. M. (1993a).

63.  The term 'currency devaluation' in the economic literature generally applies to a fixed-exchange-rate regime. Under a "managed float," a decrease in currency value is generally described as a *depreciation* of the currency.

64.  Baumol, W. J. et al. (1989, pg. 22).

65.  Dollar, D. and E. N. Wolff (1993, pg. 92).

66.  In Chapter 2, I will provide a parallel discussion of current trade patterns, based on an extension of the classical theory of comparative advantage.

67.  Sources: 1) "Going toe to toe with Big Blue and Compaq," *Business Week*, April 14, 1997, pg. 58, and 2) "Legend takes top spot in China," *Electronic Business News*, July 7, 1997, pg. 10.

68.  "Japan falls in rankings; U.S. is no. 1," *The Wall Street Journal*, April 22, 1998. For a broader perspective on U.S. competitiveness, see "Can the U.S. compete: A ten year outlook," *Chief Executive*, June 1997.

69.  There have been recent warnings voiced by scholars such as Michael Porter of Harvard University that the recent upsurge in American R&D spending and productivity

growth may cause U.S. industry to again become complacent with respect to global competitiveness. See "As innovation revives, some see complacency," *The Wall Street Journal,* March 23, 1998.

70.  Nakamura, S. (1989).

71.  Greenwood, J. (1997, pg. 13).

72.  For overviews of this topic, see 1) Willcocks, L. and S. Lester (1996), and 2) Sichel, D. E. (1997). For some practical thinking on reducing waste and improving operational productivity, see 1) Sumanth, D. J. (1998), and 2) Womack, J. P. and D. T. Jones (1996).

73.  "Disillusionment," *Information Week,* February 16, 1998.

74.  "ROI in the real world," *Information Week,* April 27, 1998.

75.  "Some firms, let down by costly computers, opt to 'de-engineer'," *The Wall Street Journal,* April 30, 1998.

76.  There is evidence from the services sector that IT implementations achieve optimal productivity only when designed around a carefully crafted downsizing plan. See "Insuring productivity," *Information Week,* November 24, 1997.

77.  "Less automation means more productivity at Sun," *Modern Materials Handling,* November 1995.

## Chapter 2: The Global Technology Marketplace

1.  The material in the first two sections of this chapter is intended to provide a general background in recent global economic integration and a foundational understanding of international trade theory. The advanced reader may wish to fast-forward to the third section, "Securing Gains from Free Trade."

2.  Bretton Woods, New Hampshire, was the venue for an international conference beginning in 1944, which established a system of global exchange rates pegged to a gold standard. Both the International Monetary Fund (IMF) and The World Bank were created during these sessions.

3.  Vernon, R. (1995).

4.  Krugman, P. R. (1989).

5.  International Monetary Fund (1997, pg. 112).

6.  Houston, T. and J. H. Dunning (1976).

7.  Wilkins, M. (1988).

8.  John, R. et al. (1997, pg. 17).

9.  This term means literally "to let do." In economic terms, it implies a policy of total nonintervention by national governments in commerce.

10.  Casson, M. (1985).

11.  Graham, E. M. (1996, pg. 26).

12.  Dunning, J. (1958).

13.  This is an excellent example of what economists refer to as a *positive externality.*

14.  John, R. et al. (1997, pg. 44).

15.  There has been a substantial push by the U.S. Department of Commerce (DOC) to expand exports to the so-called "Big Emerging Markets." This list of ten countries is projected to represent 40 percent of the growth in world import markets over the next fifteen years. Unfortunately, although the future potential of these markets may be bright, the list is dominated by developing countries (e.g., India, Brazil, Indonesia, the "Chinese Economic Area") with unstable business climates, immature financial institutions, and relatively low purchasing power. More important, the somewhat obsessive focus on exports to develop-

ing nations by the DOC sidesteps the more important economic issue for the United States: a need for rapid growth in total-factor productivity to compete effectively in global high-technology trade.

16.   Since both components and finished goods for a given product category are typically counted by the Department of Commerce as falling into the same industrial classification, this will appear statistically as North-South intra-industry trade. Note that provisions of the *Harmonized Tariff Schedule of the United States* encourage companies using foreign assembly or production operations to employ U.S.-made components, by offering a duty exemption for components that are returned to the U.S. as part of articles assembled abroad. See, for example, *Production Sharing: Use of U.S. Components and Materials in Foreign Assembly Operations, 1991–1994,* USITC Publication #2966, May 1996.

17.   Ehrlich, E. M. (1997).

18.   "Capital goes global," *The Economist,* October 25, 1997, pg. 87.

19.   It is interesting to note that full capital account convertibility has not, in the past, been a necessary condition for rapid economic growth. Both Japan and China have achieved double-digit growth rates without liberal convertibility of capital. See Bhagwati, J. (1998).

20.   "One world?" *The Economist,* October 18, 1997.

21.   The World Bank (1997).

22.   For a well-balanced discussion of the benefits and disbenefits of globalization, see Rodrik, D. (1997).

23.   UNCTAD (1995).

24.   Graham, E. M. (1996, pg. 14).

25.   This statement is descriptive of the largest MNCs. It is also relevant, however, to the majority of smaller transnational companies that participate in global industry networks.

26.   Blomstrom, M. (1990, pg. 49).

27.   The term 'rationalize' is used here to mean "make more efficient." It has been used in recent times as a euphemism for corporate downsizing.

28.   "Wintel" refers to Microsoft ( "Win," for Windows) and Intel Corporation ("tel," for Intel), and their vast network of suppliers, software producers, strategic partners, etc.

29.   See, for example, 1) Goldman, S. L. et al. (1995), and 2) Yoshino, M.Y. and U.S. Rangan (1995).

30.   Archibugi, D. and J. Michie (1997a, pg. 16).

31.   This is particularly true for technology-intensive firms, which depend heavily on the *national system of innovation* of their home countries. This structure, which includes a nation's educational, scientific, and technological environments, will be discussed in detail in Chapter 7.

32.   A tariff is simply a tax on a specific imported product. Many developing countries retain high tariffs as a source of sorely needed income.

33.   Schott, J. J. (1996, pg. 36).

34.   This is referred to as a voluntary export restraint (VER). The United States has a history of using this type of non-tariff barrier to limit imports in critical categories. Japan's voluntary restriction on the quantity of automobiles it exported to the United States in the 1980s is a well-known example.

35.   The use of local content regulations is not limited to developing countries. The European Union (EU) has employed this type of non-tariff barrier to influence other industrialized countries to build high-technology factories on EU soil.

36.   The World Bank (1997, Figure 3.1, pg. 43). For an in-depth analysis of corruption

within China, see Kwong, J. (1997).

37. For some interesting examples of how the Foreign Corrupt Practices Act can place U.S. firms at a competitive disadvantage, see Alkhafaji, A. F. (1995, pg. 335).

38. The World Bank (1997, Box 3.3, pg. 46).

39. For a thorough historical account of trade policy in the United States, see Destler, I. M. (1995).

40. This is better known as the Smoot-Hawley bill, which represented one of the most controversial, and in retrospect economically destructive, measures in American trade history.

41. Barfield, C. and D. A. Irwin (1997).

42. Iritani, E., 1997, "U.S., Kodak lose Japan trade suit," *The Los Angeles Times*, December 6.

43. Fast-track authority is granted by Congress to the president for a fixed period, to expedite the negotiation of trade agreements. It is essentially a promise by Congress to not tamper with the wording of complex agreements that may have taken years to negotiate. Congress is allowed a simple "yes" or "no" vote on the entire package.

44. For an overview of the recently adopted WTO Telecom Agreement, see Sisson, P., 1997, "The new WTO Telecom Agreement: Opportunities and challenges," *Telecommunications*, September, pp. 24–33.

45. These standards must include an equitable definition of dumping and an impartial dispute-settlement mechanism. Aside from strengthening intellectual property rights and enforcement, the issue of fair competitive practices may prove to have the greatest impact on technology-intensive industries. For additional discussion, see 1) Schott, J. J. (1996, pg. 17), and 2) Wilson, J. S. (1996).

46. Under Article XXIV of the GATT, nations are allowed to violate the principle of most-favored nation (MFN) treatment and form free-trade areas and customs unions. The primary motivation is to encourage the elimination of all trade barriers within the regional grouping, while maintaining the same level of tariffs and other barriers with respect to the outside world. In this way, other GATT signatories are not penalized, and the cause of free trade is promoted.

47. Schott, J. J. (1996, pg. 41).

48. The process of unification has applied considerable pressure to both the governments and people of several European countries, most notably Germany and France. A backlash against unification appears to be a real possibility as of this writing. For a recent discussion of the implications of the formation of the European Monetary Union on international finance, see 1) "The euro," *Business Week*, April 27, 1998, pp. 90–94, and 2) "The great money bazaar," *Business Week*, April 27, 1998, pp. 96–98.

49. Thurow, L. (1992).

50. Sager, M. A. (1997).

51. For a more complete discussion, see Bhagwati, J. and A. O. Krueger (1995).

52. Irwin, D. A. (1996a, pg. 5). Note that structural factors such as transportation and transaction costs will result in minor fluctuations in relative world prices. It is also important to note that the concept of "world market prices" assumes that buyers throughout the world have access to the same market information, a condition which rarely occurs in reality. As global communication links become stronger, however, there will be less opportunity for regional price variations, since buyers will arbitrage away any non-structural price differences.

53. Of course, there are political leaders on both sides of this issue as well, but I will make the hopeful assumption that their positions are a reflection of the constituencies they represent.

54. Interestingly, as the U.S. economy has heated up over the last several years, there has been a noticeable shift in the opinions of American workers toward a favorable position on free trade. See Greenberger, R. S., 1997, "As US exports rise, more workers benefit, and favor free trade," *The Wall Street Journal*, September 10.

55. Plato, *The Republic*, Loeb Classical Library (1930, pg. 153).

56. Irwin, D. A. (1996a, pg. 15).

57. Irwin, D. A. (1996a, pg. 83).

58. Smith, A. (1776).

59. In the mid-nineteenth century, the German economist and historian Friedrich List fiercely opposed the views of Adam Smith and his followers. His conviction was that free trade among unequal partners (from a technological standpoint) would almost always disadvantage the less-developed nation. Only under special circumstances, which included aggressive technology transfer, could free exchange be assured of benefiting both parties. List's views will surface again in Chapter 3.

60. This is in reference to the famous "invisible hand," which was said to guide each individual in such a way as to maximize the aggregate wealth of the economy. See Smith, A. (1776, pg. I-477).

61. The theory of comparative advantage is usually attributed to Ricardo; see Ricardo, D., 1817, *On the Principles of Political Economy and Taxation*. Other economists of the day, including James Mill, have been cited as originating this theory as well.

62. In this simplified model, I will use labor as a common measure of cost. In the real world, capital owners and other stakeholders share a portion of the profits from trade.

63. Another way of saying this is that Country W (and Country R as well) can purchase the same "basket of goods" more cheaply.

64. In international trade theory, this is often stated in terms of *opportunity costs*. A country is said to have a comparative advantage in product X if the opportunity cost of X in terms of product Y is less than that of its trading partner. See Markusen, J. R. et al. (1995, pg. 69).

65. For example, the gains-from-trade theorem assumes perfect competition, no taxes, constant returns to scale, and equal technologies among trading partners. At least with respect to trade in technology-intensive products, none of these assumptions is valid. The mechanism of comparative advantage always works, but the conclusion that free trade is an optimal policy choice must be carefully considered in light of real-world market conditions.

66. The complete absence of trade is referred to as *autarky*. When I suggest that people are worse off, I mean that they are worse off as compared to autarky.

67. Markusen, J. R. et al. (1995, pg. 72).

68. This is sometimes referred to as "labor arbitrage." Arbitrage is the process of buying a good or service in one market where the price is low, and selling it elsewhere where the value is high. By outsourcing semi-skilled labor to foreign markets, business owners can take advantage of low wages, while still selling into the high-profit domestic market.

69. The legendary economist John Maynard Keynes is famous for observing that "In the long run, we'll all be dead."

70. Roderik, D. (1997, pg. 1).

71. This is evident in China's recent commitment to become "fabulously rich," despite being miles apart from the Western world on human rights and democratic principles.

72.  Nivola, P. S. (1997).

73.  Bhagwati, J. (1995).

74.  The Montreal Protocol of 1987 set the first global targets for reduction of ozone-depleting gas emissions among the industrialized nations. See Cooper, R. N. (1998).

75.  There is strong evidence that *ecolabeling* has had a significant effect on U.S. market demand. The "dolphin-safe" label on cans of tuna fish, for instance, drove 98 percent of the dolphin-unsafe tuna out of the U.S. market. Moreover, this approach was ruled to be GATT-legal in 1991. See Schott, J. J. (1996, pg. 73).

76.  The recent collapse of currencies in Southeast Asia was largely due to overaggressive government spending on infrastructure and a lack of mature financial institutions to manage rapid economic development.

77.  Krugman, P. R. (1997, pg. 65).

78.  Here I use the term 'sustainable' to mean a condition of production in which all inputs can be continuously replenished, at their current rate of use, forever.

79.  Actually, since the negative experience of the Smoot-Hawley Tariff Act of 1930, American trade policy has been largely noninterventionist. A notable exception was the proposed Gephardt amendment to the Omnibus Trade and Competitiveness Act of 1988, which would have imposed import barriers against countries that failed to reduce their large bilateral trade surpluses with the United States. See Destler, I. M. (1995, pg. 92).

80.  Hufbauer, G. C. and K. A. Elliott (1994, pg. 11).

81.  Hufbauer, G. C. and K. A. Elliott (1994, pg. 15).

82.  Irwin, D. A. (1996b, pg. 11).

83.  This argument is borrowed from Irwin, D. A. (1996b, pg. 10).

84.  Jones, K. A. (1994, pg. 48).

85.  Primary inputs include labor, capital, and natural resources. Intermediate inputs have already been processed by at least one production function. Examples include such items as steel, oil (which has been refined), component parts, etc.

86.  The lack of fair access to China's domestic market has been cited as justification for the blocking by the United States and other Western countries of China's accession to the World Trade Organization.

87.  For those readers without a scorecard, trade policy in the United States is presided over by the House Ways and Means Committee and the Senate Finance Committee. In recent years, the Executive Office of the President, and in particular the U.S. Trade Representative, has played a major role in creating and influencing trade legislation.

88.  Note that there are several legitimate reasons why a company might sell products below their average cost. For example, with increasing economies of scale, the marginal cost of a product may be well below the average cost. Another reason might be as a method for damping out cyclical fluctuations in market demand, or as a forward-pricing strategy to spread out the price impact of high non-recurring development costs. See Hindley, B. and P. A. Messerlin (1996).

89.  An expanded discussion of antidumping measures is provided in Chapter 6.

90.  As with other forms of protectionism, there is a strong potential for damage to other industries that utilize protected products as inputs. The U.S.-Japan Semiconductor Agreement of 1986, which resulted from antidumping proceedings, had the effect of doubling the price of silicon memory chips in the U.S. market. IBM and other domestic computer manufacturers protested, and forced a modification to this agreement. See Destler, I. M. (1995, pg. 172).

91.  The first version of the Section 301 provision was contained in the Trade Act of

1974, and was amended to its present form in the Omnibus Trade Act of 1988.

92.  Destler, I. M. (1995, pg. 237).

93.  To be more precise, *imperfect competition* (also referred to as monopolistic competition) implies that a limited number of firms in the market are able to achieve some level of excess profits, due to insufficient threat of new entrants. An *oligopoly* is similar, except that in oligopoly, firms take into account the expected reactions of rivals in their pricing strategies. In *perfect competition*, buyers and sellers are sufficiently numerous that they can buy or sell any desired quantity without affecting the price (i.e., they can act as "price-takers").

94.  Examples of seminal work in this area include: 1) Helpman, E. and P. R. Krugman (1985), and 2) Krugman, P. R. (1986). For an in-depth analysis of the impact of strategic trade policy on the U.S. semiconductor industry, see Flamm, K. (1996).

95.  Skolnikoff, E. B. (1993, pg. 119).

96.  Krugman, P. R. (1994a).

97.  If not equalized, then made to be irrelevant. If labor cost is an issue, access to low-wage foreign labor neutralizes that imbalance. If capital is lacking from domestic sources, then presumably the world's capital markets can be tapped for investors. This picture is obviously conceptual rather than rigorously descriptive.

98.  Krugman, P. R. (1996, pg. 69).

99.  The World Bank (1997, Table 15).

100.  For a sensible tutorial, see "Trade winds," *The Economist*, November 8, 1997, pg. 85.

101.  Krugman, P. R. (1996, pg. 60).

102.  Access to foreign markets offers American firms increased economies of scale, a key factor in becoming a global leader. In addition, wages in the export sector are roughly 30 percent higher on average than those of domestic industries. Promotion of high-technology exports offers the added potential of spillover benefits to other, unrelated industries.

103.  This is true at the aggregate level. In trade theory, the Heckscher-Ohlin theorem states that countries will tend to export products that make intensive use of their relatively abundant factors. For the United States, this abundant factor is high-skilled labor. Taking this one step farther (and applying the Stolper-Samuelson theorem), the factor that is used intensively in those products will receive the gains, at the expense of other factors (in our case, unskilled labor). This is often referred to as the *redistribution effect*, meaning that if no compensation flows from the export sector to the import-competing sector, there will be an increasing welfare gap between the two groups.

104.  See Krugman, P. R. (1996, pg. 56). Actually, this is true by definition, based on the way economists calculate productivity.

105.  The factor-price-equalization theorem states, "Under identical, constant-returns-to-scale production technologies, free trade in commodities will equalize relative factor prices through the equalization of relative commodity prices, so long as both countries produce both goods." See Markusen, J. R. et al. (1995, pg.112).

106.  This is a simplified argument. In reality, there are a number of subtle issues, such as the presence of perfect competition in input markets, transportation costs, unequal natural factor endowments, etc. For the case of high-technology commodities (e.g., television sets, cellular phones, computer memory chips), these differences prove to be relatively minor.

107.  If productivity in producing a specific product were to lag equally everywhere in the world, there would be no direct effect on the terms of trade. Somewhat trivially, if productivity were to lag in a domestic sector that does not compete in world markets, our

terms of trade would not be affected.

108. For a more rigorous theoretical treatment of international trade in technology-intensive products, see Dosi, G. et al. (1990).

109. This refers to the *Linder Hypothesis*. See Linder, S. B. (1961).

110. For a recent test of the Linder Hypothesis, see Bergstrand, J. (1990).

111. Vernon, R. (1966).

112. I am specifically referring here to products that require physical manufacture and transport. Knowledge-based products, such as software and entertainment services, reside entirely on the "technology-driven" upper tier of the two-level model.

113. In Chapter 4 I will introduce a three-dimensional model of technology-intensive products, with one dimension being the "physical" content of the product. A global advantage in the physical dimension can be achieved only by finding a unique and highly productive way to utilize the essentially generic resources available in the input-driven tier of the market.

114. See, for example, Dollar, D. and E. N. Wolff (1993, pg. 188).

115. Archibugi, D. and J. Michie (1997b, pg. 172).

116. One perspective on this argument is that national governments have become even more important because of the *amplification effect* of global trade. Relatively minor changes in the domestic environment for business can have an amplified effect when firms enter the more competitive global market. See Porter, M. E. (1990).

## Chapter 3: The American Technology Enterprise

1. Link, A. N. (1987).

2. Schumpeter, J. (1943, pg. 110).

3. Smith, B. L. R. and C. E. Barfield (1996a).

4. Schmookler, J. (1966, pg. 225).

5. See, for example, 1) Arrow, K. J. (1962a), and 2) Shell, K. (1967). Note that the causality for technical progress in these early models runs from capital accumulation to innovation, rather than in the opposite (and more plausible) direction.

6. Romer, P. M. (1986).

7. See, for example, 1) Rivera-Batiz, L. A. and P. M. Romer (1991), 2) Romer, P. M. (1990), and 3) Solow, R. M. (1994). For integrated treatments, see 1) Grossman, G. M. and E. Helpman (1991), and 2) Gomulka, S. (1990).

8. Romer, P. M. (1990, pg. 74).

9. Cornes, R. and T. Sandler (1996).

10. Griliches, Z. (1973). For a comprehensive discussion of the links between technological innovation (as measured by patent data) and productivity, see Griliches, Z. (1984).

11. Gross national product (GNP) is a measure of national economic performance that includes the incomes of all residents, whether this income is generated at home or abroad, but excludes domestic income generated by non-residents. Gross domestic product (GDP), on the other hand, excludes the income generated by residents from economic activities abroad, but includes income generated by non-residents on home soil. GDP has come into favor as a more useful metric of national economic performance.

12. Link, A. N. (1987, pg. 54).

13. Rosenberg, N. (1994b).

14. This view is discussed in Mowery, D. C. and N. Rosenberg (1979).

15. Levin, R. C. et al. (1985).

16. Grossman, G. M. and E. Helpman (1991, pg. 85).

17. This is representative of the familiar "s-curve" that is characteristic of many technology-intensive products. It is important to note that real competitive markets are far more complicated than this simple model. In particular, the potential for a "hidden technology" to suddenly surface and capture an entire market is very real. This is often referred to as "jumping s-curves." See Chapters 9 and 10 for additional discussion.

18. Link, A. N. (1982).

19. See, for example, 1) Mansfield, E. R. (1965), and 2) Terleckyj, N. E. (1974).

20. Levin, R. C. et al. (1987).

21. Arrow, K. J. (1962b).

22. Levin, R. C. et al. (1987, pg. 803).

23. There is a downside to strong patent and copyright protection, particularly in high-technology industries. There have been several recent cases of contentious legal battles between aggressive high-tech firms over patent protection, involving both alleged patent infringement and accusations of antitrust violations. Although there are legitimate concerns with respect to maintaining a vital competitive environment in technology enterprise, much of this activity falls into the category of "rent-seeking," meaning that the lawsuits stem from the desire of firms to secure unjustified profits through litigation.

24. See, for example, 1) Mansfield, E. R. et al. (1977), and 2) Scherer, F. M. (1982).

25. Grossman, G. M. and E. Helpman (1991, pg. 17).

26. Roach, S. S. (1996, pg. 81).

27. In a recent study conducted by Michael E. Porter of Harvard University, he concluded that if current trends in the number of patents per million persons continue, the United States could drop to the number three position in the world by 1999, and to the number seven position by 2006. See "U.S. innovation lead imperiled," *Electronic Engineering Times*, March 23, 1998.

28. Mowery, D. C. and N. Rosenberg (1989).

29. Nelson, R. R. (1990).

30. Freeman, C. (1997, pg. 28).

31. Nelson, R. R. and G. Wright (1992).

32. The Morrill Land Grant College Act of 1862 established the initial support for agricultural and engineering education. It was supplemented by the Hatch Act of 1887, which provided each state with funds to develop agricultural-experiment stations, and to disseminate the resulting information to regional farmers. The level of support for these institutions was further increased with the passage of the Adams Act of 1906. See Nelson, R. R. and G. Wright (1992, pg. 1942).

33. In 1919, for example, the Massachusetts Institute of Technology (MIT) developed a Cooperative Course in Electrical Engineering in partnership with General Electric, which subsequently hired half of the graduating students. This program was later joined by AT&T and other firms. See Nelson, R. R. and G. Wright (1992, pg. 1949).

34. Taylor, F. W. (1947).

35. The roles of government and private investment in R&D during the postwar period will be covered in greater detail in Chapter 7.

36. It was WWII megaprojects such as the Manhattan Project that prompted both government and industry to recognize the power of science to impact our economic future. The harnessing of nuclear energy and the exploration of space were enabled by the confidence gained from these wartime projects.

37. For a current perspective on the impact and legacy of the Bush Report, see Barfield, C. E. (1997).

38. Archibugi, D. and J. Michie (1997, pg. 29, Table 2.1).

39. Tyson, L. D. (1992, pg. 23).

40. For a recent discussion of the state of trade in technology-intensive products, see Guerrieri, P. and C. Milana (1995).

41. Although U.S. exports in technology-intensive industries have steadily increased over the last several decades, the volume of high-tech imports has grown much more rapidly, resulting in an erosion of our trade balance in these critical sectors.

42. Engelbrecht, H. (1997).

43. Japanese industrial R&D expenditure as a proportion of civil industrial output surpassed that of the U.S. in the 1970s, and total civil R&D as a fraction of GNP surpassed that of the U.S. in the 1980s.

44. The linear model of innovation was perfectly compatible with the "product life-cycle" model for international diffusion of technology-intensive products, a concept that also became fashionable in the 1960s.

45. During the late 1960s, the U.S. government initiated two major studies of the innovation process, in an effort to estimate the returns from investment in R&D. Project Hindsight (1966) was administered by the Department of Defense, and focused on the scientific and technological origins of several weapons systems. This study concluded that a high payoff resulted from investment in advanced development and applied research, but that there was little evidence of a connection to a common pool of basic scientific research. The National Science Foundation's Technology in Retrospect and Critical Events in Science (TRACES) project considered five commercial innovations, over a longer time horizon. In this study, the roots of innovation in basic science were more clearly demonstrated, but more important, the highly nonlinear and variegated nature of industrial innovation was strongly suggested.

46. There have been some interesting empirical studies of the NIH syndrome. In particular, there is evidence that NIH increases with the tenure of engineers and scientists within a project group. This suggests that as team members become more inwardly focused on their individual project, they begin to close their minds to outside information. Clearly, this can result in a less than optimal outcome, particularly in industries characterized by rapid and discontinuous technological change. See Katz, R. and T. J. Allen (1982).

47. The topic of Japanese industrial policy and its impact on global competitiveness will be discussed more extensively in Chapter 7.

48. See 1) Mansfield, E. R. (1988), and 2) Mansfield, E. R. (1996).

49. In a different study of 1,119 manufacturing firms over the period from 1979 to 1985, American companies reported that 81 percent of R&D went toward product rather than process innovations. During the same period, a Japanese survey reported that 72 percent of its firms' R&D investment went to process improvements and the adaptation of the advanced technologies of other nations. See Caravatti, M. (1992).

50. Many Japanese firms rotate their key personnel between R&D, marketing, and manufacturing for extended periods, to give these individuals an understanding of the needs of each function. This also serves to improve communication and networking between these groups, thereby facilitating the diffusion and accumulation of knowledge by the firm. The opposite situation has proven to be a severe liability for centrally planned economies such as the former Soviet Union's, in which R&D activities were intentionally held separate from manufacturing operations. See Cohen, W. M. and D. A. Levinthal (1990).

51. Much of this section follows Cohen, W. M. and D. A. Levinthal (1990).

52. Anderson, J. R. et al. (1984).

53. This limitation could be overcome by outsourcing the technology selection process to "expert" consultants from industry or the university system. Although the hiring of consultants can be a valuable means to absorb frontier technology, using "outsiders" to make critical strategic decisions is poor business practice.

54. Cohen, W. M. and D. A. Levinthal (1990, pg. 129). Note that there is a real danger associated with the popular concept of a *virtual company*, in which many operations are outsourced. Without a sufficient level of internal R&D experience and expertise, a "virtual" firm may not have an adequate ability to evaluate new technologies for products or to qualify the manufacturing processes of suppliers.

55. Bell, M. and K. Pavitt (1997, pg. 116).

56. See, for example, Senge, P. M. (1990).

57. In the military, it is common for strategists to assume that if a specific weapons system is technologically feasible, the "enemy" must also be developing it. This somewhat paranoid "mirror" logic can be extended to commercial competition. If a new, low-power integrated circuit technology has just become available, for example, it is reasonable to assume that your competition will exploit it to the same extent that you could. This proactive strategy protects a firm from finding itself months behind its competition by the time a "leak" alerts it to a threat.

58. Mansfield, E. R. (1985). Note that this empirical study is more than twelve years old. One must assume that with increased technological integration and improved communication, the time frame for leakage must be even shorter today.

59. Almeida, P. (1996).

60. For a very recent empirical study, see Florida, R. (1997). Generally, MNCs establish R&D facilities in foreign countries to access capabilities that are not available at home, rather than to perform the same work as their domestic counterparts.

61. Paci, R. et al. (1997).

62. For a somewhat dated discussion of this topic, see Reich, R. B., 1987, "The rise of techno-nationalism," *Atlantic*, May Issue, pp. 63–69.

63. Note that the gap between the ability to operate equipment and the technological knowledge required to design and improve that equipment is steadily widening. A distinction should be made between the "learning by doing" associated with recurring manufacturing activities and the intensive learning derived through research and development on groundbreaking new technologies. See Bell, M. and K. Pavitt (1997).

64. Bell, M. and K. Pavitt (1997, pg. 117).

65. It is important to recognize that advanced training and education are not the exclusive domain of scientists and engineers. Every employee in a company has the potential to innovate, and should be provided with the best possible education within his or her job classification. The accumulated technical knowledge of a firm extends well beyond the R&D laboratory. At every stage of the business process, from marketing to customer service, there are opportunities to increase productivity and enhance competitive advantage.

66. The topic of knowledge management within the firm is discussed in detail in Chapter 10.

67. A *point design* implies that the design team's activities were focused on meeting the narrow requirements of the project, without regard to broader learning opportunities or to future extensions or enhancements.

68. This is an important reason why technology-intensive firms should retain a reasonable proportion of scientists and engineers in executive positions. Accountants and business-school graduates all too often consider technical experts as interchangeable parts

("a Ph.D. is a Ph.D."). Having spent much of my professional career leading technical teams, I cannot overemphasize how valuable a mature and experienced team can be to retaining technological leadership.

69.   Since some level of attrition is inevitable, I recommend that firms protect themselves by ensuring that critical competencies are not embodied in a single individual. It might even be worth the extra expense, for example, to add nonessential members to a development team, as both a training opportunity and to ensure sufficient redundancy of vital tacit experience.

70.   There is some important recent evidence that membership in the "convergence club" of nations is highly dependent on levels of accumulated technology. This is a reasonable extension of earlier evidence that human capital development was a strong factor in productivity growth. See 1) Bell, M. and K. Pavitt (1997, pg. 120), and 2) Daniels, P. L. (1997).

71.   Daniels, P. L. (1997, pg. 1190).

72.   "Just how big is high tech?" *Business Week*, March 31, 1997, pg. 68.

73.   "Companies drive research with giant-sized investments," *R&D Magazine*, October 1997.

74.   The reason why these questions are of concern to policymakers (and American society in general) is that there is a strong correlation between the locations where R&D is performed and the direct and spillover benefits accrued to the regional economy. If R&D is being performed overseas, it is reasonable to assume that foreign economies will benefit from the value-added gained in their region, independent of the home nation of firms. See Fors, G. (1997).

75.   See, for example, Ohmae, K. (1990). For an overview of national specialization in technology, see 1) Amable, B. (1993), and 2) Archibugi, D. and M. Pianta (1992). For the U.S. government's perspective on R&D globalization, see Office of Technology Policy (1995).

76.   This discussion follows Archibugi, D. and J. Michie (1997b).

77.   Patents are an imperfect proxy for R&D investment, which is itself an unreliable measure of productive innovation. The intensity of patenting is highly industry-dependent, and numbers of patents do not necessarily reflect economic benefits. For an excellent discussion of the limitations of various metrics of innovation, see Griliches, Z. (1990).

78.   Patel, P. (1997). For early work on this topic, see Patel, P. and K. Pavitt (1991).

79.   For a more detailed discussion of global R&D strategies and the "polyp model," see Chapter 5.

80.   For recent studies, see 1) Daniels, P. L. (1997), and 2) Cantwell, J. (1997).

81.   Doremus, P. N. et al. (1998, pg. 100).

82.   Doremus, P. N. et al. (1998, pg. 143).

## Chapter 4: A Framework for Sustainable Advantage

1.   For a discussion of the impact of high concentrations of skilled technical professionals on the economic growth of nations, see Romer, P. M. (1989a).

2.   List, F. (1841).

3.   Freeman, C. (1997).

4.   The literature often uses the alternative term *codified knowledge*. I choose to use the term *explicit knowledge* because, well, it seems more explicit.

5.   Paul M. Romer has referred to the three facets of knowledge as "hardware," "software" (what I'm calling explicit knowledge), and "wetware" (representing tacit knowledge).

Although these terms are descriptive, I prefer not to be reminded so vividly that human knowledge is carried around in what amounts to lumpy bags of protoplasm. See Nelson, R. R. and P. M. Romer (1996).

6. Tacit knowledge can be thought of as "the things that we don't know we know."

7. Gibbons, M., et al. (1994).

8. Fransman, M. (1997, pg. 74).

9. Nye, W. W. (1996).

10. In recent years there have been attempts to bridge the gap between tacit and explicit knowledge through rules-based expert systems and other artificial-intelligence algorithms. Du Pont, for example, has made extensive use of expert systems to capture the heuristic knowledge of senior chemical technicians. For an overview, see Dutta, S. (1997).

11. Barro, R. J. (1989).

12. For a popular overview, see Stewart, T. A. (1997).

13. An excellent comparative definition is offered by Gibbons, M. et al. (1994, pg. 13): "Knowledge is produced by configuring human capital. However, unlike physical capital, human capital is potentially more malleable. Human resources can be configured again and again to generate new forms of specialized knowledge."

14. Mowery, D. C. and N. Rosenberg (1989).

15. For an interesting discussion of the role of tacit knowledge in group innovation, see Leonard, D. and S. Sensiper (1998).

16. According to the originators of the concept, a core competency must 1) make a disproportionate contribution to perceived customer value, 2) be competitively unique, and 3) be extendible to a range of products or services. See Hamel, G. and C. K. Prahalad (1994).

17. Hu, Y. (1995).

18. One way to think about this is to consider a new model of personal computer. At the moment of its release into the public domain, the value of its embodied knowledge content (another way of saying explicit knowledge) is at its peak. As time passes, competitors imitate aspects of the product, and technology evolves beyond its capabilities. After only a year or two, the explicit portion of this now-obsolete computer is almost all in the public domain and has essentially zero value. The team of scientists and engineers who created it, however, have continued to learn from their experiences, and are even more valuable than before.

19. Louisville Slugger is a registered trademark of Hillerich & Bradsby Company. For more on the low-tech/high-tacit production of baseball bats, see Morrison, J., 1997, "Feel-good wood," *Spirit Magazine*, October Issue, pg. 39.

20. The term 'orthogonal' is used here to mean that these three aspects of a product can be treated independently of each other. Although this might not be rigorously true, it is a reasonable simplifying assumption.

21. Along the commodity axis of this model, markets approximate the perfect competition of classical economics.

22. Software and entertainment products, for example, are reproduced and distributed with essentially zero tacit content. This makes them embarrassingly easy to pirate, even in the most technologically backward nations.

23. The work of Schumpeter has received increased attention in recent years. A new school of economic thought, referred to as *evolutionary economics*, has formed around the concept of "disequilibrium" analysis of markets and competition. For overview discussions, see 1) Freeman, C. (1994), and 2) Metcalfe, S. (1997).

24.  Schumpeter, J. (1943, pg. 82).

25.  For a more detailed discussion, see Rosenberg, N. (1994a, pp. 47–61).

26.  In ecological systems, the analogue to rate of innovation would be a higher rate of reproduction. Species that reproduce rapidly have increased opportunities to adapt to changes in their environment. This is the reason why cockroaches will very likely rule the Earth in another few hundred million years.

27.  See, for example, 1) Porter, M. E. (1980), 2) Porter, M. E. (1985), 3) Bartlett, C. A. and S. Ghoshal (1989), 4) Reich, R. B. (1991), and 5) Yip, G. S. (1992). For a comparative overview of these strategies, see Porter, M. E. and R. E. Wayland (1995).

28.  Porter, M. E. (1990).

29.  There are typically a number of ways in which a complex system can be analyzed. By grouping relevant factors in different ways, these alternative frameworks can offer unique insights. A topographical map, for example, provides essential information to hikers, whereas a roadmap is more useful to motorists.

30.  Balachandra, R., and J. H. Friar (1997).

31.  Bettis, R. A. and M. A. Hitt (1995).

32.  Noyce, R. (1977, pg. 68).

33.  Freeman, C. (1997, pg. 30).

34.  A fast-second strategy can be effective in industries where first-mover advantage is overcome by the difficulties in "working the bugs" out of a design. In this case, a technically capable firm can rapidly duplicate the first-mover's product, and by learning from its mistakes (particularly with respect to meeting customer needs) can capture a substantial market share.

35.  It is becoming increasingly apparent that some areas of technology development will be so costly that risk-sharing among several countries is the only viable option.

36.  Branscomb, L. M. (1993, pg. 21).

37.  Porter, M. E. (1990).

## Chapter 5: Industry and Market Structure

1.  "How Motorola lost its way," *Business Week*, May 4, 1998, pp. 140–148.

2.  Saxenian, A. (1994, pg. 119).

3.  I am assuming for this discussion that first-mover innovations are not protected by effective patent laws or other legal means. This is often the case in technology industries, given the cost and difficulty of establishing broad patent coverage.

4.  For an in-depth case study of this industry segment, see Trajtenberg, M. (1990).

5.  Scherer, F. M. (1992, pg. 78).

6.  Unfortunately, the number of rivals tends to fall off as an industry matures. As markets shift from innovation-based competition to price-based competition, a few large firms will tend to dominate the market, due to scale advantages. This further contributes to the stagnation of mature industries.

7.  This, in turn, mandates the development of a sophisticated infrastructure of technical standards. The shift toward open architectures for information systems in the last decade provides an excellent example of the critical role of standards. This topic is discussed in the final section of this chapter.

8.  For a detailed case study that demonstrates the interrelationships between rivals in a rapidly changing technology sector, see Bowonder, B. et al. (1996).

9.  Cohan, P. S. (1997).

10.  National Academy of Engineering (1993, pg. 76).

11. For example, the possibilities for embedding automation and control into virtually every aspect of our lives have just begun to be exploited. Likewise, the application of advanced bioengineering techniques to chemical and pharmaceutical production is likely to produce explosive advances over the next few decades.

12. Patel, P. and K. Pavitt (1991).

13. Archibugi, D. and J. Mitchie (1997a, pg. 15).

14. For an overview of three alternative strategies, see Porter, M.E. and R.E. Wayland (1995).

15. Incidentally, one of the more significant scientific breakthroughs of the last two decades took place at an IBM research laboratory outside the United States. The first demonstrations of high-temperature superconductivity were performed by a German scientist working in a Swiss R&D affiliate of IBM. Mowery, D. C. and N. Rosenberg (1989, pg. 210).

16. For an expanded discussion, see 1) Kuemmerle, W. (1997), and 2) Chiesa, V. (1996).

17. For a good overview of networked R&D organizations and strategies, see Chiesa, V. (1996).

18. For a detailed case study of a firm that has achieved high R&D efficiency levels through a networked innovation structure, see Bowonder, B. and T. Miyake (1997).

19. For several fine discussions of collaboration in R&D, see 1) Katz, M. L. and J. A. Ordover (1990), 2) Foray, D. (1991), and 3) Quintas, P. and K. Guy (1995).

20. Mowery, D. C. and N. Rosenberg (1989, pg. 239).

21. Pavitt, K. (1991).

22. For a detailed analysis of R&D collaboration strategies in the biotechnology industry, see Greis, N. P. et al. (1995).

23. Sykes, A. O. (1995).

24. For a discussion of the antitrust issues associated with industry-standards consortia, see Moore, H. P., Smith, W. S., and R. Jensen, 1996, "Antitrust implications in the adoption of industry standards," *Intellectual Property Magazine*, Winter Issue.

25. Cohan, P. S. (1997, pg. 64).

26. Mowery, D. C. and N. Rosenberg (1989, pg. 263).

27. Mowery, D. C. (1989).

28. The propensity of firms to enter into collaborative agreements appears to depend on the phase in the technological life cycle of a given industry segment. Thus, cooperative agreements are more likely to occur in the early stages of development and product introduction, and rapidly decline as the industry reaches maturity. See Cainarca, G. C. et al. (1992).

29. Tacit collusion should not be confused with tacit knowledge; the terms are unrelated. For an expanded discussion, see Barney, J. B. (1997, pg. 255).

30. See, for example, 1) Tezuka, H. (1997), and 2) Weinstein, D. E. and Y. Yafeh (1995).

31. Barney, J. B. (1997, pg. 258).

32. Lux, H., 1995, "An economists' supergroup will review NASDAQ charges," *Investment Dealers Digest*, Vol. 61, July 3, pg. 5.

# Chapter 6: Trade and Competition Policy

1. References for the general topic of trade and competition in high-technology industries include: 1) National Research Council (1997), 2) National Research Council

(1996), and 3) Graham, E. M. and J. D. Richardson (1997).

    2.  See 1) Fagerberg, J. (1988), and 2) Amendola, G. et al. (1994).

    3.  After years of being ignored by both the public and Congress, trade policy has recently received disproportionate attention. What was once a relatively obscure segment of American policy has risen to prominence, due to a steadily eroding balance of trade and the loss of domestic technology industries. With respect to bilateral trade balance, it is interesting to note that while both Japan and Germany run large trade surpluses in high-technology products, they are the world's leading *importers* of disembodied technical knowledge. The United States, on the other hand, is the world's largest *exporter* of such knowledge, through licensing and other transfer agreements.

    4.  Tyson, L. D. (1992, pg. 276).

    5.  It has been suggested that a monitoring rule could be contrived to signal that a "critical" industry had reached unacceptable levels of concentration in foreign nations. One such criterion is the so-called 4-4-50 rule, in which a danger flag is raised if more than 50 percent of the global market is held by either four firms or four nations. I believe that arbitrary standards such as this will likely prove to be unworkable in a more technically specialized future. See Moran, T. H. (1990).

    6.  For an extended discussion of the issue of American dependency on foreign sources of semiconductors, see Flamm, K. (1993).

    7.  For an overview of the liquid crystal display (LCD) antidumping case and its predictably negative consequences, see Tyson, L. D. (1992, pp. 141–142).

    8.  Federal Interagency Staff Working Group (1987, pg. 31).

    9.  Dick, A. (1995, pp. 7–8). The real credit for the success of Japanese firms must go to their people's entrepreneurial spirit, a willingness to take risks, and a consistently high domestic savings and investment rate. Compared with these factors, government subsidy in Japan has been of minor significance.

    10.  Tyson, L. D. (1992, pg. 2).

    11.  Hindley, B. and P. A. Messerlin (1996).

    12.  This does not inherently require high production volumes, however, depending on the degree of monopoly power and level of appropriability that a firm can achieve for its product. Clearly, high volume is relative; in memory chips, for example, sufficient volume is measured in millions of units, whereas in commercial aircraft or supercomputers, sales goals are typically in the tens or hundreds.

    13.  National Science Board, *Science and Engineering Indicators–1991*, pg. 401.

    14.  Sophisticated markets can evolve anywhere; some examples include consumer electronics products, 35mm cameras, and energy-saving devices (Japan), sophisticated kitchen appliances (France and Italy), and precision machine tools (Germany). In each case, conditions in the domestic market allowed firms in these countries to anticipate a market need and gain a first-mover advantage as global demand grew.

    15.  There have been several studies providing evidence of the positive impact that international diversification of products can have on competitive advantage. See, for example, Hitt, M. A. et al. (1997).

    16.  Henderson, J. (1989, pg. 45). Duties on semiconductor imports to the EU are frequently suspended. Quotas have had a greater impact, particularly with respect to relatively low-tech manufactured products such as color TVs, VCRs, and microwave ovens. Japan and Korea have been singled out for market-share ceilings and voluntary "arrangements" by the EU, prompting a substantial inflow of foreign direct investment from these countries.

    17.  By exercising monopsonistic (i.e., single-buyer) power with respect to its domes-

tic aircraft market, Brazil was able to negotiate a very favorable technology-transfer agreement with Piper Corporation, which contributed significantly to the growth of its indigenous aircraft industry. See Mowery, D. C. (1993, pg. 50).

18. World Trade Organization (1995, Tables IV.31 and IV.36).

19. To make matters worse, Japan had imposed official restrictions on inward foreign direct investment until the mid-1980s, resulting in a conspicuously low level of foreign investment; the ratio of FDI to total capital formation in Japan in 1985 was only 0.1 percent. Although inward flows have increased significantly in recent years, there are still informal barriers to foreign investment, including cross-ownership of stock in Japanese firms. See Graham, E. M. (1996, pp. 16–17).

20. For several perspectives on this classic story of managed trade, see 1) Tyson, L. D. (1992), 2) Dick, A. (1995), 3) Flamm, K. (1996), and 4) Howell, T. and D. Ballantine (1992).

21. Dynamic random-access memories (DRAMs) and erasable programmable read-only memories (EPROMs) are extensively used in many commercial electronics products, most notably in personal computers.

22. While there is merit to the assertion that U.S. semiconductor manufacturers were disadvantaged in world markets by Japan's predatory actions, there is another side to the decline of our firms. Japanese manufacturers achieved consistently higher yields from their production processes during the 1980s, along with fewer latent defects. In other words, American firms were disadvantaged by poor process controls and product quality, factors well within their own control.

23. The Semiconductor Industry Association was formed in 1977 to serve as a political-action group. A number of leading American firms, including Micron Technologies, Advanced Micro Devices, Intel, and National Semiconductor were charter members.

24. Pressure for a trade agreement in commodity semiconductors was fueled by the public disclosure of an infamous memo sent by Hitachi America to its distributors, instructing them to undercut all rival price quotes on EPROM chips by 10 percent, at a guaranteed 25 percent profit margin. Interestingly, the memo instructed dealers to target not only American firms, but also a Japanese rival, Fujitsu, with predatory pricing. See Flamm, K. (1993, pg. 267).

25. Tyson, L. D. (1992, pg. 116).

26. No such agreement was ever signed. This was not due to recognition of the negative effects of restricting global competition, but rather to a decision by the instigator of the antidumping action against the South Koreans, who felt that they could benefit more from the imposition of an antidumping penalty. See Dick, A. (1995, pg. 62).

27. The story of managed trade in semiconductors continues to this day. In 1991 a new Semiconductor Trade Agreement was reached, which included explicit reference to a 20 percent market-share commitment. Negotiations in 1996 to again renew the trade agreement was marked by an encouraging emphasis on cooperative activities between the U.S. and Japanese governments and between industries in both countries, including the formation of a Semiconductor Council made up of representatives from the United States, Japan, South Korea, and the European Union. See the Semiconductor Industry Association Website for various press releases within its on-line database. For a general discussion of market dynamics and the effects of the learning curve on the DRAM and EPROM markets, see Gruber, H. (1996).

28. The trade conflict between the United States and Japan has by no means been limited to semiconductor markets. Since the early 1980s, the United States has engaged in market-oriented, sector-specific (MOSS) talks dealing with telecommunications, electronics, medical equipment, and drugs. These negotiations also tend to be outcome-directed, as

was the case with actions to improve access to Japanese markets for supercomputers by Cray Research. See Tyson, L. D. (1992, pg. 58).

29.   General references on the protection of intellectual property include: 1) International Chamber of Commerce (1996), and 2) Warshofsky, F. (1994).

30.   *The Constitution of the United States of America*, Article 1, Section 8, Clause 8.

31.   Abraham Lincoln was the only American president to hold a U.S. patent. He noted in 1861 that "the patents system added the fuel of interest to the fire of genius." See Lehman, B. A. (1996).

32.   Thurow, L. C. (1997).

33.   See, for example, Horowitz, A. W. and E. L. C. Lai (1996, pg. 785).

34.   The reader may recall the famous lawsuit filed by Apple Computer against Microsoft, claiming that the Windows operating system was a blatant infringement on Apple's Macintosh operating-system patents. The courts found in favor of Microsoft, establishing a powerful precedent in software patent law. For a general discussion, see Nelson, R. R. and P. M. Romer (1996, pg. 64).

35.   Flaherty, K., 1997, "Calling the patent suit bluff," *Intellectual Property Magazine*, on-line edition, November Issue.

36.   As an example of how intellectual property has become a product in itself, it is now common in the semiconductor industry to refer to modular and reusable integrated-circuit design elements as "intellectual property." Motorola, for example, has introduced a vast portfolio of mutually compatible circuit design "blocks," which can be mixed and matched into highly integrated devices. This was touted as a "Lego" approach to intellectual property building blocks. See "Motorola semi stacks IP blocks to recast image," *Electronic Engineering Times*, September 8, 1997, pg. 26.

37.   See 1) UNTCMD (1993), 2) Seyoum, B. (1996), and 3) Mansfield, E. R. (1995).

38.   OECD (1989).

39.   Mossinghoff, G. J. and T. Bombelles (1996). Some Latin American countries have expressed a willingness to improve IPRs in return for increased trading opportunities with advanced economies. See Sood, J. and G. L. Miller (1996).

40.   Grindley, P. C. and D. J. Teece (1997, pg. 21).

41.   The United States has run a trade surplus in intellectual property for most of the postwar period. International sales of American firms' intellectual property increased from $8.1 billion in 1986 to $27 billion by 1995. This resulted in a positive U.S. technology trade balance of $20.6 billion in 1995. See Doremus, P. N. et al. (1998).

42.   See 1) Nill, A. and C. J. Shultz II (1996, pg. 37), and 2) Lehman, B. A. (1996).

43.   Section 337 of the Tariff Act of 1930 provides some protection against counterfeit products entering the United States from foreign sources. Unfortunately, it is often hard to distinguish pirated merchandise from the genuine article. In one case, a shipment of Liz Claiborne purses intercepted by U.S. Customs was suspected of being counterfeit. Company experts, however, were unable to verify that the purses were not their own. The possibility of "gray-market" shipments of genuine products further reduces the effectiveness of border enforcement. See Mutti, J. and B. Yeung (1996). For more on gray markets in copyrighted goods, see 1) Buffon, C. E. and R. G. Dove, Jr., 1997, "A not-so-gray area," *Intellectual Property Magazine*, April Issue, and 2) Buffon, C. E. and C. A. Bradley, 1995, "Using copyright as a barrier to gray-market imports," *Intellectual Property Magazine*, Winter Issue.

44.   It is interesting to note that Taiwan has for many years been a center for the pirating of fake first editions of books by collectable authors. These are commonly known in the antiquarian book trade as "Taiwan piracies."

45. Thurow, L. C. (1997, pg. 100).

46. Sources: 1) Alford, W. P. (1995), 2) Snee, A., 1997, "To catch a Chinese pirate," *AsiaWeek*, Vol. 22, No. 23, and 3) Cox, R. E. 1996, "Checkered Chinese policy," *Intellectual Property Magazine*, on-line edition, Summer Issue.

47. "Listed company raided for 'software piracy'," *The Straits Times*, August 14, 1997, on-line edition.

48. Business Software Alliance, 1997, *Overview: Global Software Piracy Report*, Business Software Alliance Website. Note that poor enforcement of IPRs in Singapore has resulted in action by the U.S. Trade Representative under Special 301 provisions of the Omnibus Trade and Competitiveness Act of 1988. Special 301 is designed specifically to address intellectual property protection, by allowing the USTR to take aggressive action, including trade penalties, against countries that fail to meet multilateral treaty provisions for IPR protection.

49. WIPO, 1997, *International Protection of Industrial Property*, WIPO Website.

50. WTO, 1997, *Intellectual Property*, WTO Website.

51. Primo Braga, C. A. (1995).

52. Mossinghoff, G. J. and T. Bombelles (1996, pg. 44).

53. Even among advanced trading partners there can be substantive differences in IPR regimes. The Japanese system, for example, requires less "originality" and does not require disclosure of "prior art." The approval period can also be quite different; in the United States a patent can take up to three years for approval, whereas in Japan the wait can be far longer (up to seven years). See Baughn, C. C. et al. (1997).

54. "New penalties for Net software pirates," *BBC News*, on-line edition, December 18, 1997.

55. Kirk, M. K., 1997, "Remaking a PTO in crisis," *Intellectual Property Magazine*, on-line edition, September Issue.

56. The two examples of recent antitrust action which follow have been covered extensively in the popular press, and therefore will not be discussed in detail here. The reader who is interested in the Microsoft case is referred to the following references: 1) "What to do about Microsoft?" *Business Week*, April 20, 1998, pg. 112, 2) "Ready to drop the big one on Bill?" *Business Week*, February 23, 1998, pg. 36, 3) "Justice vs. Microsoft: What's the big deal?" *Business Week*, December 1, 1997, pg. 159, 4) "Microsoft's future," *Business Week*, January 19, 1998, pg. 58, 5) "Microsoft and Justice end a skirmish, yet war could escalate," *The Wall Street Journal*, January 23, 1998, 6) "Is antitrust relevant in the digital age? Watch Microsoft's case," *The Wall Street Journal*, October 22, 1997, and 7) "Why software and antitrust law make an uneasy mix," *The Wall Street Journal*, October 22, 1997. For two interesting and contrasting points of view, see 1) Arquit, K. J., 1998, "Cracking down on Microsoft," *Intellectual Property Magazine*, January Issue, and 2) Siskind, L. J., 1998, "Cracking down on Microsoft," *Intellectual Property Magazine*, January Issue.

57. The reader who is interested in the Intel case is referred to the following references: 1) "Intel: The feds are loaded for bear," *Business Week*, October 13, 1997, pg. 36, 2) Takahashi, D. and E. Ramstad, 1998, "Court ruling in Intel case may give its chip rivals antitrust ammunition," *The Wall Street Journal*, April 14, and 3) "The smell of blood at Intel," *Business Week*, April 27, 1998, pg. 45.

58. AT&T was the subject of antitrust investigation for an extended period, leading up to the monumental breakup of the company in 1983. The Modified Final Judgment in the case of *United States vs. AT&T* opened the domestic U.S. market to both foreign and domestic competitors, but also unleashed AT&T to compete in markets outside of telecommunications.

59. For a discussion of Japan's antimonopoly law, see Doremus, P. N. et al. (1998, pg. 77).

60. For detailed discussions of the pro-competitive behavior of *keiretsu* in Japan, see 1) Tezuka, H. (1997), and 2) Weinstein, D. E. and Y. Yafeh (1995).

61. Additional legislation was passed in 1990 to relax penalties on the formation of joint-venture partnerships among U.S. firms. See Mowery, D. C. (1993, pp. 50–53). Recently the Justice Department has been getting tough on monopolistic behavior, particularly in the telecommunications industry. See "Up against the wall, monopolist," *Business Week*, March 23, 1998, pg. 35.

62. Cohen, P. S. (1997).

63. Brunetti, A. C., 1997, "Wading into patent pooling," *Intellectual Property Magazine*, on-line edition, November Issue. For a more advanced treatment of the topics of pooling and cross-licensing in high technology, see 1) Grindley, P. C. and D. J. Teece (1997), and 2) Brod, A. and R. Shivakumar (1997).

64. Another important area of policy that has a significant impact on American competitiveness is liability laws. The U.S. judiciary is virtually unique among the justice systems of the industrialized world in its strict interpretation of product liability. This strictness can represent a significant risk to firms that are developing and commercializing a new product. See Huber, P. (1992).

## Chapter 7: The American System of Innovation

1. Mowery, D. C. (1993, pg. 3).

2. For early work on this subject, see Andersen, E. and B. Lundvall (1988).

3. A formal definition for national systems of innovation is elusive. It has been suggested, for example, that a successful NSI represents the kind of environment that Thomas Edison, if he were alive today, could exploit in the execution of his research. For a comprehensive discussion, see Nelson, R. R. (1993). The OECD initiated a program to study national systems of innovation in 1994. This program seeks to compare knowledge flows in various national innovation systems, with the goal of highlighting new approaches for technology policy. See OECD, 1997, *National Innovation Systems,* Directorate for Science, Technology and Industry, OECD Website.

4. The laboratories of the National Institute of Standards and Technology (NIST) have reported aggregate economic rates of return ranging from 63 to 428 percent on investment in technology. The reader should note that statistics of this sort can be highly influenced by subjective factors, so take them with a grain of salt. See Executive Office of the President (1996, pg. 39).

5. Note that improved relationships between industry and government would tend to improve the appropriability of returns from technological spillovers for the domestic economy.

6. Pre-competitive research is defined by the Department of Commerce as "Research and development activities up to the stage where technical uncertainties are sufficiently reduced to permit preliminary assessment of commercial potential and prior to development of application-specific commercial prototypes." See Branscomb, L. M. and G. Parker (1993, pg. 83).

7. Branscomb, L. M. (1993b, pg. 3).

8. For a more detailed discussion, see Brooks, H. (1996, pg. 16).

9. During 1944 and 1945, the Manhattan Project received substantially more research funding than the entire R&D budget for the rest of the war effort. See Mowery, D. C. (1989, pg. 124).

10.  Branscomb, L. M. and R. Florida (1998, pg. 14).

11.  As the requirements for military hardware became more specialized, the potential for costless transfer into the commercial sector declined. A military communications satellite, for example, must have componentry which can withstand very high levels of radiation. This requirement alone created an entire subdiscipline within the semiconductor industry to produce so-called "radiation-hardened" devices. Despite the highly sophisticated technology required to produce such devices, there is little demand for such expensive and specialized components in a peaceful world.

12.  For a short list of important U.S. technology policy acts and laws, see "Technology innovation legislation highlights," The Federal Laboratory Consortium Website, www.fedlabs.org.

13.  Ham, R. M. and D. C. Mowery (1995).

14.  Clinton, W. J. and A. Gore, Jr. (1993).

15.  See 1) U.S. Department of Commerce (1994, pg. 47), and 2) Branscomb, L. M. (1993, pg. 20).

16.  For recent overviews of U.S. R&D investment strategies, see 1) Council on Competitiveness (1996), and 2) Office of Technology Policy (1997).

17.  The European subsidy of the Airbus consortium over the last twenty years included government contracts, loan guarantees on favorable terms, equity infusions, tax breaks, debt forgiveness, and bailouts. For an in-depth discussion, see Tyson, L. D. (1992, pg. 172).

18.  There is a recurring challenge conducted by *The Wall Street Journal*, in which the editors ask several respected investment analysts to pick a "virtual portfolio" of growth stocks. The *Journal* editors then create their own portfolio by throwing darts at the stock exchange listings. Surprisingly, the dartboard has yielded a higher rate of return than the professionally selected stocks in a significant number of contests.

19.  Proponents of a proactive technology policy often cite the benefits of first-mover advantage in advanced technology markets as a motivation for early subsidy of promising industries. It is assumed that an early advantage will "lock in" control of a long stream of follow-on products, while a late start may make entry for "followers" far more difficult. See Borrus, M. and J. Stowsky (1998, pg. 44).

20.  Both Ford and General Motors had established a dominant presence in Japan by 1927. Similarly, Toshiba was affiliated with General Electric, Westinghouse was a 10 percent shareholder in Mitsubishi Electric, and Fuji Electric was formed as a joint venture between Siemens and Furukawa Mine. Only Hitachi, among Japan's technological giants, has never been affiliated with a foreign company. See Hiroyuki, O. and A. Goto (1993, pg. 94).

21.  One interesting example involves the rate of improvement of fighter planes during the war years. Evidently, over the six-year duration of World War II, the horsepower of Japanese airplane engines increased by only 20 percent, while American and European fighter engines more than doubled in horsepower. This technology gap rendered the Japanese Zero fighter plane obsolete by the end of the war. See Hiroyuki, O. and A. Goto (1993, pg. 111).

22.  For an in-depth analysis of the role of MITI in Japanese economic development, see Okimoto, D. I. (1989).

23.  MITI officials stress the criterion of "market failure" as a justification for their involvement in commercial markets. They expend considerable energy in identifying areas in which either the risks are too high or the appropriability of returns is too low for firms to adequately invest in R&D. While this criterion may have been adequate in the past, today one can hardly find an industry that does not suffer from some form of market failure. The current challenge is to identify the points of greatest leverage for government investment, rather than

simply reacting to insufficient private-sector investment. See Fransman, M. (1997, pg. 64).

24. For an overview of Japan's Fifth Generation Computer Project, see Nakamura, Y. and M. Shibuya (1996).

25. Japanese drug firms, for example, have made major strides in the development of commercially viable processes using advanced biotechnology. They may be in a position to leapfrog American firms in the next decade, in much the same way as our computer and semiconductor firms were overtaken in the 1980s. See Lansing, P. and J. Gabriella (1995).

26. Fransman, M. (1997, pg. 66).

27. For a concise overview of future challenges to U.S. technology policy, see Branscomb, L. M. et al. (1997).

28. There is growing evidence that "localized" systems of innovation, such as the unique environment that has spontaneously formed in Silicon Valley, play a central role in providing advantages and opportunities for technology industries. This topic will be covered in more detail in Chapter 8. See Antonelli, C. (1994).

29. It is interesting and somewhat disturbing to note that the U.S. government did not actually take stock of public funding for science and technology until the early 1990s. It was discovered that funding was so dispersed among various mission-oriented agencies that there was virtually no coordination of either goals or strategies.

30. Branscomb, L. M. (1993c, pg. 282).

31. One option that has been proposed in recent years is the formation of a Civilian Technology Corporation, which would receive a one-time endowment of funding. This government-chartered corporation would then strategically invest its resources, and, hopefully, sustain itself through return on investment. This may be an effective supplement to the U.S. venture-capital system and could promote increased rates of innovation. See Committee on Science Engineering and Public Policy (1992).

32. From the perspective of this book at least, the impact of government investment on technology-intensive industries is of paramount importance. Others could argue that the funding of basic scientific research or strategic-defense technology is more critical to our society's well-being. Of course, I would be forced to disagree.

33. The only exceptions that should be made to this rule would be for programs that provide small amounts of "seed money" for start-up firms and entrepreneurs, such as the Small Business Innovation Research (SBIR) and Advanced Technology Program (ATP).

34. National Academy of Engineering (1993, pg. 9).

35. Branscomb, L. M. (1993, pg. 12).

36. Gore, A. Jr. (1997, pg. 26).

37. Executive Office of the President (1996, pp. 16–17).

38. In inflation-adjusted (real) terms, total R&D spending increased by 3.8 percent in 1997, 3.2 percent in 1996, and 5.9 percent in 1995. In contrast, the entire U.S. economy grew at only 2.4 percent, 2.1 percent, and 2.1 percent respectively. See National Science Foundation, 1997, *Data Brief: R&D Exceeds Expectations Again, Growing Faster than the U.S. Economy During the Last Three Years*, National Science Foundation, November 5, 1997. As of this writing, it appears that the federal R&D budget for 1998 will include a 4.1 percent increase over 1997, according to the American Association for the Advancement of Science (AAAS).

39. Ibid. (1997, pg. 3).

40. As noted earlier, improvement in the budget deficit over the last several years may cause a change in this trend. It appears, for example, that federal R&D spending will increase in real terms in 1998.

41. Nelson, S. D. and K. Koizumi (1997, pg. 82).

42. One proposed method for prioritizing government R&D suggests that the time frame for investment could be used to determine the appropriateness of public funding. In this scheme, activities that are intended to yield short-term commercial benefits would not be supported by government funds, while longer-term (six-to-ten-year) projects would be candidates for subsidy. See Council on Competitiveness (1996).

43. The concept of a federal science and technology (FS&T) budget has been proposed in a recent influential report. See Committee on Criteria for Federal Support of Research and Development (1995).

44. Domenici, P. V. (1997, pg. 60).

45. Gore, A. Jr. (1997, pg. 27).

46. "Japan's plan to double Its R&D budget is on track," *SSTI Weekly Digest*, July 18, 1997, State Science and Technology Institute.

47. Branscomb, L. M. (1993, pg. 73).

48. Smith, B. L. R. and C. E. Barfield (1996b, pg. 7).

49. There is somewhat dated evidence that the American scientific community shares the unfortunate "not invented here" syndrome common within U.S. industry. If still prevalent, this "self-preoccupation" may be limiting their openness to foreign scientific advances, thereby reinforcing our already poor national absorptive abilities. See Frame, J. D. and F. Narin (1988).

50. Nelson, R. R. (1993, pg. 12).

51. Mowery, D. C. and N. Rosenberg (1989, pg. 226).

52. Branscomb, L. M. (1993, pg. 93).

53. Universities often represent the "seed crystal" for high-technology industrial clusters, as was the case with Stanford in Silicon Valley and MIT in Massachusetts' Route 128 region. Stanford has actively sponsored industry collaborative efforts for several decades, including the Stanford University Center for Integrated Systems (CIS).

54. Committee on Criteria for Federal Support of Research and Development (1995, pg. 24).

55. A study of twenty-eight early ATP award winners determined that half of the companies surveyed estimated that they reduced their technology-development cycle by 50 percent. The overwhelming majority (twenty-seven out of twenty-eight) estimated that ATP participation reduced cycle times anywhere from 30 to 66 percent. See "New study finds ATP speeds technology development," *SSTI Weekly Digest*, November 7, 1997, State Science and Technology Institute. For a good summary of the statistical distribution of ATP funds among various research entities, see Hill, C. T. (1998).

56. Branscomb, L. M. and G. Parker (1993, pg. 70).

57. Simons, G. R. (1993, pg. 169).

58. In a sincere effort to encompass all the opportunities for government to support technology enterprise, some authors have begun to blur the lines between policy functions. This is not simply a semantic issue. Each aspect of technology policy, including knowledge creation, diffusion, etc., has a specific place in the innovation system and performs a reasonably well-defined function. Without a common and precise language for discussing these activities, we cannot begin to make tradeoffs or attempt to optimize the system. I have chosen rather narrow definitions for the purposes of this book, and look forward to a better consensus in the open literature in the future. See, for example, Justman, M. and M. Teubal (1995).

59. In other words, infrastructure should support increased rates of innovation and the ability of firms to build a platform of accumulated knowledge, two of the three elements of the innovation cycle. See Tassey, G. (1991).

60. U.S. Department of Commerce (1994, pg. 58).

61. Howell, T. and D. Ballantine (1992, pg. xii).

62. This discussion expands upon Kahin, B. (1993, pg. 139). For a broad perspective on the infrastructure required to effectively manage knowledge within firms and industries, see Brown, J. S. and P. Duguid (1998).

63. Some have argued that investment in technology infrastructure provides as much benefit to foreign firms located in the United States as it does to domestic industry. While this is most likely the case, I fail to see how this diminishes the value of infrastructure investment to our economy. Foreign technology firms with U.S. subsidiaries provide high-paying jobs, offer opportunities for tacit skill development and technology spillovers, inject revenue into both local and national economies, and help support our domestic firms with whom they cooperate. The fact that some profits leave the country seems to be a minor consideration when compared with the substantial advantages of attracting technology-intensive foreign direct investment.

64. U.S. Department of Commerce (1996, pg. 13).

65. At the first World Telecommunications Development conference in Buenos Aires in 1994, the United States promoted a vision for the GII that incorporated five basic principles: private investment, competition, universal service, open access, and flexible regulations. For an overview of the Clinton administration's perspective on U.S. involvement in the GII, see 1) Executive Office of the President (1995), and 2) Gore, A. Jr. (1996).

66. For an exhaustive overview of the impact of electronic commerce on global business, see Choi, S. et al. (1997).

67. In addition to protecting the safety of individuals, there is growing concern over "cyber threats" to our national information infrastructure. To address these issues, the current administration has created the President's Commission on Critical Infrastructure Protection. This group is tasked with assessing threats to national security vis-à-vis willful damage to our power, transportation, water, telecommunications, and data systems, and recommending a strategy for protecting this critical infrastructure. See Executive Office of the President, 1997, *Critical Foundations: Thinking Differently*, President's Commission on Critical Infrastructure Protection Website.

68. Kahin, B. (1993, pg. 154).

69. Sources: 1) Clinton, W. J. and A. Gore Jr. (1997), 2) "Talking Internet commerce: IC interview with Ira Magaziner," *Intellectual Capital*, July 17, 1997, 3) "U.S. rejects levying tax on Internet," *International Herald Tribune*, July 1, 1997, and 4) "Clinton backs Internet free-trade zone," *International Herald Tribune*, July 2, 1997.

70. For more on trade restrictions on strong encryption technology, see Vesely, R., 1997, "The generation gap," *Wired*, October Issue.

71. In the 1980s, two parallel programs were launched by the federal government to create strategic plans for development of enabling and infrastructural technology. The Research Applied to National Needs (RANN) program was managed by the NSF, while a similar activity, the Experimental Technology Incentives Program (ETIP), was initiated by the National Bureau of Standards (now NIST). Although some valuable scientific soul-searching was accomplished, neither activity took root as a continuing program. See Brooks, H. (1996, pg. 25).

72. Branscomb, L. M. (1993d, pg. 56).

73. Industry associations and consortia must play an increasing role in the development of industry-specific infratechnology in the future. Groups such as the American National Standards Institute (ANSI), the Society of Automotive Engineers (SAE), the Institute for Interconnecting and Packaging Electronic Circuits (IPC), and the Institute of Electrical and

Electronic Engineers (IEEE) all fill a critical void in our national technology infrastructure.

74. For a broad discussion of the role of SMEs in the development of the U.S. economy, see National Academy of Engineering (1995).

75. Total venture-capital investment in the United States has been at record levels in recent years. In the third quarter of 1997, total investments exceeded $3.5 billion, up from $3.2 billion in the second quarter of the same year. A total of 675 companies received financing during the third quarter, representing primarily the biotechnology, communications, computer, semiconductor, and software industries. See "Venture capital investments at record level for second straight quarter," *SSTI Weekly Digest*, December 12, 1997, State Science and Technology Institute.

76. Early in the development of television, the United States and Europe established different and totally incompatible standards for broadcast and display formats. This has caused incalculable waste and inefficiency in television markets for the last half-century. For a detailed discussion of the early stages of the HDTV debate, see Farrell, J. and C. Shapiro (1992).

77. In the United States, the current allocation of cellular phone standards is split three ways: the GSM standard has 24 percent of the market, a newer digital standard called TDMA has 14 percent, and most of the remainder has gone to an even newer technology known as CDMA. See "Uncle Sam, please pick a cell-phone standard," *Business Week*, February 24, 1997, pg. 34.

78. For a comprehensive discussion of international commercial standards, see 1) Office of Technology Assessment (1992), and 2) Sykes, A. O. (1995).

79. "ANSI, partners launch standards network on Web," *Electronic Business Times*, March 3, 1997, pg. 26.

80. Henke, C. (1997).

81. McDaniels, I. K. and M. G. Singer (1997, pg. 22).

82. See, for example, 1) Mansfield, E. R. (1993), and 2) Beede, D. N. and K. H. Young (1998).

83. Irwin, S. M. (1993, pg. 62).

84. Council on Competitiveness (1996, pg. 23).

85. National Academy of Engineering (1993, pg. 78).

86. Simons, G. R. (1993, pg. 189).

87. These corporations include the Japan Electronic Computer Company, established in 1961, and the Japan Robot Leasing Company, established in 1980. See Mowery, D. C. (1993, pg. 37).

88. Nakamura, Y. and M. Shibuya (1996).

89. The formation of cooperative relationships to share technical information is not restricted to the advanced economies. China, for example, has established cooperation and exchange relations in science and technology with 134 countries and regions. See Mitchell, G. R. (1997, pg. 244).

90. "NSF announces funds for new science and technology centers," *SSTI Weekly Digest*, November 7, 1997.

91. Simons, G. R. (1993, pg. 179).

92. Huttner, S. L. and C. Yarkin (1998).

93. For an overview of the evolution of technology alliances from 1970 to the present, see Doremus, P. N. et al. (1998, pp. 111–115).

94. Hemphill, T.A. (1997).

95. As I mentioned earlier, the transfer of dual-use technology has proved to be

problematic rather than "automatic." To assist in the commercialization of dual-use technology, the government enacted the 1980 Stevenson-Wydler Innovation Act and the Technology Transfer Act of 1986. See Branscomb, L. M. (1993e, pg. 113).

96. National Science Foundation (1996).

97. Packard, D. (1983, pg. 9).

98. For a more detailed discussion of CRADAs in the fields of medicine and biotechnology, see Guston, D. H. (1998).

99. See, for example, 1) Brod, A. and R. Shivakumar (1997), and 2) Teichert, T. A. (1997).

100. Quintas, P. and K. Guy (1995).

101. For a firm-specific (Westinghouse Electric Corporation) perspective on the rising importance of collaborative R&D, see Melissaratos, A. (1998).

102. Olk, P. and K. Xin (1997).

103. Mowery, D. C. (1993, pg. 41).

104. Executive Office of the President (1996, pg. 50).

105. Rea, D. G. et al. (1997).

106. One of the unique aspects of government funding for SEMATECH was that it was provided with very few strings attached. This caused some concern on the part of policymakers, but allowed member firms the flexibility to adapt their original mission to the tasks that could yield the greatest benefit to their industry.

107. Browning, L. D. et al. (1995).

108. The need for SEMI/SEMATECH was driven by the high price of entry into the main consortium. The minimum buy-in was set at $1 million at the founding of the group, and has not changed to this day. This was highly criticized by small supplier firms, but was perceived to be necessary to avoid "free riding" on the more substantial investments of larger members. For an overview of SEMI, see Davis, J., 1997, "SEMI: The growth of a global association," Solid State Technology, May Issue.

109. Dick, A. (1995, pg. 69).

110. More important, American semiconductor equipment manufacturers now enjoy a 55 percent global market share, as compared to Japan's 35 percent and Europe's 10 percent. See Hackman, S. (1997).

111. For a balanced analysis of the economic benefits of SEMATECH, see Irwin, D. A. and P. J. Klenow (1996).

112. The Sortec consortium consists of thirteen large Japanese electronics firms. The majority of funding for this collaborative organization comes from industry; public funding was set at $100 million for the entire ten-year duration of the agreement. See Skolnikoff, E. B. (1993, pg. 124).

113. For a broader discussion of "free riding," see Henriques, I. (1994).

114. SEMATECH is not unique in its exclusion of foreign participants. The U.S. Consortium for Automotive Research (USCAR) also excludes foreign membership. In some cases, the U.S. government has used rules similar to "local content" requirements to determine qualification for membership, or has demanded reciprocity from foreign nations with respect to their domestic consortia. See Ham, R. M. and D. C. Mowery (1995).

115. For additional discussions of the issues surrounding international consortia, see 1) Ohba, S. (1996), and 2) Solomond, J. P. (1996).

116. Tripsas, M. et al. (1995).

117. Zinberg, D. S. (1993, pg. 247).

118. In 1995, China adopted a new policy document, *The Decision to Accelerate the Development of Science and Technology*, which, among other initiatives, calls for upgrading

one hundred of the nation's best universities to international standards of excellence by early in the twenty-first century. See Suttmeier, R. P. (1997).

119. Zinberg, D. S. (1993, pg. 242).

120. For several years, the Information Technology Association has released annual figures on the shortage of skilled labor in the information technology industry. In 1997, it estimated that 190,000 jobs in this field were going unfilled, and increased its estimate for 1998 to a shortfall of over 300,000 jobs. See "U.S. to foster training of high-tech workers," *The International Herald Tribune*, October 1, 1997, on-line edition. For a recent update, see Lerman, R. I. (1998).

121. Johnston and Packer, *Workforce 2000: Work and Workers for the Twenty-First Century*.

122. National Academy of Engineering (1993, pg. 14).

123. "Filling high-tech jobs is getting very tough," *The Wall Street Journal*, December 1, 1997.

124. There seems to be a growing pattern of reverse brain drain, particularly in the newly industrialized nations. After losing many of their best and brightest technologists and managers to foreign opportunities, countries such as the Philippines, Taiwan, China, India, South Korea, and Malaysia are attracting these expatriates back with offers of high-paying jobs and bright futures. See Hagerty, B. (1996).

125. National Science Foundation (1996, pp. 2–14).

126. In Japan, it is common for employees to be members of a single company-wide union. These organizations perform a different function in Japanese industry than that of American unions, by serving as coordinators for the movement of employees, providing labor-relations services, and managing training needs. See Odagiri, H. and A. Goto (1993, pg. 108).

127. There is a growing trend among technically savvy high school graduates to skip college in favor of opportunities in high-technology industry. These individuals often find themselves in dead-end positions, however, due to their lack of formal education. See 1) Silverstein, S., 1997, "They're trading higher education for high tech—but at what cost?" *The Los Angeles Times*, November 9, and 2) Valverde, G. A. and W. H. Schmidt (1997).

128. Terdiman, D., 1997, "Post boomer boom times: Silicon Valley's roaring economy has firms in a frenzy to hire law school graduates," *Intellectual Property Magazine*, November Issue, on-line edition.

129. "Everyone knows E=MC²–now who can explain it?" *Business Week*, October 6, 1997, pg. 66.

130. Several government agencies began awarding fellowships to doctoral students in the sciences after World War II. The National Institutes of Health, for example, introduced traineeships, and the National Science Foundation offered university fellowships. The National Defense Education Act of 1958 provided for increased government funding for education in science and engineering. See Zinberg, D. S. (1993, pg. 255).

131. Wiggenhorn, W. (1990).

132. Hibbard, J., 1998, "The learning revolution," *Information Week*, March 9, pp. 44–60.

133. Several foreign firms have taken up the challenge of improving levels of human capital within the United States. The Korean automotive giant, Hyundai Motors, established an eighteen-week training program for auto mechanics in an ethnically mixed neighborhood of South Central Los Angeles. One of the goals of this program was to reduce racial tensions between African Americans and Korean Americans in the region. See Alkhafaji, A. F. (1995, pg. 312).

## Chapter 8: Regional Advantage

1. Marshall, A. (1890).

2. See, for example, 1) Porter, M. E. (1990), 2) Porter, M. E. (1998), 3) Feldman, M. P. (1993), and 4) Sabourin, V. and I. Pinsonneault (1997).

3. GEST (1986).

4. Sabourin, V. and I. Pinsonneault (1997, pg. 166).

5. A number of firms have naively assumed that professional workers would be willing to relocate at the whim of their employers. In the late 1980s, Hughes Aircraft Company announced plans to relocate several thousand research and engineering professionals from Southern California to Tucson, Arizona. The response to this relocation plan was so negative that the company was forced to cancel the move after losing a number of its top employees to competing firms.

6. The competition for excellent employees is no different than the competition for market share; if firms wish to gain leadership in global markets, they must be prepared to expend Herculean efforts to retain their core of creative talent. Note that despite the fact that firms are often attracted to technology clusters for their rich pool of human capital, these regions can suffer from periodic shortages of skilled labor. See, for example, Thuermer, K. E. (1998).

7. Henderson, J. (1989, pg. 34).

8. Hagedoorn, J. and J. Schakenraad (1992).

9. For a recent account of the rising fortunes of the Route 128 region, see Auerbach, J. G. and R. Kerber, 1998, "Despite Digital deal, Massachusetts rises again," *The Wall Street Journal*, January 30.

10. Irwin, S. M. (1993, pg. 179).

11. The Stanford Industrial Park began life as a means for Stanford University to capture some economic benefit from its vast land holdings. Initial tenants included Varian Associates, General Electric, Eastman Kodak, Admiral Corporation, and Hewlett-Packard. By 1961, the park had grown to 650 acres and was home to more than 11,000 high-technology employees. See Saxenian, A. (1994, pp. 23–24).

12. Suttmeier, R. P. (1997, pg. 386).

13. Sources: 1) Mathews, J. A. (1997), and 2) Xue, L. (1997).

14. Castells, M. and P. Hall (1994, pg. 37).

15. Sources: 1) Bullis, D. (1997), 2) *Malaysia - The Star*, various issues, and 3) Greenwald, J., 1997, "Thinking big," *Wired*, August Issue, pp. 95–144.

16. See, for example, 1) Porter, M. E. (1998, pg. 12), and 2) Debresson, C. (1989).

17. See 1) Sabourin, V. and I. Pinsonneault (1997), and 2) Audretsch, D. B. and M. P. Feldman (1996).

18. See for example, Jaffe, A. B. et al. (1993).

19. Saxenian, A. (1994, pg. 67).

20. Feldman, M. P. (1993).

21. Saxenian, A. (1994, pg. 175).

22. Ferdows, K. (1997).

23. Simons, G. R. (1993, pg. 178).

24. "The rich ecosystem of Silicon Valley," *Business Week*, August 25, 1997, pg. 202.

25. "In the Valley," *Contract Professional*, January/February 1998, pg. 35.

26. For an overview of the economic history of Silicon Valley, see Khanna, D. M. (1997).

27. Saxenian, A. (1994, pg. 26).

28. Henderson, J. (1989, pg. 41).
29. Saxenian, A. (1994, pg. 115).
30. Saxenian, A. (1994, pg. 33). Recent trade-theft suits filed by major competitors in Silicon Valley have raised concerns over the region's "freewheeling ways." See "A nest of software spies?" *Business Week,* May 19, 1997, pg. 100.
31. Saxenian, A. (1991).
32. Saxenian, A. (1994, pg. 59).
33. Saxenian, A. (1994, pg. 126).
34. The location of facilities in these regions offers another important advantage: Foreign firms can circumvent the tariff barriers and local-content requirements imposed on many high-technology imports to the European Union. See 1) Macleod, G. (1996), 2) McCann, P. (1997), and 3) Morgan, K. (1997).
35. Bindra, A., 1997, "Software savvy draws firms to India," *Electronic Engineering Times,* March 31.
36. Porter, M. E. (1990, pg. 232).
37. Siegel, J., 1997, "The world is coming to Silicon Wadi," *The Jerusalem Post - Internet Edition,* June 16.
38. For more on the development of high-technology industry in Israel, see Sandler, N., 1998, "Mazeltech on the Mediterranean," *Electronic Business,* February Issue.
39. Walker, K., 1997, "The UK's science parks are attracting high-tech U.S. companies," *Expansion Management,* November/December Issue, pp. 60–63.
40. Scott, A. J. (1991).
41. See 1) Robinson, R., 1996, "The Telecom Corridor: Linking the world to the twenty-first century," *Business Horizons,* Vol. 5, No. 2, pp. 18–20, and 2) Orr, D. E., 1996, "Global opportunities for telecom companies," *Business Horizons,* Vol. 5, No. 2, pp. 21–23.
42. For a discussion of the locational links among scientists and firms in the biotechnology industry, see Audretsch, D. B. and P. E. Stephan (1996).
43. Monahan, K., 1997, "Alabama: Silicon Valley South," *Expansion Management,* November/December Issue, pp. 48–50.
44. Saxenian, A. (1994, pg. 158).
45. Antonelli, C. (1994).
46. Cooney, S. L. Jr. (1997, pg. 375).
47. Henderson, J. (1989, pg. 51).
48. Porter, M. E. and R. E. Wayland (1995, pg. 101).
49. Powers, W. F. (1997, pg. 368).
50. Kuemmerle, W. (1997).
51. Kuemmerle, W. (1997, pg. 62).
52. Grove, A., 1993, "How Intel makes spending pay off," *Fortune,* February 22, pg. 58.
53. The contract manufacturing industry evolved out of a need for specialized assemblers of printed circuit boards. One of the earliest entrants was Flextronics, Inc. of Silicon Valley, which started life as a "board-stuffing" house. As with many other firms in this industry, Flextronics worked its way steadily up the value-added food chain until today it produces sophisticated "turnkey" products, and even provides competent product design services. See Saxenian, A. (1994, pg. 151).
54. Until recently, the United States stood alone in the size and depth of its venture-capital markets. The recognition that start-ups play a disproportionate role in the advance of commercial technology, however, has prompted the development of a dynamic "small-cap" market in Europe as well. Since 1996, Europe has created six secondary capital mar-

kets, including the EASDAQ, London's AIM, and four "new market" exchanges. See "Europe finally wakes up to high-tech start-ups," *Business Week*, May 26, 1997, pg. 76.

55. "Tales from spin-off city," *Business Week*, February 23, 1998, pg. 112.

56. Saxenian, A. (1994, pg. 29).

57. It is interesting to note that since his embarrassing dismissal as CEO of DEC in 1992, Ken Olsen has gone on to found his own start-up, Advanced Modular Solutions. See "Ken Olsen gets small," *Your Company*, February/March 1998, pg. 48.

58. For an interesting discussion of the impact of both formal and informal networks on the learning ability of firms, see Powell, W. W. (1998).

59. Bettis, R. A. and M. A. Hitt (1995, pg. 9).

60. Saxenian, A. (1991).

61. See 1) Florida, R. (1997), and 2) Howells, J. (1990).

62. Saxenian, A. (1994, pg. 164).

# Chapter 9: Negotiating the Technology Rapids

1. For a broad-based analysis of competition in information technology, including a number of excellent case studies, see 1) Yoffie, D. B. (1994), and 2) Yoffie, D. B. and B. Gomes-Casseres (1994).

2. Rumelt, R. P. et al. (1994, pg. 26).

3. For an overview, see Hammond, T. H. (1994). For the seminal discussion of "structure follows strategy," see Chandler, A. D. Jr. (1962).

4. See Rumelt, R. P. et al. (1994, pg. 39). Other references related to competition in knowledge-intensive industries include: 1) Quinn, J. B. (1992), and 2) Montgomery, C. A. and M. E. Porter (1991).

5. Kanz, J. and D. Lam (1996).

6. Porter, M. E. (1980). For a comparison of several analysis tools, see Porter, M. E. (1994). For a discussion of strategy development in rapidly changing environments, see Chakravarthy, B. (1997).

7. For some sincere efforts (which provide valuable, but often contradictory insights), see Gaynor, G. H. (1996), Betz, F. (1993), and Burgelman, R. A. et al. (1996).

8. It is beyond the scope of this book to provide a thorough survey of the literature in strategic management, management of technology, industrial organization theory, and other relevant fields of study. Instead, I will suggest a "most likely to succeed" list, and include some possible ways to organize these insights into a practical form.

9. One of the formulations of resource-based strategy is the "VRIO framework," which addresses value, rareness, imitability, and organization. This model, with the effects of change over time duly noted, will form the basis for much of what is presented in subsequent discussions of sustainable advantage. See Barney, J. B. (1997, pg. 145).

10. Hammond, T. H. (1994, pg. 151).

11. Williams, J. R. (1994, pg. 238).

12. Burgelman, R. A. et al. (1996, pg. 22).

13. See, for example, 1) Day, G. S. (1997, pg. 72), and 2) Barney, J. B. (1997, pg. 171).

14. Saloner, G. (1994, pg. 178).

15. Saloner, G. (1994, pg. 167).

16. Von Neumann, J. and O. Morgenstern (1944).

17. Buchanan, J. M. (1978).

18. Barney, J. B. (1994). This article contrasts game theory to a competing field of study, known as prospect theory, which treats firms as collections of biased decision-makers.

19. Perhaps a more accurate description would be that firms suffer from an organizational version of "multiple personality disorder." See Barney, J. B. (1994, pg. 68).

20. For a useful description of the complex nature of competition in the electronics display industry, see Bowonder, B. et al. (1996). A less rigorous but otherwise interesting analysis of competition in the information-technology sector is provided in Moschella, D. C. (1997). For a theoretical treatment of the application of game theory to changing technology in economic systems, see Hall, P. (1994).

21. Saloner, G. (1994, pg. 174).

22. Rumelt, R. P. et al. (1994, pg. 51).

23. Kretschmer, M. (1998).

24. Sources: 1) Camerer, C. F. (1994), 2) Saloner, G. (1994), and 3) Barney, J. B. (1997).

25. Ackoff, R. (1981).

26. Juhasz, J. E. (1994, pg. 16.5).

27. Juhasz, J. E. (1996, pg. 16.5). For a sampler of software tools for business-strategy development, see Baker, S. and K. Baker (1998).

28. Barney, J. B. (1994).

29. See, for example, Kahneman, D. and A. Tversky (1979).

30. Kahneman, D. and D. Lovallo (1994, pg. 81).

31. Barney, J. B. (1994, pg. 58).

32. The strategic choices of Generals Grant and Lee during the American Civil War provide an excellent example. Once Grant determined that he had the war won, he embarked on a conservative "war of attrition" strategy. Lee, on the other hand, recognizing the desperation of his situation, attempted brilliant but risky maneuvers to save his cause.

33. This statistical impossibility brings to mind the famous opening to Garrison Keillor's "Tales of Lake Wobegon," where "all the children are above average."

34. Kahneman, D. and D. Lovallo (1994, pg. 90).

35. These criteria, along with much of this section, follow Dawes, R. M. (1988). This is an excellent and highly readable book, which I recommend to all readers facing the challenges of decision-making under change and uncertainty.

36. Dawes, R. M. (1988, pg. 9).

37. There is at least one rational justification for honoring sunk costs: the saving of a firm's reputation in the marketplace. If killing a project would make a firm (or its management team) appear to be indecisive or weak to customers or shareholders, it may be worthwhile investing in a "scaled-down version" or contriving some other graceful exit.

38. One of the most important fields of research into the decision process involves the systematic skewing of people's perceptions of monetary value and utility as scale increases. One of the central principles of prospect theory is that the utility of both positive and negative consequences of decisions exhibits a diminishing-returns characteristic. See, for example, Kahneman, D. and A. Tversky (1979).

39. Dawes, R. M. (1988, pg. 122).

40. DeGroot, A. D. (1965).

41. Humorist James Thurber once noted that we often fall on our face because we have previously fallen on our ass.

42. For some "decision shortcuts" that might (or might not) improve the degree of rationality in decision-making, see Meyer, R. J. and D. Banks (1997).

43. Simon, H. A. (1979).

44. Dawes, R. M. (1988, pg. 51).

45. See, for example, Kelly, S. (1996).

46. One way of focusing the decision processes of firms is through the formation of a *strategic attitude*, or alternatively a *strategic intent*, that captures the long-term goals and near-term directions of the organization. The strategic attitude represents a first-level filter for new concepts and ideas, and provides some assurance that alternatives will synergize with the desired evolution of the company. See Betz, F. (1993, pg. 38).

47. Dawes, R. M. (1988, pg. 87).

48. "Is this the end of sticky prices?" *The Economist*, May 16, 1998.

49. Barney, J. B. (1997, pg. 105).

50. Porter, M. E. and R. E. Wayland (1995, pg. 85).

51. "The art of innovation," *Information Week*, December 1, 1997.

52. For a broader discussion, see Reibstein, D. J. and M. J. Chussil (1997, pg. 395).

53. "Virtual chips: The next step up," *Upside*, December 1997.

54. These "virtual semiconductor companies" have not been warmly received by venture-capital investors, who tend to feel more comfortable with some capital assets on the balance sheet. Nonetheless, this concept has been well received by Wall Street in several recent initial public offerings (IPOs).

55. "The changing landscape of contract electronics manufacturing," *Circuitree*, May 1997. For an in-depth discussion of the evolution of outsource manufacturing among the "Asian tigers," see Hobday, M. (1995).

56. For a detailed discussion of the use of "building-block" design cores in the design of electronic circuits, see "Accelerate system designs by leveraging intellectual property," *Electronic Design*, January 21, 1998, and "Gazing at the new age of intellectual property," *Electronic Design*, January 12, 1998.

57. The concept of mass customization was initially developed in Pine, B. J., II (1993). For a more traditional discussion of product-line strategy, see McGrath, M. E. (1995).

58. Anderson, D. M. (1997, pg. 25).

59. For an interesting perspective on the relationship between mass customization and tacit knowledge, see Kotha, S. (1995).

60. Gilmore, J. H. and B. J. Pine II (1997).

61. For an example of how a class of software products could benefit from collaborative mass customization, see Dietrich, G. B. et al. (1997).

62. See 1) Lee, H. L. et al.(1997), and 2) Feitzinger, E. and H. L. Lee (1997).

63. Sources: 1) "Design reuse: From a wish to a reality," *Engineering Design News*, May 8, 1997, 2) "Taking different approaches to design reuse," *Computer Design*, February 1998, and 3) "Board designers get into the 'reuse' act," *Electronic Engineering Times*, May 1, 1995.

## Chapter 10: The Tools of the Oracle

1. Cerf, C. and V. Navasky, 1984, *The Experts Speak*, Pantheon Books.

2. For a parallel development of learning strategies that uses Motorola University as a case study, see Baldwin, T. T. et al. (1998).

3. Hibbard, J., 1998, "The learning revolution," *Information Week*, March 9.

4. Source: American Society for Training and Development.

5. Hibbard, J., 1998, "The learning revolution," *Information Week*, March 9, pg. 48.

6. For an expanded discussion of collaborative tools in R&D, see King, E., 1997, "Collaboratories are poised to take the next step," *Scientific Computing and Automation*, August Issue.

7.  A recent "summit" convened by the Boeing Company concluded that the shape and structure of higher education must change in the next decade to become both geographically and intellectually aligned with industry. Such a convergence between the two primary sources of human-capital formation could enable tremendous gains in productivity. See Condit, P. and R. B. Pipes, 1997, "The global university," *Issues in Science and Technology*, Fall Issue.

8.  For a useful overview of organizational learning methods, see Barabaschi, S. (1992).

9.  Cohan, P. S. (1997, pg. 33).

10.  Cohan, P. S. (1997, pg. 59). Note that traditional methods of "arm's-length" licensing of new technologies might not be a productive solution in this age of increased specialization. To ensure a high return from a licensing investment, it might be advisable to contract for the exchange of tacit knowledge as well, through the provision of technical services by the innovating firm. For a discussion of this topic as it applies to North-South technology transfer, see Arora, A. (1996).

11.  For the case of the biotechnology industry, see Shan, W. and J. Song (1997).

12.  For a discussion of the advantages and disadvantages of international participation in high-technology consortia, see Solomond, J. P. (1996).

13.  "Tech watching becomes the industry to watch," *The Wall Street Journal*, March 19, 1998.

14.  Romer, P. M. (1993b).

15.  See 1) "For the tech industry, market in Washington is toughest to crack," *The Wall Street Journal*, March 4, 1998, and 2) "Capital spending," *Electronic Business*, December 1997.

16.  For an expanded discussion of the practical application of knowledge management, including results from a survey of 431 firms, see Ruggles, R. (1998). For some warnings of pitfalls on the pathway to effectively managing knowledge within the firm, see Fahey, L. and L. Prusak (1998). An excellent overview of value creation through knowledge assets is provided in Teece, D. J. (1998).

17.  Sources: 1) "Knowing what we know," *Information Week*, October 20, 1997, and 2) "Leverage your knowledge base," *Internet World*, February 1998. For further reading, see 1) Nonaka, I. and H. Takeuchi (1995), and 2) Leonard-Barton, D. (1995).

18.  See, for example, 1) Bhalla, S. K. (1987), and 2) Millett, S. M. and E. J. Honton, (1991). For a "logic tree" guide to the literature on technology forecasting, see Worlton, J., 1988, "Some patterns of technological change in high performance computers," *Proceedings Supercomputing 1988*, IEEE Computer Society Press.

19.  See, for example, 1) Betz, F. (1996), 2) Burgelman, R. A. et al. (1996, pg. 143), 3) Betz, F. (1993, pp. 305–326), and 4) "Jumping the technology s-curve," *IEEE Spectrum*, June 1995, pp. 49–54.

20.  Here I am borrowing the technology-evolution model described in Moore, G. A. (1991).

21.  Cohen, P. S. (1997, pg. 55).

22.  Campbell, A. and M. Alexander (1997).

23.  For an interesting description of the method used by a pioneering global high-technology firm, see Stillman, H. M. (1997).

24.  Prokesch, S. E. (1994, pg. 363).

25.  Let me caveat this "rule" by noting that not all visions of the future are used in a comparative way. If, however, a firm is determining a choice of direction (as is often the case with forecasting activities), it is important that even a very rough estimate be made of

probabilities. One simple approach is to consider two or three possible future outcomes, and add a catchall possibility that captures "everything else." More on the assignment of (admittedly subjective) probabilities is found in later sections of this chapter.

26.   Millett, S. M. and E. J. Honton (1991, pg. 43).

27.   Betz, F. (1993, pg. 32).

28.   For some additional discussion on this topic, see Weil, E., 1997, "The future is younger than you think," *Fast Company*, April/May Issue.

29.   "Science's dream team," *The Los Angeles Times*, December 22, 1997.

30.   Moore's Law states (roughly paraphrased) that the density (i.e., the performance) of semiconductor devices will increase by four times every three years, while the cost to the consumer will not change. Thus far, this "law" has been remarkably accurate over a period of nearly two decades, and forms the foundation for such respected long-range industry forecasts as the SIA's National Technology Roadmap for Semiconductors. Such a reliable long-run projection is rare indeed, and in this case may be a self-fulfilling prophecy: since most of the U.S. semiconductor industry marches to the aggressive "beat" of the SIA roadmap, who is to say whether progress in this industry would have remained as rapid had Gordon Moore been more conservative in his early predictions?

31.   See 1) Millett, S. M. and E. J. Honton (1991, pg. 7), and 2) Burgelman, R. A. et al. (1996, pg. 146).

32.   See 1) Burgelman, R. A. et al (1996, pg. 148), and 2) Millett, S. M. and E. J. Honton (1991, pg. 51).

33.   For a somewhat arcane treatment of some of the irrationalities in R&D decision-making, see Malueg, D. A. and S. O. Tsutsui (1997).

34.   For a small sampling, see 1) Stillman, H. M. (1997), 2) Thomas, C. W. (1996), 3) Burgelman, R. A. et al. (1996, pg. 532), 4) LeBlanc, L. J. et al. (1997), 5) Giget, M. (1997), and 6) Bacon, G. et al. (1994).

35.   See, for example, Calvin, W. H., 1989, *The Cerebral Symphony*, Bantam Books.

36.   Schwartz, P. (1991, pg. 4).

37.   Millett, S. M. and E. J. Honton (1991, pg. 66).

38.   Wack, P. (1985).

39.   The BASICS system utilizes cross-impact analysis to generate multiple possible outcomes for each causal factor, and subsequently cross-matches them to determine conditional probabilities. For a more complete discussion, see Millett, S. M. and E. J. Honton (1991, pg. 68).

40.   Here I am adopting the formalism described in Schwartz, P. (1991).

41.   This is a somewhat simplified version of the story presented in Schwartz, P. (1991, pg. 100).

42.   Dawes, R. M. (1988, pg. 132).

43.   Schwartz, P. (1991, pg. 50).

44.   Gates, W. F.,1997, "The Web lifestyle," *The World in 1998*, The Economist Publications, pg. 104.

45.   Dawes, R. M. (1988, pg. 137).

46.   This scenario has already occurred as of this writing, and the jump-shift described herein is almost certainly representative of the future direction of the optical-data-storage industry. This case highlights one of the advantages of creating scenario examples: you can use hindsight to give the impression of foresight. For more on the technology behind this case example, see 1) Whipple, C. T., 1998, "Blue diode lasers: Are laser manufacturers blue with envy?" *Photonics Spectra*, May Issue, and 2) Nakamura, S., 1998, "Blue lasers meet commercial requirements," *Photonics Spectra*, same issue.

47.  Grove, A. (1996).

48.  For more on this topic, see 1) *Electronic Engineering Times*, May 18, 1998 and June 1, 1998 issues, and 2) "Tantalum, copper and damascene: The future of interconnects," *Semiconductor International*, Vol. 21, No. 6, June 1998. A basic discussion is provided in "The road to Damascus," *The Economist*, June 6, 1998, pp. 77–78.

49.  For an even broader perspective on "roadmapping" that organizes the knowledge inputs to innovation, by means of a process known as *knowledge mapping*, see Gaynor, G. H. (1996, pg. 13.9).

50.  "The SIA's 1997 National Technology Roadmap for Semiconductors," *Solid State Technology*, January 1998.

51.  "The National Technology Roadmap for Semiconductors - 1994," The Semiconductor Industry Association.

52.  "Technology roadmaps: A path to success," *Circuitree*, May 1995. A growing number of technology-roadmapping initiatives are being performed by industry/government/university consortia. The Next-Generation Manufacturing (NGM) project has begun the development of the Integrated Manufacturing Technology Roadmap, which is sponsored by several government departments and led by the Oak Ridge Centers for Manufacturing Technology. See "A roadmap to the future of manufacturing," *APICS Magazine*, April 1998. Similarly, the U.S. Display Consortium (USDC) is developing a National Technology Roadmap for Flat Panel Displays, which was scheduled to be available for review by members in mid-1998. See "National technology roadmap for flat panel displays," *Solid State Technology*, January 1998.

53.  The SIA has decided, at least for now, to open its document up to inputs from foreign contributors, but to retain overall control and coordination at the domestic level. See 1) "SIA road map goes global," *Electronic Buyer's News*, April 20, 1998, and 2) "SIA ponders whether to take its road map global," *Electronic Engineering Times*, March 16, 1998.

54.  Willyard, C. H. and C. W. McClees (1987).

55.  Groenveld, P. (1997).

# Chapter 11: Models for Sustained Market Leadership

1.  Moore, G. A. (1995).

2.  It is important to recognize that case studies represent single events rather than statistically valid samplings of the behavior of firms. The study of competitive strategy is burdened by the inability of researchers to "stage" controlled, statistically meaningful experiments.

3.  Yoshida, J., 1996, "Untold fortunes: How the cocky consumer startups went bust," *OEM Magazine*, October Issue.

4.  Rumelt, R. P. (1994, pg. 229).

5.  As any good gardener knows, dying roses should be trimmed quickly from the bush so that the maximum amount of energy is available to the rosebush for making new blossoms. This is clearly analogous to the need for managers to terminate mature and dying product lines promptly, to avoid their draining precious resources from the firm.

6.  See, for example, Maidique, M. A. and R. H. Hayes (1984).

7.  Gruner, S., 1998, "Start up. Cash out. Repeat.," *Inc. Magazine - State of Small Business 1998*.

8.  Useem, J., 1997, "The start-up factory," *Inc. Magazine*, February Issue.

9.  Cohen, P. S. (1997, pg. 39).

10. "R&D 'assembly line' revs up Genentech," *The Wall Street Journal*, March 12, 1998.

11. "Intel holds sway as a venture capitalist," *Electronic Engineering Times*, February 16, 1998.

12. For a brief overview, see Porter, M. E. (1985, pg. 286).

13. Burgleman, R. A. et al. (1996, pg. 309).

14. Note that this model focuses on revenues rather than profits. In technology-intensive industries, however, the share of revenue is not necessarily a good indicator of the competitiveness of a firm. Low-price commodity producers, for example, can enjoy a dominant share of revenues in a given market segment, but might not receive sufficient returns on their R&D investments to retain their edge into the next generation of products. In technology-intensive industries, profits rather than revenues are often the driving force behind the cycle of innovation.

15. For more details on this familiar case study, see Grindley, P., 1989, "Product standards and market development: The case of video-cassette recorders," London Business School Centre for Business Strategy working paper.

16. There is a format battle being waged overseas among several firms, each hoping to install its proprietary products as the accepted standard in the potentially gigantic Chinese market. See "China braces for video-DC format battle," *Electronic Engineering Times*, June 22, 1998.

17. See 1) "A 'disposable' videodisk threatens to undercut nascent market for DVDs," *The Wall Street Journal*, September 9, 1997, and 2) "Rival formats offer two views of videodisc future," *Los Angeles Times*, January 12, 1998.

18. "Vying to be a site for more eyes," *Business Week*, May 18, 1998.

19. "The rebirth of IBM," *The Economist*, June 6, 1998.

20. For a discussion of "distributed core competencies" in multi-disciplinary firms, see Granstand, O. et al. (1997).

21. "To sell a world phone, play to executives' fears of being out of touch," *The Wall Street Journal*, June 4, 1998.

22. "Microprocessors are for wimps," *Business Week*, December 15, 1997. The Japanese have embraced the system-on-a-chip strategy as their next opportunity to gain leadership in the global semiconductor industry. See "Japan looks to system ICs for a fourth miracle," *Electronic Engineering Times*, June 15, 1998.

23. "The incredible shrinking OEM," *Electronic Business Today*, April 1997.

24. "At last, telecom unbound," *Business Week*, July 6, 1998.

25. For a classic description of the bundling strategy, see Porter, M. E. (1985, pp. 425–436).

26. Porter, M. E. (1985, pg. 430).

27. "Netscape uses browser to beef up Web business," *The Wall Street Journal*, June 1, 1998.

28. I am oversimplifying a bit for clarity. All businesses have fixed costs that must be defrayed, and even the purest non-rival good (i.e., a software product available for download on the Internet) has some variable costs associated with it as well. From a strategic standpoint, these costs are typically not significant, however.

29. "For Microsoft, the phone is a potent weapon," *The Wall Street Journal*, May 27, 1998. For yet another overview of this case, see "The battle for the cyber future," *Business Week*, June 1, 1998.

30. "Gateway offers choice on Web software," *The Wall Street Journal*, May 28, 1998.

31. "Chip case revives 'essential' doctrine," *Interactive Week*, May 4, 1998.

32. Sources: 1) "To sell a world phone, play to executives' fears of being out of touch," *The Wall Street Journal*, June 4, 1998, 2) "Soap powder, with added logic," *The Economist*, December 6, 1997, and 3) "Battle of the brand," *The Economist*, June 13, 1998.

33. For an interesting discussion of the distinction between "hard" and "soft" strategies, see Thomas, L. A. (1997).

34. "Gateway loses the folksy shtick," *Business Week*, July 6, 1998. As an interesting side note, Gateway has decided to move its headquarters from the cornfields of South Dakota to the growing technology cluster surrounding San Diego, California. This move was in response to the need for high levels of technical and managerial expertise, which Gateway found difficult to attract in its farmbelt location.

35. "Netscape pulls out stops with browser offer," *The Los Angeles Times*, January 23, 1998.

36. "Buzz: In search of the most elusive force in all of marketing," *Inc. Magazine*, May 1998.

37. Fisher, L. M. (1997).

38. Rifkin, G. (1998).

39. Dyson, F. J. (1998).

40. "Model business," *The Economist*, April 18, 1998.

41. "Art and Technology," *IEEE Spectrum*, July 1998.

## Conclusion

1. Esty, D. C. (1994, pg. 10, footnote no. 2).

2. For further discussion, see OECD (1997b).

3. Esty, D. C. (1994, pg. 4).

4. For an overview of the economic complications of assigning costs to environmental assets and damage, see "An invaluable environment," *The Economist*, April 18, 1998.

5. "Greenhouse common sense: Why global-warming economics matters more than science," *U.S. News and World Report*, December 1, 1997, pg. 60.

6. Elkington, J. (1994, pg. 91).

7. Cline, W. R. (1992).

8. For a broad overview of the issue of global warming, see Read, P. (1994).

9. Schelling, T. C. (1997).

10. "Greenhouse common sense: Why global-warming economics matters more than science," *U.S. News and World Report*, December 1, 1997.

11. "Exxon urges developing nations to shun environmental curbs hindering growth," *The Wall Street Journal*, October 12, 1997.

12. See, for example, 1) OECD (1997c), and 2) Cooper R. N. (1994).

13. For additional discussion of the Kyoto Protocol, see 1) Coppock, R. (1998), 2) Jacoby, H. D. et al. (1998), and 3) Cooper, R. N. (1998).

14. Swift, B. (1998, pg. 75).

15. "Global-warming debate gets no consensus in industry," *The Wall Street Journal*, April 16, 1998.

16. Swift, B. (1998, pg. 77). The U.S. Acid Rain Program is a notable model for a broader multilateral credit-trading scheme for another reason. The use of high-quality monitoring, the implementation of a public Allowance Tracking System, and the imposition of steep penalties have led to 100 percent compliance among U.S. utilities. Similar tough enforcement regimes will be needed to avoid rampant cheating on any multilateral global-warming agreement.

17. Office of Technology Assessment (1992b, pg. 37).

18. *The World in 1998*, The Economist Publications, pg. 27. For more information, see "Green party grows strong on soil peculiar to Germany," *The Wall Street Journal*, April 20, 1998.

19. Office of Technology Assessment (1992b, pg. 67).

20. Elkington, J. (1994, pg. 97).

21. OECD (1997c, pg. 67).

22. For more information on "environmentally critical technologies," see World Resources Institute (1992).

23. "Hot air treaty," *U.S. News and World Report*, December 22, 1997.

# References

Abelson, P. H., 1996, "The changing frontiers of science and technology," in Teich, A. H. et al. (eds.) (1997).

Ackoff, R., 1981, *Creating the Corporate Future*, Wiley.

Alford, W. P., 1995, *To Steal a Book is an Elegant Offense: Intellectual Property Law in Chinese Civilization*, Stanford University Press.

Alkhafaji, A. F., 1995, *Competitive Global Management: Principles and Strategies*, St. Lucie Press.

Almeida, P., 1996, "Knowledge sourcing by foreign multinationals: Patent citation analysis in the U.S. semiconductor industry," *Strategic Management Journal*, Vol. 17, Winter Special Issue, pp. 155–165.

Amable, B., 1993, "National effects of learning, international specialization and growth paths," in Foray, D. and C. Freeman (eds.) (1993).

Amendola, G., G. Dosi, and E. Papagni, 1994, "The dynamics of international competitiveness," *Weltwirtschaftliches Archiv*, Vol. 129, pp. 451–471.

Andersen, E. and B. Lundvall, 1988, "Small national systems of innovation facing technological revolutions," in C. Freeman and B. Lundvall, (eds.), *Small Nations Facing Technological Revolutions*, Pinter.

Anderson, D. M., 1997, *Agile Product Development for Mass Customization*, Irwin.

Anderson, J. R., R. Farrel, and R. Sauers, 1984, "Learning to program in LISP," *Cognitive Science*, Vol. 8, pp. 87–129.

Antonelli, C., 1994, "Technological districts, localized spillovers and productivity growth: The Italian evidence on technological externalities in the core regions," *International Review of Applied Economics*, Vol. 8, pp. 18–30.

Archibugi, D. and J. Michie, (eds.), 1997a, *Technology, Globalisation and Economic Performance*, Cambridge University Press.

Archibugi, D. and J. Michie, 1997b, "The globalisation of technology: A new taxonomy," in Archibugi, D. and J. Michie (eds.) (1997a).

Archibugi, D. and M. Pianta, 1992, *The Technological Specialization of Advanced Countries*, Kluwer Academic Publishers.

Arora, A., 1996, "Contracting for tacit knowledge: The provision of technical services in technology licensing contracts," *Journal of Development Economics*, Vol. 50, pp. 233–256.

Arrow, K. J., 1962a, "The economic implications of learning by doing," *Review of Economic Studies*, Vol. 29, pp. 155–173.

Arrow, K. J., 1962b, "Economic welfare and the allocation of resources for invention," in R. R. Nelson (ed.), *The Rate and Direction of Inventive Activity*, Princeton University Press, pp. 609–625.

Audretsch, D. B. and M. P. Feldman, 1996, "R&D spillovers and the geography of innovation and production," *American Economic Review*, Vol. 83, No. 3, pp. 630–640.

Audretsch, D. B. and P. E. Stephan, 1996, "Company-scientist locational links: The case of biotechnology," *American Economic Review*, Vol. 86, No. 3, pp. 641–652.

Bacon, G. et al., 1994, "Managing product definition in high-technology industries: A pilot study," *California Management Review*, Vol. 36, No. 3, pp. 32–56.

Baker, S. and K. Baker, 1998, "New software for business strategists," *IEEE Engineering Management Review*, Spring 1998, pp. 100–113.

Balachandra, R. and J. H. Friar, 1997, "Factors for success in R&D projects and new product innovation: A contextual framework," *IEEE Transactions on Engineering Management*, Vol. 44, No. 3, pp. 276–287.

Baldwin, T. T., C. Danielson, and W. Wiggenhorn, 1998, "The evolution of learning strategies in organizations: From employee development to business redefinition," *IEEE Engineering Management Review*, Spring 1998, pp. 50–57.

Barabaschi, S., 1992, "Managing the growth of technical information," in Rosenberg, N. et al. (eds.) (1992).

Barfield, C. E. (ed.), 1997, *Science for the 21st Century: The Bush Report Revisited*, The American Enterprise Institute.

Barfield, C. E. and D. A. Irwin, 1997, "The future of free trade," *Business Economics*, Vol. 32, No. 2, pp. 26–31.

Barney, J. B., 1994, "Beyond individual metaphors in understanding how firms behave: A comment on game theory and prospect theory models of firm behavior," in Rumelt, R. P. et al. (eds.) (1994).

Barney, J. B., 1997, *Gaining and Sustaining Competitive Advantage*, Addison-Wesley Publishing Company.

Barro, R. J., 1989, "A cross-country study of growth, saving, and government," NBER Working Paper No. 2855, Cambridge, MA.

Barro, R. J., 1991, "Economic growth in a cross section of countries," *Quarterly Journal of Economics*, May, Vol. 105, pp. 407–443.

Bartlett, C. A. and S. Ghoshal, 1989, *Managing Across Borders: The Transnational Solution*, Harvard Business School Press.

Baughn, C. C., M. Bixby, and L. S. Woods, 1997, "Patent laws and the public good: IPR protection in Japan and the United States," *Business Horizons*, July - August, pp. 59–65.

Baumol, W. J., S. A. B. Blackman, and E. N. Wolff, 1989, *Productivity and American Leadership: The Long View*, The MIT Press.

Beede, D. N. and K. H. Young, 1998, "Patterns of advanced technology adoption and manufacturing performance," *Business Economics*, Vol. 33, No. 2, pp. 43–48.

Bell, M. and K. Pavitt, 1997, "Technological accumulation and industrial growth," in Archibugi, D. and J. Michie (eds.) (1997a).

Bergstrand, J., 1990, "The Heckscher-Ohlin-Samuelson Model, the Linder Hypothesis and the determinants of bilateral intra-Industry trade," *Economic Journal*, Vol. 100, pp. 850–868.

Bettis, R. A. and M. A. Hitt, 1995, "The new competitive landscape," *Strategic Management Journal*, Vol. 16, Special Summer Issue, pp. 7–19.

Betz, F., 1993, *Strategic Technology Management*, McGraw-Hill.

Betz, F., 1996, "Forecasting and planning technology," in Gaynor, G. H. (ed.) (1996).

Bhagwati, J.,1995, "Trade liberalisation and 'fair trade' demands: Addressing the environmental and labour standards issues," *The World Economy*, Vol. 18, No. 6, pp. 745–759.

Bhagwati, J., 1998, "The capital myth: The difference between trade in widgets and dollars," *Foreign Affairs*, Vol. 77, No. 3, pp. 7–12.

Bhagwati, J. and A. O. Krueger, 1995, *The Dangerous Drift to Preferential Trade Agreements*, American Enterprise Institute.

Bhalla, S. K., 1987, *The Effective Management of Technology*, Battelle Press.

Blomstrom, M., 1990, *Transnational Corporations and Manufacturing Exports from Developing Countries*, United Nations Centre on Transnational Corporations.

Borrus, M. and J. Stowsky, 1998, "Technology policy and economic growth," in Branscomb, L. M. and J. H. Keller (eds.) (1998).

Bowonder, B. and T. Miyake, 1997, "R&D and business strategy: Analysis of practices at Canon," *International Journal of Technology Management*, Vol. 13, Nos. 7/8, Special Issue on R&D Management, pp. 833–852.

Bowonder, B., S. L. Sarnot, M. S. Rao, and T. Miyake, 1996, "Competition in the global electronics display industry: Strategies of major players," *International Journal of Technology Management*, Vol. 12, Nos. 5/6, Special Issue, pp. 551–576.

Brandenburger, A. M. and B. J. Nalebuff, 1996, *Co-opetition*, Currency Doubleday.

Branscomb, L. M. (ed.), 1993a, *Empowering Technology: Implementing a U.S. Strategy*, The MIT Press.

Branscomb, L. M., 1993b, "The national technology policy debate," in Branscomb, L. M. (ed.) (1993a).

Branscomb, L. M., 1993c, "Empowering technology policy," in Branscomb, L. M. (ed.) (1993a).

Branscomb, L. M., 1993d, "Targeting critical technologies," in Branscomb, L. M. (ed.) (1993a).

Branscomb, L. M., 1993e, "National laboratories: The search for new missions and new structures," in Branscomb, L. M. (ed.) (1993a).

Branscomb, L. M. and R. Florida, 1998, "Challenges to technology policy in a changing world economy," in Branscomb, L. M. and J. H. Keller (eds.) (1998).

Branscomb, L. M. and J. H. Keller (eds.), 1996, *Converging Infrastructure*, The MIT Press.

Branscomb, L. M. and J. H. Keller (eds.), 1998, *Investing in Innovation: Creating a Research and Innovation Policy that Works*, The MIT Press.

Branscomb, L. M. and G. Parker, 1993, "Funding civilian and dual-use industrial technology," in Branscomb, L. M. (ed.) (1993a).

Branscomb, L. M. et al., 1997, *Investing in Innovation: Toward a Consensus Strategy for Federal Technology Policy*, John F. Kennedy School of Government Working Paper.

Brod, A. and R. Shivakumar, 1997, "R&D cooperation and the joint exploitation of R&D," *Canadian Journal of Economics*, Vol. 30, No. 3, pp. 673–684.

Brooks, H., 1996, "The evolution of U.S. science policy," in Smith, B. L. R and C. E. Barfield (eds.) (1996a).

Brown, J. S. and P. Duguid, 1998, "Organizing knowledge," *California Management Review*, Vol. 40, No. 3, pp. 90–111.

Brown, S. L. and K. M. Eisenhardt, 1998, *Competing on the Edge: Strategy as Structured Chaos*, Harvard Business School Press.

Browning, L. D., J. M. Beyer, and J. C. Shetler, 1995, "Building cooperation in a competitive industry: SEMATECH and the semiconductor industry," *Academy of Management Journal*, Vol. 38, No. 1, pp. 113–151.

Buchanan, J. M., 1978, *Cost and Choice: An Inquiry in Economic Theory*, University of Chicago.

Bullis, D., 1997, "Why Malaysia's MSC and Singapore One are missing the market," *Asia Pacific Economic Review*, Vol. 5, No. 2, pp. 16–18.

Burgelman, R. A., M. A. Maidique, and S. C. Wheelwright, 1996, *Strategic Management of Technology and Innovation*, 2nd Edition, Irwin.

Cainarca, G. C., M. G. Colombo, and S. Mariotti, 1992, "Agreements between firms and the technological life cycle model: Evidence from information technologies," *Research Policy*, Vol. 21, pp. 45–62.

Camerer, C. F., 1994, "Does strategy research need game theory?" in Rumelt, R. P. et al. (eds.) (1994).

Campbell, A. and M. Alexander, 1997, "What's wrong with strategy?" *Harvard Business Review*, November - December, pp. 42–51.

Cantwell, J., 1997, "The globalization of technology: What remains of the product cycle model?" in Archibugi, D. and J. Michie (eds.) (1997a).

Caravatti, M., 1992, "Why the United States must do more process R&D," *Research Technology Management*, September - October, pp. 8–9.

Casson, M., 1985, "Multinational monopolies and international cartels," in *The Economic Theory of the Multinational Enterprise*, Macmillan.

Castells, M., 1996, *The Rise of the Network Society*, Blackwell Publishers.

Castells, M., 1997, *The Power of Identity*, Blackwell Publishers.

Castells, M., 1998, *End of Millennium*, Blackwell Publishers.

Castells, M. and P. Hall, 1994, *Technopoles of the World: The Making of 21st Century Industrial Complexes*, Routledge.

Chakravarthy, B., 1997, "A new strategy framework for coping with turbulence," *Sloan Management Review*, Vol. 38, No. 2, Winter, pp. 69–82.

Chandler, A. D., Jr., 1962, *Strategy and Structure: Chapters in the History of the Industrial Enterprise*, The MIT Press.

Chapman, Clark, and Dobson, 1990, *Technology-Based Economic Development*, National Institute of Standards and Technology.

Chawla, S. and J. Renesch (eds.), 1995, *Learning Organizations: Developing Cultures for Tomorrow's Workplace*, Productivity Press.

Chen, E. K. Y., 1997, "The total factor productivity debate: Determinants of economic growth in East Asia," *Asian Pacific Economic Literature*, Vol. 11, No. 1, pp. 18–38.

Chiesa, V., 1996, "Strategies for global R&D," *Research - Technology Management*, September - October, pp. 19–25.

Choi, S., D. O. Stahl, and A. B. Whinston, 1997, *The Economics of Electronic Commerce*, Macmillan Technical Publishing.

Clark, K. B. and S. C. Wheelwright, 1994, *The Product Development Challenge: Competing Through Speed, Quality, and Creativity*, Harvard Business Review Press.

Cline, W. R., 1992, *The Economics of Global Warming*, Institute for International Economics.

Clinton, W. J. and A. Gore, Jr., 1993, *Technology for America's Economic Growth: A New Direction to Build Economic Strength*, Executive Office of the President.

Clinton, W. J. and A. Gore Jr., 1997, *A Framework for Global Electronic Commerce*, Executive Office of the President.

Cohan, P. S., 1997, *The Technology Leaders*, Jossey-Bass Publishers.

Cohen, W. M. and D. A. Levinthal, 1989, "Innovation and learning: The two faces of R&D," *The Economic Journal*, Vol. 99, pp. 569–596.

Cohen, W. M. and D. A. Levinthal, 1990, "Absorptive capacity: A new perspective on learning and innovation," *Administrative Science Quarterly*, Vol. 35, pp. 128–152.

Committee on Criteria for Federal Support of Research and Development, 1995, *Federal Funds for Science and Technology*, National Academy of Sciences.

Committee on Science Engineering and Public Policy, 1992, *The Government Role in Civilian Technology: Building a New Alliance*, The National Academy Press.

Cooney, S. L., Jr., 1997, "Government policies and the siting of private-sector R&D facilities: Views from foreign firms," in Teich, A.H. et al. (eds.) (1997).

Cooper, R. N., 1994, *Environment and Resource Policies for the World Economy*, The Brookings Institution.

Cooper, R. N., 1998, "Toward a real global warming treaty," *Foreign Affairs*, Vol. 77, No. 2, pp. 66–79.

Coppock, R., 1998, "Implementing the Kyoto Protocol," *Issues in Science and Technology*, Vol. 14, No. 3, pp. 66–74.

Cornes, R. and T. Sandler, 1996, *The Theory of Externalities, Public Goods and Club Goods*, 2nd Edition, Cambridge University Press.

Costello, D. M., 1993, "A cross-country, cross-industry comparison of productivity growth," *Journal of Political Economy*, Vol. 101, No. 2, pp. 207–222.

Council on Competitiveness, 1996, *Endless Frontier, Limited Resources: U.S. R&D Policy for Competitiveness*, Council on Competitiveness.

Daniels, P. L., 1997, "National technology gaps and trade - an empirical study of the influence of globalisation," *Research Policy*, Vol. 25, pp. 1189–1207.

Dawes, R. M., 1988, *Rational Choice in an Uncertain World*, Harcourt Brace College Publishers.

Day, G. S., 1997, "Maintaining the competitive edge: Creating and sustaining advantages in dynamic competitive environments," in Day, G. S. and D. J. Reibstein (eds.) (1997).

Day, G. S. and D. J. Reibstein (eds.), 1997, *Wharton on Dynamic Competitive Strategy*, John Wiley & Sons.

Dean, E. (ed.), 1984, *Education and Economic Productivity*, Ballinger Publishing Company.

Debresson, C., 1989, "Breeding innovation clusters: A source of dynamic development," *World Development*, Vol. 17, No. 1, pp. 1–16.

DeGroot, A. D., 1965, *Thought and Choice in Chess*, Morton.

Denison, E. F., 1985, *Trends in American Economic Growth, 1929 - 1982*, The Brookings Institution.

Denison, E. F., 1989, *Estimates of Productivity Change by Industry*, The Brookings Institution.

Destler, I. M., 1995, *American Trade Politics*, 3rd Edition, Institute for International Economics.

Dick, A., 1995, *Industrial Policy and Semiconductors: Missing the Target*, American Enterprise Institute.

Dietrich, G. B., D. B. Walz, and J. L. Wynekoop, 1997, "The failure of SDT diffusion: A case for mass customization," *IEEE Transactions on Engineering Management*, Vol. 44, No. 4, pp. 390–398.

Dollar, D., 1992, "Outward-oriented developing economies really do grow more rapidly: Evidence from 95 LDCs, 1976-85," *Economic Development and Cultural Change*, April, Vol. 40, pp. 523–544.

Dollar, D. and E. N. Wolff, 1993, *Competitiveness, Convergence, and International Specialization*, The MIT Press.

Domenici, P. V., 1997, "Science and technology and the balanced budget," in Teich, A. H. et al. (eds.) (1997).

Doremus, P. N., W. W. Keller, L. W. Pauly, and S. Riech, 1998, *The Myth of the Global Corporation*, Princeton University Press.

Dornbusch, R. and S. Fischer, 1994, *Macroeconomics*, 6th Edition, McGraw Hill.

Dosi, G., K. Pavitt, and L. Soete, 1990, *The Economics of Technical Change and International Trade*, New York University Press.

Dowrick, S. and N. Gemmell, 1991, "Industrialisation, catching up and economic growth: A comparative study across the world's capitalist economies," *The Economic Journal*, Vol. 101, pp. 263–275.

Dunne, T., 1994, "Plant age and technology use in U.S. manufacturing industries," *RAND Journal of Economics*, Vol. 25, No. 3, Autumn pp. 488–499.

Dunning, J., 1958, *American Investment in British Manufacturing Industry*, Allen & Unwin.

Dutta, S., 1997, "Strategies for implementing knowledge-based systems," *IEEE Transactions on Engineering Management*, Vol. 44, No. 1, pp. 79–90.

Dyson, F. J., 1998, "Science as a craft industry," *Science*, Vol. 280, pp. 1014–1015.

Ehrlich, E. M., 1997, "Notes on a borderless world," *Business Economics*, Vol. 32, No. 3, pp. 32–36.

Elkington, J., 1994, "Towards the sustainable corporation: Win-win-win business strategies for sustainable development," *California Management Review*, Vol. 36, No. 2, pp. 90–100.

Engelbrecht, H., 1997, "International R&D spillovers, human capital and productivity in OECD economies: An empirical investigation," *European Economic Review*, Vol. 41, pp. 1479–1488.

Englander, Evenson and Hanazaki, 1988, "R&D, Innovation and the Total Factor Productivity Slowdown," *OECD Economic Studies*, No. 11, Autumn.

Esty, D. C., 1994, Greening the GATT: *Trade, Environment, and the Future*, Institute for International Economics.

Executive Office of the President, 1995, *Global Information Infrastructure: An Agenda for Cooperation*, Executive Office of the President.

Executive Office of the President, 1996, *Technology in the National Interest*, Executive Office of the President.

Fagerberg, J., 1987, "A technology gap approach to why growth rates differ," *Research Policy*, Vol. 16, pp. 87–99.

Fagerberg, J., 1988, "International competitiveness," *The Economic Journal*, Vol. 98, pp. 355–374.

Fagerberg, J., 1994, "Technology and international differences in growth rates," *Journal of Economic Literature*, Vol. 32, No. 3, pp. 1147–1175.

Fahey, L. and L. Prusak, 1998, "The eleven deadliest sins of knowledge management," *California Management Review*, Vol. 40, No. 3, pp. 265–276.

Farrell, J. and C. Shapiro, 1992, "Standard setting in high-definition television," *Brookings Papers: Microeconomics*, pp. 1–93.

Federal Interagency Staff Working Group, 1987, *The Semiconductor Industry*. National Science Foundation.

Feitzinger, E. and H. L. Lee, 1997, "Mass customization at Hewlett-Packard: The power of postponement," *Harvard Business Review*, January - February, pp. 116–121.

Feldman, M. P., 1993, "An examination of the geography of innovation," *Industrial and Corporate Change*, Vol. 2, No. 3, pp. 451–470.

Ferdows, K., 1997, "Making the most of foreign factories," *Harvard Business Review*, March–April, pp. 73–88.

Fisher, L. M., 1997, "How to manage creative people: The case of Industrial Light and Magic," *Strategy and Business*, Issue 7, Second Quarter, pp. 79–86.

Flamm, K., 1993, "Semiconductor dependency and strategic trade policy," *Brookings Papers: Microeconomics.*

Flamm, K., 1996, *Mismanaged Trade? Strategic Policy and the Semiconductor Industry,* Brookings Institution Press.

Florida, R., 1997, "The globalization of R&D: Results of a survey of foreign-affiliated R&D laboratories in the USA," *Research Policy,* Vol. 26, pp. 85–103.

Foray, D., 1991, "The secrets of industry are in the air: Industrial cooperation and the organizational dynamics of the innovative firm," *Research Policy,* Vol. 20, pp. 393–405.

Foray, D. and C. Freeman (eds.), 1993, *Technology and the Wealth of Nations,* Pinter.

Fors, G., 1997, "Utilization of R&D results in the home and foreign plants of multinationals," *The Journal of Industrial Economics,* Vol. 45, No. 2, pp. 341–358.

Frame, J. D. and F. Narin, 1988, "The national self-preoccupation of American scientists: An empirical view," *Research Policy,* Vol. 17, pp. 203–212.

Fransman, M., 1997, "Is national technology policy obsolete in a globalized world? The Japanese response," in Archibugi, D. and J. Michie (eds.) (1997a).

Freeman, C., 1994, "The economics of technical change," *Cambridge Journal of Economics,* Vol. 18, pp. 463–514.

Freeman, C., 1997, "The 'national system of innovation' in historical perspective," in Archibugi, D. and J. Michie (eds.) (1997a).

Gaynor, G. H. (ed.), 1996, *Handbook of Technology Management,* McGraw-Hill.

Gerschenkron, A., 1952, "Economic Backwardness in Historical Perspective," in Hoseletz, B. F. (ed.), *The Progress of Underdeveloped Areas,* University of Chicago Press.

GEST, 1986, Grappes Technologiques. *Les Nouvelles Strategies d'entreprise,* McGraw-Hill.

Gibbons, M., et al., 1994, *The New Production of Knowledge,* Sage Publications.

Giget, M., 1997, "Technology, innovation and strategy: Recent developments," *International Journal of Technology Management,* Vol. 14, No. 6/7/8, pp. 613–634.

Gilmore, J. H. and B. J. Pine II, 1997, "The four faces of mass customization," *Harvard Business Review,* January - February, pp. 91–101.

Goldman, S. L., R. N. Nagel, and K. Preiss, 1995, *Agile Competitors and Virtual Organizations,* Van Nostrand Reinhold.

Gomulka, S., 1990, *The Theory of Technological Change and Economic Growth,* Routledge.

Gore, A. Jr., 1996, "Bringing information to the world: The Global Information Infrastructure," *Harvard Journal of Law & Technology,* Vol. 9, Winter.

Gore, A. Jr., 1997, "What is the role of science in American society?" in Teich, A. H. et al. (eds.) (1997).

Graham, E. M., 1996, *Global Corporations and National Governments,* Institute for International Economics.

Graham, E. M. and J. D. Richardson (eds.), 1997, *Global Competition Policy,* Institute for International Economics.

Granstand, O., P. Patel, and K. Pavitt, 1997, "Multi-technology corporations: Why they have 'distributed' rather than 'distinctive core' competencies," *California Management Review,* Vol. 39, No. 4, pp. 8–25.

Greenwood, J., 1997, *The Third Industrial Revolution: Technology, Productivity and Income Inequality,* American Enterprise Institute.

Greider, W., 1997, *One World, Ready or Not: The Manic Logic of Global Capitalism,* Simon & Schuster.

Greis, N. P., M. D. Dibner, and A. S. Bean, 1995, "External partnering as a response to innovation barriers and global competition in biotechnology," *Research Policy,* Vol. 24, No. 4, pp. 609–630.

Griliches, Z., 1973, "Research expenditures and growth accounting," in R. R. Williams (ed.), 1973, *Science and Technology in Economic Growth*, John Wiley.

Griliches, Z. (ed.), 1984, *R&D, Patents, and Productivity*, The University of Chicago Press.

Griliches, Z., 1988, "Productivity puzzles and R&D: Another nonexplanation," *Journal of Economic Perspectives*, Vol. 2, No. 4, pp. 9–21.

Griliches, Z., 1990, "Patent statistics as economic indicators: A survey," *Journal of Economic Literature*, Vol. 28, pp. 1661–1707.

Griliches, Z., 1994, "Productivity, R&D, and the data constraint," *American Economic Review*, Vol. 84, No. 1, pp. 1–23.

Grindley, P. C. and D. J. Teece, 1997, "Managing intellectual capital: Licensing and cross-licensing in semiconductors and electronics," *California Management Review*, Vol. 39, No. 2.

Groenveld, P., 1997, "Roadmapping integrates business and technology," *Research -Technology Management*, Vol. 40, No. 5, pp. 48–55.

Grossman, G. M. and E. Helpman, 1991, *Innovation and Growth in the Global Economy*, The MIT Press.

Grove, A., 1996, *Only the Paranoid Survive*, Currency Doubleday.

Gruber, H., 1996, "Trade policy and learning by doing: The case of semiconductors," *Research Policy*, Vol. 25, pp. 723–739.

Guerrieri, P. and C. Milana, 1995, "Changes and trends in the world trade in high-technology products," *Cambridge Journal of Economics*, Vol. 19, pp. 225–242.

Guston, D. H., 1998, "Technology transfer and the use of CRADAs at the National Institutes of Health," in Branscomb, L. M. and J. H. Keller (eds.) (1998).

Hackman, S., 1997, "Winning through cooperation: An interview with William Spencer," *Technology Review*, Vol. 100, No. 1, pp. 22–27.

Hagedoorn, J. and J. Schakenraad, 1992, "Leading companies and networks of strategic alliances in information technology," *Research Policy*, Vol. 21, pp. 163–190.

Hagerty, B., 1996, "Booming economies lure Asian executives home," *Business Horizons*, Vol. 5, No. 2, pp. 60–61.

Hall, B. H., 1993a, "The stock market's valuation of R&D investment during the 1980's," *American Economic Review*, Vol. 83, No. 2, pp. 259–264.

Hall, B. H., 1993b, "Industrial research during the 1980s: Did the rate of return fall?" *Brookings Papers: Microeconomics 2*, pp. 289–343.

Hall, P., 1994, *Innovation, Economics & Evolution*, Harvester Wheatsheaf.

Ham, R. M. and D. C. Mowery, 1995, "Enduring dilemmas in U.S. technology policy," *California Management Review*, Vol. 37, No. 4, pp. 89–107.

Hamel, G. and C. K. Prahalad, 1994, *Competing for the Future*, Harvard Business School Press.

Hammond, T. H., 1994, "Structure, strategy, and the agenda of the firm," in Rumelt, R. P. et al. (eds.) (1994).

Heaton, G. R., Jr., R. Repetto, and R. Sobin, *Backs to the Future: U.S. Government Policy Toward Environmentally Critical Technology*, World Resource Institute.

Helpman, E. and P. R. Krugman, 1985, *Market Structure and Foreign Trade: Increasing Returns, Imperfect Competition, and the International Economy*, The MIT Press.

Hemphill, T. A., 1997, "U.S. technology policy, intraindustry joint ventures, and the National Cooperative Research and Production Act of 1993," *Business Economics*, October, pp. 48–54.

Henderson, J., 1989, *The Globalisation of High Technology Production*, Routledge.

Henke, C., 1997, "International electronics standards: Compatibility or interference?" *Medical Device & Diagnostic Industry*, November, pp. 8–13.

Henriques, I., 1994, "Do firms free-ride on rivals' R&D expenditure? An empirical analysis," *Applied Economics,* Vol. 26, pp. 551–561.

Hill, C. T., 1998, "The Advanced Technology Program: Opportunities for enhancement," in Branscomb, L. M. and J. H. Keller (eds.) (1998).

Hindley, B. and P. A. Messerlin, 1996, *Antidumping Industrial Policy: Legalized Protectionism in the WTO and What to Do About It,* American Enterprise Institute.

Hiroyuki, O. and A. Goto, 1993, "The Japanese system of innovation: Past, present, and future," in Nelson, R. R. (ed.) (1993).

Hitt, M. A., R. E. Hoskisson, and H. Kim, 1997, "International diversification: Effects on innovation and firm performance in product-diversified firms," *Academy of Management Journal,* Vol. 40, No. 4, pp. 767–798.

Hobday, M., 1995, *Innovation in East Asia: The Challenge to Japan,* Edward Elgar Publishing.

Horowitz, A. W. and E. L. C. Lai, 1996, "Patent length and the rate of innovation," *International Economic Review,* Vol. 37, No. 4.

Houston, T. and J. H. Dunning, 1976, *UK Industry Abroad,* Financial Times Publications.

Howell, T. and D. Ballantine, 1992, *Creating Advantage,* The Semiconductor Industry Association.

Howells, J., 1990, "The location and organisation of research and development: New horizons," *Research Policy,* Vol. 19, pp. 133–146.

Hu, Y., 1995, "The international transferability of the firm's advantages," *California Management Review,* Vol. 37, No. 4, pp. 73–88.

Huber, P., 1992, "Liability and insurance problems in the commercialization of new products: A perspective from the United States and England," in Rosenberg, N. et al. (eds.) (1992).

Hufbauer, G. C. and K. A. Elliott, 1994, *Measuring the Costs of Protection in the United States,* Institute for International Economics.

Huttner, S. L. and C. Yarkin, 1998, "California's R&D partnerships for a knowledge-based economy," in Teich, A. H. et al. (eds.) (1998).

International Chamber of Commerce, 1996, *Intellectual Property & International Trade: A Guide to the Uruguay Round TRIPS Agreement,* ICC Publishing SA.

International Monetary Fund, 1997, *World Economic Outlook - May 1997,* International Monetary Fund.

Irwin, D. A., 1996a, *Against the Tide: An Intellectual History of Free Trade,* Princeton University Press.

Irwin, D. A., 1996b, *Three Simple Principles of Trade Policy,* American Enterprise Institute.

Irwin, D. A. and P. J. Klenow, 1996, "High-tech R&D subsidies: Estimating the effects of Sematech," *Journal of International Economics,* Vol. 40, pp. 323–344.

Irwin, S. M., 1993, *Technology Policy and America's Future,* St. Martin's Press.

Jacoby, H. D., R. G. Prinn, and R. Schmalensee, 1998, "Kyoto's unfinished business," *Foreign Affairs,* July - August, pp. 54–66.

Jaffe, A. B., M. Trajtenberg, and R. Henderson, 1993, "Geographic localization of knowledge spillovers as evidenced by patent citations," *The Quarterly Journal of Economics,* August, pp. 577–598.

John, R. and G. Ietto-Gillies (eds.), 1997, *Global Business Strategy,* International Thompson Business Press.

Jones, C. I., 1995, "Time series tests of endogenous growth models," *Quarterly Journal of Economics,* Vol. 110, No. 2, pp. 495–525.

Jones, K. A., 1994, *Export Restraint and the New Protectionism: The Political Economy of Discriminatory Trade Restrictions,* The University of Michigan Press.

Juhasz, J. E., 1996, "Enterprise engineering in the systems age," in Gaynor, G. H. (ed.) (1996).

Justman, M. and M. Teubal, 1995, "Technology infrastructure policy (TIP): Creating capabilities and building markets," *Research Policy,* Vol. 24, pp. 259–281.

Kahin, B., 1993, "Information technology and information infrastructure," in Branscomb, L. M. (ed.) (1993a).

Kahneman, D. and D. Lovallo, 1994, "Timid choices and bold forecasts: A cognitive perspective on risk taking," in Rumelt, R. P. et al. (eds.) (1994).

Kahneman, D. and A. Tversky, 1979, "Prospect theory: An analysis of decision under risk," *Econometrica,* Vol. 47, pp. 263–291.

Kanz, J. and D. Lam, 1996, "Technology, strategy, and competitiveness: An institutional-managerial perspective," in Gaynor, G. H. (ed.) (1996).

Katz, M. L. and J. A. Ordover, 1990, "R&D cooperation and competition," *Brookings Papers: Microeconomics,* pp. 137–203.

Katz, R. and T. J. Allen, 1982, "Investigating the not invented here (NIH) syndrome: A look at the performance, tenure, and communication patterns of 50 R&D project groups," *R&D Management,* Vol. 12, pp. 7–12.

Kelly, S., 1996, *Data Warehousing: The Route to Mass Customization,* John Wiley & Sons.

Kendrick, J. W., 1973, *Postwar Productivity Trends in the United States, 1948-1969,* National Bureau of Economic Research.

Khanna, D. M., 1997, *The Rise, Decline, and Renewal of Silicon Valley's High Technology Industry,* Garland Publishing.

Kotha, S., 1995, "Mass customization: Implementing the emerging paradigm for competitive advantage," *Strategic Management Journal,* Vol. 16, pp. 21–42.

Kretschmer, M., 1998, "Game theory: The developer's dilemma, Boeing vs. Airbus," *Strategy & Business,* Issue 11, Second Quarter, pp. 4–6.

Krugman, P. R. (ed.), 1986, *Strategic Trade Policy and the New International Economics,* The MIT Press.

Krugman, P. R., 1989, *Exchange Rate Instability,* The MIT Press.

Krugman, P. R., 1994a, "Proving My Point," *Foreign Affairs,* July–August.

Krugman, P. R., 1994b, "The Myth of Asia's Miracle," *Foreign Affairs,* November–December, pg. 62.

Krugman, P. R., 1996, *Pop Internationalism,* The MIT Press.

Kuemmerle, W., 1997, "Building effective R&D capabilities abroad," *Harvard Business Review,* March - April, pp. 61–70.

Kwong, J., 1997, *The Political Economy of Corruption in China,* M. E. Sharp.

Lansing, P. and J. Gabriella, 1995, "Is history repeating itself? A comparative analysis of technology 'policies' in the U.S. and Japanese pharmaceutical industries," *Journal of Asia-Pacific Business,* Vol. 1, No. 3, pp. 57–79.

LeBlanc, L. J. et al., 1997, "A comparison of US and Japanese technology management and innovation," *International Journal of Technology Management,* Vol. 13, No. 5/6, pp. 601–614.

Lee, H. L., E. Feitzinger, and C. Billington, 1997, "Getting ahead of your competition through design for mass customization," *Target,* Vol. 13, No. 2, pp. 8–17.

Lehman, B. A., 1996, "Intellectual property: America's competitive advantage in the 21st century," *The Columbia Journal of World Business,* Spring, pp. 6–16.

Leonard, D. and S. Sensiper, 1998, "The role of tacit knowledge in group innovation," *California Management Review,* Vol. 40, No. 3, pp. 112–132.

Leonard-Barton, D., 1995, *Wellsprings of Knowledge,* Harvard Business School Press.

Lerman, R. I., 1998, "Is there a labor shortage in the information technology industry?" *Issues in Science and Technology,* Vol. 14, No. 3, Spring, pp. 82–83.

Levin, R. C., W. M. Cohen and D. C. Mowery, 1985, "R&D appropriability, opportunity, and market structure: New evidence on some Schumpeterian hypotheses," *American Economic Review,* Vol. 75, No. 2.

Levin, R. C. et al., 1987, "Appropriating the returns from industrial research and development," *Brookings Papers on Economic Activity,* Vol. 3, pp. 783–831.

Linder, S. B., 1961, *An Essay on Trade and Transformation,* Almqvist & Wiksel.

Link, A. N., 1982, "A disaggregated analysis of industrial R&D: Product versus process innovation," in D. Sahal (ed.), 1982, *The Transfer and Utilization of Technical Knowledge,* D. C. Heath.

Link, A. N., 1987, *Technological Change and Productivity Growth,* Harwood Academic Publishers.

List, F., 1841, *The National System of Political Economy,* English edition, Longman, ([1841],1904).

Macleod, G., 1996, "The cult of enterprise in a networked, learning region? Governing business and skills in lowland Scotland," *Regional Studies,* Vol. 30, No. 8, pp. 749–755.

Maddison, A., 1982, *Phases of Capitalist Development,* Oxford University Press.

Maddison, A., 1987, "Growth and slowdown in advanced capitalist economies: Techniques of quantitative assessment," *Journal of Economic Literature,* Vol. 25, June, pp. 649–706.

Maidique, M. A. and R. H. Hayes, 1984, "The art of high-technology management," *Sloan Management Review,* Vol. 25, Winter Issue.

Malerba, F., 1992, "Learning by firms and incremental technical change," *The Economic Journal,* Vol. 102, July.

Malueg, D. A. and S. O. Tsutsui, 1997, "Dynamic R&D competition with learning," *RAND Journal of Economics,* Vol. 28, No. 4, pp. 751–772.

Mansfield, E. R., 1965, "Rates of return from industrial research and development," *American Economic Review,* Vol. 55, pp. 310–322.

Mansfield, E. R., 1985, "How rapidly does new industrial technology leak out?" *The Journal of Industrial Economics,* Vol. 34, No. 2., pp. 217–223.

Mansfield, E. R., 1988, "Industrial R&D in Japan and the United States: A comparative study," *American Economic Review,* Vol. 78, No. 2, pp. 223–228.

Mansfield, E. R., 1993, "The diffusion of flexible manufacturing systems in Japan, Europe and the United States," *Management Science,* Vol. 39, No. 2, pp. 149–159.

Mansfield, E. R., 1995, *Intellectual Property Protection, Direct Investment, and Technology Transfer,* Discussion Paper Number 27, International Finance Corporation—The World Bank.

Mansfield, E. R., 1996, "Contributions of new technology to the economy," in Smith, B. L. R. and C. E. Barfield (eds.) (1996a).

Mansfield, E. R. et al., 1977, "Social and private rates of return from industrial innovation," *Quarterly Journal of Economics,* Vol. 91, pg. 211–240.

Markusen, J. R., J. R. Melvin, W. H. Kaempfer, and K. E. Maskus, 1995, *International Trade: Theory and Evidence,* McGraw-Hill.

Marshall, A., 1890, *Principles of Economics,* Macmillan.

Mathews, J. A., 1997, "A Silicon Valley of the East: Creating Taiwan's semiconductor industry," *California Management Review,* Vol. 39, No. 4, pp. 26–54.

McCann, P., 1997, "How deeply embedded is Silicon Glen? A cautionary note," *Regional Studies,* Vol. 31, No. 7, pp. 695–703.

McDaniels, I. K. and M. G. Singer, 1997, "Standard fare," *The China Business Review,* May– June.

McGrath, M. E., 1995, *Product Strategy for High-Technology Companies,* Irwin.

Melissaratos, A., 1998, "R&D in an era of cooperation," in Teich, A. H., et al. (1998).

Metcalfe, S., 1997, "Technology systems and technology policy in an evolutionary framework," in Archibugi, D. and J. Michie (eds.) (1997a).

Meyer, R. J. and D. Banks, 1997, "Behavioral theory and naïve strategic reasoning," in Day, G. S. and D. J. Reibstein (eds.) (1997).

Millett, S. M. and E. J. Honton, 1991, *A Manager's Guide to Technology Forecasting and Strategy Analysis Methods,* Battelle Press.

Mitchell, G. R., 1997, "International science and technology: Emerging trends in government policies and expenditures," in Teich, A. H., et al. (eds.) (1997).

Montgomery, C. A. and M. E. Porter (eds.), 1991, *Strategy: Seeking and Securing Competitive Advantage,* Harvard Business Review Books.

Moore, G. A., 1991, *Crossing the Chasm,* HarperBusiness.

Moore, G. A., 1995, *Inside the Tornado,* HarperBusiness.

Moran, T. H., 1990, "The globalization of America's defense industries: Managing the threat of foreign dependence," *International Security,* Vol. 15, pp. 57–100.

Morgan, K., 1997, "The learning region: Institutions, innovation and regional renewal," *Regional Studies,* Vol. 31, No. 5, pp. 491–503.

Moschella, D. C., 1997, *Waves of Power: The Dynamics of Global Technology Leadership 1964 - 2010,* American Management Association.

Mossinghoff, G. J. and T. Bombelles, 1996, "The importance of intellectual property protection to the American research-intensive pharmaceutical industry," *The Columbia Journal of World Business,* Spring Issue, pp. 38–48.

Mowery, D. C., 1989, "Collaborative ventures between U.S. and foreign manufacturing firms," *Research Policy,* Vol. 18, pp. 19–32.

Mowery, D. C., 1993, *Survey of Technology Policy,* Working Paper No. 93-7, University of California at Berkeley.

Mowery, D. C. and N. Rosenberg, 1979, "The influence of market demand on innovation: A critical review of some empirical studies," *Research Policy,* Vol. 8, pp. 103–153.

Mowery, D. C. and N. Rosenberg, 1989, *Technology and the Pursuit of Economic Growth,* Cambridge University Press.

Mutti, J. and B. Yeung, 1996, "Section 337 and the protection of intellectual property in the United States: The complainants and the impact," *The Review of Economics and Statistics,* Vol. 78, No. 3, pp. 510–520.

Nakamura, S., 1989, "Productivity and factor prices as sources of differences in production costs between Germany, Japan, and the U.S.," *The Economic Studies Quarterly,* Vol. 40, pp. 701–715.

Nakamura, Y. and M. Shibuya, 1996, "Japan's technology policy: A case study of the R&D of the Fifth Generation computer systems," *International Journal of Technology Management,* Vol. 12, No. 5/6, Special Issue, pp. 509–533.

National Academy of Engineering, 1993, *Prospering in a Global Economy: Mastering a New Role,* National Academy Press.

National Academy of Engineering, 1995, *Risk and Innovation: The Role and Importance of Small High-Tech Companies in the U.S. Economy,* National Academy Press.

National Center on Education and the Economy, 1990, *America's Choice: High Skills or Low Wages. The Report of the Commission on the Skills of the American Workforce.*

National Research Council, 1996, *Conflict and Cooperation in National Competition for High-Technology Industry,* National Academy Press.

National Research Council, 1997, *International Friction and Cooperation in High-Technology Development and Trade,* National Academy Press.

National Science Foundation, 1996, *Industry / University Cooperative Research Centers Program,* The National Science Foundation, publication No. NSF 93–97.

National Science Foundation, 1996, *Science & Engineering Indicators 1996,* National Science Board.

Nelson, R. R., 1990, "U.S. technological leadership: Where did it come from and where did it go?" *Research Policy,* Vol. 19, pp. 117–132.

Nelson, R. R. (ed.),1993, *National Innovation Systems: A Comparative Analysis,* Oxford University Press.

Nelson, R. R. and P. M. Romer, 1996, "Science, economic growth, and public policy," in Smith, B. L. R. and C. E. Barfield (eds.) (1996a).

Nelson, R. R. and G. Wright, 1992, "The rise and fall of American technological leadership: The postwar era in historical perspective," *Journal of Economic Literature,* December 1992, pp. 1931–1964.

Nelson, S. D. and K. Koizumi, 1997, "Overview of R&D in the President's FY 1997 budget proposals," in Teich, A. H. et al. (eds.) (1997).

Nill, A. and C. J. Shultz II, 1996, "The scourge of global counterfeiting," *Business Horizons,* November - December.

Nivola, P. S. (ed.), 1997, *Comparative Disadvantages?: Social Regulations and the Global Economy,* Brookings Institution Press.

Nonaka, I. and H. Takeuchi, 1995, *The Knowledge-Creating Company,* Oxford University Press.

Nordhaus, W. D., 1997, "Traditional productivity estimates are asleep at the (technological) switch," *The Economic Journal,* Vol. 107, September, pp. 1548–1559.

Noyce, R., 1977, "Microelectronics," *Scientific American,* Vol. 237, No. 3, September.

Nye, W. W., 1996, "Firm-specific learning-by-doing in semiconductor production: Some evidence from the 1986 Trade Agreement," *Review of Industrial Organization,* Vol. 11, No. 3, pp. 383–394.

Odagiri, H. and A. Goto, 1993, "The Japanese system of innovation: Past, present, and future," in Nelson, R. R. (ed.) (1993).

OECD, 1989, *The Internationalization of Software and Computer Services.* Organization for Economic Co-operation and Development.

OECD, 1997a, *Science, Technology and Industry: Scoreboard of Indicators 1997,* OECD Publications.

OECD, 1997b, *Sustainable Development: OECD Policy Approaches for the 21st Century,* Organization for Economic Co-operation and Development.

OECD, 1997c, *Economic Globalization and the Environment,* Organization for Economic Co-operation and Development.

Office of Technology Assessment, 1992a, *Global Standards: Building Blocks for the Future,* Congress of the United States.

Office of Technology Assessment, 1992b, *Green Products by Design,* U.S. Government Printing Office.

Office of Technology Policy, 1995, *Globalizing Industrial Research and Development,* U.S. Department of Commerce.

Office of Technology Policy, 1997, *The Global Context for U.S. Technology Policy,* Office of Technology Policy.

Ohba, S., 1996, "Critical issues related to international R&D programs," *IEEE Transactions on Engineering Management,* Vol. 43, No. 1, pp. 78–87.

Ohmae, K., 1990, *The Borderless World,* Harper.

Okimoto, D. I., 1989, *Between MITI and the Market: Japanese Industrial Policy for High Technology,* Stanford University Press.

Olk, P. and K. Xin, 1997, "Changing the policy on government-industry cooperative R&D arrangements: Lessons from the U.S. effort," *International Journal of Technology Management,* Special Issue, pp. 710–728.

Paci, R., A. Sassu, and S. Usai, 1997, "International patenting and national technological specialization," *Technovation,* Vol. 17, No. 1, pp. 25–38.

Pack, H., 1994, "Endogenous growth theory: Intellectual appeal and empirical shortcomings," *Journal of Economic Perspectives,* Vol. 8, No. 1, pp. 55–72.

Packard, D., 1983, *Report of the White House Science Council: Federal Laboratory Review Panel,* Office of Science and Technology Policy.

Patel, P., 1997, "Localized production of technology for global markets," in Archibugi, D. and J. Michie (eds.) (1997a).

Patel, P. and K. Pavitt, 1991, "Large firms in the production of the world's technology: An important case of 'non-globalisation,'" *Journal of International Business Studies,* Vol. 22, No. 1, pp. 1–21.

Patel, P. and K. Pavitt, 1997, "The technological competencies of the world's largest firms: Complex and path-dependent, but not much variety," *Research Policy,* Vol. 26, pp. 141–156.

Pavitt, K., 1991, "What makes basic research economically useful?" *Research Policy,* Vol. 20, pp. 109–119.

Pine, B. J. II, 1993, *Mass Customization: The New Frontier in Business Competition,* Harvard Business School Press.

Porter, M. E., 1980, *Competitive Strategy,* The Free Press.

Porter, M. E., 1985, *Competitive Advantage: Creating and Sustaining Superior Performance,* The Free Press.

Porter, M. E., 1990, *The Competitive Advantage of Nations,* The Free Press.

Porter, M. E., 1994, "Toward a dynamic theory of strategy," in Rumelt, R. P. et al. (eds.) (1994).

Porter, M. E., 1998, "The Adam Smith Address: Location, clusters, and the 'new' microeconomics of competition," *Business Economics,* Vol. 33, No. 1, pp. 7–13.

Porter, M. E. and R. E. Wayland, 1995, "Global Competition and the Localization of Competitive Advantage," *Advances in Strategic Management,* Vol. 11A, pg. 101.

Powell, W. W., 1998, "Learning from collaboration: Knowledge and networks in the biotechnology and pharmaceutical industries," *California Management Review,* Vol. 40, No. 3, pp. 228–240.

Powers, W. F., 1997, "Government policies and the siting of private-sector R&D facilities: A view from U.S. industry," in Teich, A. H. et al. (eds.) (1997).

Primo Braga, C. A., 1995, "Protection on a global scale," *The China Business Review,* March - April, pp. 25–27.

Prokesch, S. E., 1994, "Mastering chaos at the high-tech frontier: An interview with Silicon Graphics' Ed McCracken," in Clark, K. B. and S. C. Wheelwright (1994).

Quinn, J. B., 1992, *Intelligent Enterprise,* The Free Press.

Quintas, P. and K. Guy, 1995, "Collaborative, pre-competitive R&D and the firm," *Research Policy,* Vol. 24, pp. 325–348.

Rea, D. G., H. Brooks, R. M. Burger, and R. LaScala, 1997, "The semiconductor industry -model for industry/university/government cooperation," *Research-Technology Management,* Vol. 40, No. 4, pp. 46–54.

Read, P., 1994, *Responding to Global Warming,* Zed Books Limited.

Reibstein, D. J. and M. J. Chussil, 1997, "Putting the lesson before the test: Using simulation to analyze and develop competitive strategies," in Day, G. S. and D. J. Reibstein (eds.) (1997).

Reich, R. B., 1991, *The Work of Nations: Preparing Ourselves for $21^{st}$- Century Capitalism,* Alfred A. Knopf.

Ricardo, D., 1817, *On the Principles of Political Economy and Taxation.*

Rifkin, G., 1997, "Growth by acquisition: The case of Cisco Systems," *Strategy and Business,* Issue 7, Second Quarter, pp. 91–102.

Rifkin, G., 1998, "Competing through innovation: The case of Broderubund," *Strategy and Business,* Issue 11, Second Quarter, pp. 48–58.

Rifkin, J., 1995, *The End of Work: The Decline of the Global Labor Force and the Dawn of the Post-Market Era,* G.P. Putnam's Sons.

Rivera-Batiz, L. A. and P. M. Romer, 1991, "Economic integration and endogenous growth," *Quarterly Journal of Economics,* Vol. 106, pp. 531–555.

Roach, S. S., 1996, "The hollow ring of the productivity revival," *Harvard Business Review,* November - December, pp. 81–89.

Rodrik, D., 1997, *Has Globalization Gone Too Far?,* Institute for International Economics.

Romer, P. M., 1986, "Increasing returns and long-run growth," *Journal of Political Economy,* Vol. 94, No. 5, pp. 1002–1037.

Romer, P. M., 1989a, "What determines the rate of growth and technical change?" The World Bank Policy, Planning and Research Working Paper No. WPS 279, Washington, D. C.

Romer, P. M., 1989b, "Capital accumulation in the theory of long run growth," in R. Barro (ed.), *Modern Business Cycle Theory,* Harvard University Press.

Romer, P. M., 1990, "Endogenous technological change," *Journal of Political Economy,* Vol. 98, No. 5 , Part 2, pp. S71–S102.

Romer, P. M., 1993a, "Idea gaps and object gaps in economic development," *Journal of Monetary Economics,* Vol. 32, pp. 543–573.

Romer, P. M., 1993b, "Implementing a national technology strategy with self-organizing industry investment boards," *Brookings Papers: Microeconomics 2,* pp. 345–399.

Romer, P. M., 1994, "The origins of endogenous growth," *Journal of Economic Perspectives,* Vol. 8, No. 1, pp. 3–22.

Rosenberg, N. (ed.), 1994a, *Exploring the Black Box: Technology, Economics, and History,* Cambridge University Press.

Rosenberg, N., 1994b, "Critical issues in science policy research," in Rosenberg, N. (ed.) (1994a).

Rosenberg, N., R. Landau, and D. C. Mowery (eds.), 1992, *Technology and the Wealth of Nations,* Stanford University Press.

Ruggles, R., 1998, "The state of the notion: Knowledge management in practice," *California Management Review,* Vol. 40, No. 3, pp. 80–89.

Rumelt, R. P., D. E. Schendel, and D. J. Teece (eds.), 1994, *Fundamental Issues in Strategy: A Research Agenda,* Harvard Business School Press.

Sabourin, V. and I. Pinsonneault, 1997, "Strategic formation of competitive high technology clusters," *International Journal of Technology Management,* Vol. 13, No. 2, pp. 165–178.

Sager, M. A., 1997, "Regional trade agreements: Their role and the economic impact on trade flows," *The World Economy*, Vol. 20, No. 2, pp. 239–252.

Saloner, G., 1994, "Game theory and strategic management: Contribution, applications, and limitations," in Rumelt, R. P. et al. (eds.) (1994).

Saxenian, A., 1991, "The origins and dynamics of production networks in Silicon Valley," *Research Policy*, Vol. 20, pp. 423–437.

Saxenian, A., 1994, *Regional Advantage: Culture and Competition in Silicon Valley and Route 128*, Harvard University Press.

Schelling, T. C., 1997, "The cost of combating global warming," *Foreign Affairs*, November–December, pp. 8–14.

Scherer, F. M., 1982, "Inter-industry technology flows and productivity growth," *Review of Economics and Statistics*, Vol. 64, pp. 627–634.

Scherer, F. M., 1992, *International High-Technology Competition*, Harvard University Press.

Schmookler, J., 1966, *Invention and Economic Growth*, Harvard University Press.

Schott, J. J. (ed.), 1996, *The World Trading System: Challenges Ahead*, Institute for International Economics.

Schumpeter, J., 1943, *Capitalism, Socialism and Democracy*, Second edition, Allen & Unwin.

Schwartz, P., 1991, *The Art of the Long View*, Currency Doubleday.

Scott, A. J., 1991, "The aerospace-electronics industrial complex of Southern California: The formative years, 1940 - 1960," *Research Policy*, Vol. 20, pp. 439–456.

Senge, P. M., 1990, *The Fifth Discipline: The Art and Practice of the Learning Organization*, Currency Doubleday.

Seyoum, B., 1996, "The impact of intellectual property rights on foreign direct investment," *The Columbia Journal of World Business*, Spring, pp. 50–59.

Shan, W. and J. Song, 1997, "Foreign direct investment and the sourcing of technological advantage: Evidence from the biotechnology industry," *Journal of International Business Studies*, Vol. 28, No. 2, pp. 267–284.

Shell, K., 1967, "A model of inventive activity and capital accumulation," in K. Shell (ed.), *Essays on the Theory of Optimal Economic Growth*, The MIT Press.

Sichel, D. E., 1997, *The Computer Revolution: An Economic Perspective*, The Brookings Institution.

Simon, H. A., 1979, "Rational decision making in business organizations," *American Economic Review*, Vol. 69, pp. 493–513.

Simons, G. R., 1993, "Industrial extension and innovation," in Branscomb, L. M. (ed.) (1993a).

Skolnikoff, E. B., 1993, *The Elusive Transformation: Science, Technology, and the Evolution of International Politics*, Princeton University Press.

Smith, A., 1776, *An Inquiry into the Nature and Causes of the Wealth of Nations*, The University of Chicago Press ([1776] 1976).

Smith, B. L. R. and C. E. Barfield (eds.), 1996a, *Technology, R&D, and the Economy*, The Brookings Institution.

Smith, B. L. R. and C. E. Barfield, 1996b, "Contributions of research and technical advance," in Smith, B. L. R. and C. E. Barfield (eds.) (1996a).

Solomond, J. P., 1996, "International high technology cooperation: Lessons learned," *IEEE Transactions on Engineering Management*, Vol. 43, No. 1, pp. 69–77.

Solow, R. M., 1956, "A contribution to the theory of economic growth," *Quarterly Journal of Economics*, Vol. 70, pp. 65–94.

Solow, R. M., 1957, "Technical change and the aggregate production function," *Review of Economics and Statistics*, August.

Solow, R. M., 1994, "Perspectives on growth theory," *Journal of Economic Perspectives*, Vol. 8, No. 1, pp. 45–54.

Sood, J. and G. L. Miller, 1996, "Intellectual property rights and trade expansion," *The International Executive*, Vol. 38, No. 2, pp. 243–254.

Stewart, T. A., 1997, *Intellectual Capital: The New Wealth of Organizations*, Currency Doubleday.

Stillman, H. M., 1997, "How ABB decides on the right technology investments," *Research - Technology Management*, Vol. 40, No. 6, pp. 14–22.

Sumanth, D. J., 1998, *Total Productivity Management*, St. Lucie Press.

Suttmeier, R. P., 1997, "The role of science and technology policies in the economic growth of East Asia / Pacific Rim nations," in Teich, A. H., et al. (eds.) (1997).

Swift, B., 1998, "A low-cost way to control climate change," *Issues in Science and Technology*, Vol. 14, No. 3, pp. 75–81.

Sykes, A. O., 1995, *Product Standards for Internationally Integrated Goods Markets*, The Brookings Institution.

Tassey, G., 1991, "The functions of technology infrastructure in a competitive economy," *Research Policy*, Vol. 20, pp. 345–361.

Taylor, F. W., 1947, *Scientific Management*, Harper.

Teece, D. J., 1998, "Capturing value from knowledge assets: The new economy, markets for know-how, and intangible assets," *California Management Review*, Vol. 40, No. 3, pp. 55–79.

Teich, A. H., S. D. Nelson, and C. M. McEnaney (eds.), 1997, *AAAS Science and Technology Policy Yearbook 1996/97*, American Association for the Advancement of Science.

Teich, A. H., Nelson, S. D., and C. M. McEnaney, (eds.), 1998, *AAAS Science and Technology Policy Yearbook - 1998*, American Association for the Advancement of Science.

Teichert, T. A., 1997, "Success potential of international R&D co-operations," *International Journal of Technology Management*, Vol. 14, Nos. 6/7/8, pp. 804–821.

Terleckyj, N. E., 1974, *Effects of R&D on the Productivity Growth of Industries: An Exploratory Study*, National Planning Association.

Tezuka, H., 1997, "Success as the source of failure? Competition and cooperation in the Japanese economy," *Sloan Management Review*, Vol. 38, No. 2, pp. 83–93.

Thomas, C. W., 1996, "Strategic technology assessment, future products and competitive advantage," *International Journal of Technology Management*, Special Publication on Technology Assessment, pp. 651–666.

Thomas, L. A., 1997, "Commitment: How narrowing options can improve competitive positions," in Day, G. S. and D. J. Reibstein (eds.) (1997).

Thuermer, K. E., 1998, "Supply and demand: Growing high-tech clusters often face shortages of adequately skilled workers," *World Trade*, April, pp. 84–88.

Thurow, L., 1992, *Head to Head: The Coming Economic Battle Among Japan, Europe, and America*, Warner Books.

Thurow, L. C., 1997, "Needed: A new system of intellectual property rights," *Harvard Business Review*, September–October, pp. 95–103.

Tilton, J. E., 1971, *International Diffusion of Technology: The Case of Semiconductors*, The Brookings Institution.

Trajtenberg, M., 1990, *Economic Analysis of Product Innovation: The Case of CT Scanners*, Harvard University Press.

Tripsas, M., Schrader, S. and M. Sobrero, 1995, "Discouraging opportunistic behavior in collaborative R&D: A new role for government," *Research Policy*, Vol. 24, No. 3, pp. 367–389.

Tyson, L. D., 1992, *Who's Bashing Whom? Trade Conflict in High-Technology Industries,* Institute for International Economics.

UNCTAD, 1994, *World Investment Report 1994,* The United Nations.

UNCTAD, 1995, *World Investment Report 1995,* The United Nations.

UNCTAD, 1997, *World Investment Report 1997: Transnational Corporations, Market Structure and Competition Policy,* The United Nations.

UNTCMD, 1993, *Intellectual Property Rights and Foreign Direct Investment,* The United Nations.

U.S. Department of Commerce, 1994, *Competing to Win in a Global Economy,* United States Department of Commerce.

U.S. Department of Commerce, 1996, *A Nation of Opportunity: Realizing the Promise of the Information Superhighway,* United States Department of Commerce.

Valverde, G. A. and W. H. Schmidt, 1997, "Refocusing U.S. math and science education," *Issues in Science and Technology,* Vol. 14, No. 2, pp. 60–66.

Vernon, R., 1966, "International investment and international trade in the product cycle," *Quarterly Journal of Economics,* Vol. 80, pp. 190–207.

Vernon, R., 1995, "Passing through regionalism: The transition to global markets," John F. Kennedy School of Government Working Paper, Harvard University.

Von Neumann, J. and O. Morgenstern, 1944, *Theory of Games and Economic Behavior,* Wiley.

Wack, P., 1985, "Scenarios: Shooting the rapids," *Harvard Business Review,* Vol. 63, No. 5, pp. 139–150.

Warshofsky, F.,1994, *The Patent Wars: The Battle to Own the World's Technology,* John Wiley and Sons.

Weinstein, D. E. and Y. Yafeh, 1995, "Japan's corporate groups: Collusive or competitive? An empirical investigation of *keiretsu* behavior," *The Journal of Industrial Economics,* Vol. 43, No. 4, pp. 359–376.

Wiegner, K., 1997, "Cisco's rock-solid purchase," *Wired,* Issue 5.03, March.

Wiggenhorn, W., 1990, "Motorola U: When training becomes an education," *Harvard Business Review,* July - August, pp. 71–83.

Wilkins, M., 1988, "The free-standing company, 1870-1914; an important type of British foreign direct investment," *Economic History Review, 2nd Series,* Vol. 41, No. 2, pp. 259–282.

Willcocks, L. and S. Lester, 1996, "Beyond the IT productivity paradox," *European Management Journal,* Vol. 14, No. 3, pp. 279–290.

Williams, J. R., 1994, "Strategy and the search for rents: The evolution of diversity among firms," in Rumelt, R. P. et al. (eds.) (1994).

Willyard, C. H. and C. W. McClees, 1987, "Motorola's technology roadmap process," *Research Management,* September–October.

Wilson, J. S., 1996, "The new trade agenda: Technology, standards, and technical barriers," *SAIS Review,* Winter - Spring, pp. 67–91.

Womack, J. P. and D. T. Jones, 1996, *Lean Thinking,* Simon & Schuster.

The World Bank, 1996, *Global Economic Prospects and the Developing Countries,* The World Bank.

The World Bank, 1997, *World Development Report 1997,* The World Bank.

World Trade Organization, 1995, *International Trade: Trends and Statistics 1995,* World Trade Organization.

Xue, L., 1997, "Promoting industrial R&D and high-tech development through science parks: The Taiwan experience and its implications for developing countries," *International Journal of Technology Management,* Vol. 13, No. 7/8, pp. 744–761.

Yip, G. S., 1992, *Total Global Strategy: Managing for Worldwide Competitive Advantage,* Prentice-Hall.

Yoffie, D. B., 1994, *Strategic Management in Information Technology,* Prentice-Hall.

Yoffie, D. B. and B. Gomes-Casseres, 1994, *International Trade and Competition: Cases and Notes in Strategy and Management,* 2nd edition, McGraw-Hill.

Yoshino, M. Y. and U. S. Rangan, 1995, *Strategic Alliances: An Entrepreneurial Approach to Globalization,* Harvard Business School Press.

Young, A., 1992, "A tale of two cities: Factor accumulation and technical change in Hong Kong and Singapore," *NBER Macroeconomic Annual.*

Young, A., 1995, "The tyranny of numbers: Confronting the statistical realities of the East Asian growth experience," *Quarterly Journal of Economics,* Vol. 110, No. 3.

Zinberg, D. S., 1993, "Putting people first: Education, jobs, and economic competitiveness," in Branscomb, L. M. (ed.) (1993a).

# Index

# The Growth Warriors

Creating Sustainable Global Advantage for
America's Technology Industries

ISBN: 0-9662697-0-5
Item Number: 1015

To order additional copies, please contact Technology
Perspectives at the numbers below, or visit our Website at
www.techper.com

| | | |
|---|---|---|
| Single Copy Price – | U.S.A. | $34.95 |
| | Canada | $52.95 |
| Five or More– | U.S.A. | $27.95 |
| | Canada | $42.95 |

**Technology Perspectives**
P.O. Box 8539, Northridge, CA 91327
Phone Orders:
Toll-free within the U.S. – (888) 366-7488
Outside the U.S. – (818) 366-7488

Fax Orders:
(818) 366-6085